Management and Industrial Engineering

Series Editor

J. Paulo Davim, Department of Mechanical Engineering, University of Aveiro, Aveiro, Portugal

This series fosters information exchange and discussion on management and industrial engineering and related aspects, namely global management, organizational development and change, strategic management, lean production, performance management, production management, quality engineering, maintenance management, productivity improvement, materials management, human resource management, workforce behavior, innovation and change, technological and organizational flexibility, self-directed work teams, knowledge management, organizational learning, learning organizations, entrepreneurship, sustainable management, etc. The series provides discussion and the exchange of information on principles, strategies, models, techniques, methodologies and applications of management and industrial engineering in the field of the different types of organizational activities. It aims to communicate the latest developments and thinking in what concerns the latest research activity relating to new organizational challenges and changes world-wide. Contributions to this book series are welcome on all subjects related with management and industrial engineering. To submit a proposal or request further information, please contact Professor J. Paulo Davim, Book Series Editor, pdavim@ua.pt

More information about this series at https://link.springer.com/bookseries/11690

Jitesh J. Thakkar

Project Management

Strategic and Operational Planning

Springer

Jitesh J. Thakkar
School of Management and Leadership,
National Rail and Transportation Institute
Vadodara, Gujarat, India

ISSN 2365-0532 ISSN 2365-0540 (electronic)
Management and Industrial Engineering
ISBN 978-981-15-3694-6 ISBN 978-981-15-3695-3 (eBook)
https://doi.org/10.1007/978-981-15-3695-3

This Springer imprint is published by the registered company Springer Nature Singapore Pte Ltd.
The registered company address is: 152 Beach Road, #21-01/04 Gateway East, Singapore 189721, Singapore

Dedicated to
my guru
Prof. S. G. Deshmukh

Preface

The book on *Project Management: Strategic and Operational Planning* intends to provide a detailed understanding on various concepts and techniques useful in executing real-life projects. Project management is offered as one of the core courses in engineering and management education. There is also a huge demand of this course from professionals working in manufacturing and service industry. It is necessary to develop insights into the key areas of project management as specified by internationally recognized body Project Management Institute (PMI) and Project Management Body of Knowledge (PMBOK).

This book will enable undergraduate, graduate, and industry executives to develop in-depth understanding on both soft and hard issues of the project management. The present book intends to address both strategic and operational issues in executing the projects. The book covers the basic concepts and project management and specifically includes THE following topics to broaden the knowledge domain of university students and industry executives on project management.

- Focus on 10 key PMBOK knowledge areas of project management
- Imparts both qualitative (case studies) and quantitative (mathematical models and heuristics) learning
- Simulation-based scheduling of project
- Graphical evaluation and review technique (GERT) for project scheduling
- Theory of constraint (TOC) and critical chain project management (CCPM)
- Project procurement and contracts management
- Project risk and quality management
- Project communication management.

This book includes illustrative applications of the project management concepts and details select cases on various issues of project management to develop an interpretive understanding on the subject. It is expected that the book will immensely benefit the reader for developing a holistic and systems perspective on project management.

Vadodara, India Jitesh J. Thakkar

Acknowledgements

It gives me an immense pleasure to deliver the book on *Project Management: Strategic and Operational Planning* to the students. The book will benefit the students/researchers in engineering, and management fields.

The field of project management has received immense attention in the private and public organizations because of complexities associated in executing the mega and mission critical projects. The researchers and practitioners have contributed significantly in the field of project management which has helped to shape this book. The leading international organizations such as Project Management Institute (PMI) and Project Management Body of Knowledge (PMBOK) have developed a deep understanding on critical aspects of the project management and defined the key areas in executing the projects. This has helped to decide the overall content and coverage of this book.

I express my deep gratitude toward Late Professor Arun Kanda, Ex-Professor, IIT Delhi. He has always inspired and motivated me to appreciate the various concepts on project management with its real-life relevance. His unique teaching style and deliberation on the various concepts of project management have immensely helped me in writing this book.

I have always derived inspiration and energy for executing academic projects from my teacher and Guru Prof. S. G. Deshmukh. I express my deep gratitude for his valuable inputs in improving the quality of this book. This book could not have attained its present form, both in content and in presentation without his active interest, direction, and help. I feel proud by mentioning that whatever I am today, it is because of his inspiration, affection, and personal care. Professor Deshmukh has always inspired me for academic excellence and contributions.

My father, Jayprakashbhai Thakkar, and mother, Ushaben Thakkar, have provided a constant moral support and motivation for this work.

I deeply express my love and affection for my wife, Amee, daughter, Prachi, and son, Harshit, for giving me a freedom and moral support in timely completion of this book.

I am thankful to the publisher Springer Nature for supporting my book proposal.

At present, the coronavirus (COVID-19) pandemic is causing large-scale loss of life and severe human suffering. I express my deep gratitude to all the selfless people protecting us like our doctors, nurses, hospital staff and our police, municipal corporation employees, and our able government. I would like to express my sincere condolences for the people who have died in COVID-19.

Dr. Jitesh J. Thakkar

Contents

About the Author

Dr. Jitesh J. Thakkar is a Dean (Academics) and Professor, School of Management and Leadership at National Rail and Transportation Institute (NRTI), Vadodara—India's first University dedicated in the field of Rail and Transportation education. He has served as a faculty at Department of Industrial and Systems Engineering, IIT Kharagpur for 10 years. He has professional experience of more than 23 years. He holds a Ph.D in Supply Chain Management from IIT Delhi, M.Tech. in Industrial Engineering from IIT Delhi and Bachelors of Engineering in Mechanical Engineering conferred with Gold Medal from one of the oldest Government Engineering College—Birla Vishvakarma Mahavidyalaya Engineering College, Sardar Patel University. His areas of interests include Logistics & Supply Chain Management, Transportation Management, Project Management, Service Operations Management, Six Sigma, Lean Manufacturing, Optimization, Statistical Modeling and System Dynamics. He has supervised 13 Ph.D and 80 B.Tech./M.Tech. projects at IIT Kharagpur. He is a productive researcher with H-index 33 and 80 research papers published in SCI/SCOPUS listed journals. The research papers are published in high impact factor and A* & A grade journals of ABDC (Australian Business Deans Council) such as International Journal of Production Economics, International Journal of Operations and Production Management, Transportation Research (Part-E), International Journal of Production Research, Computers and Industrial Engineering, Production Planning and Control, Expert Systems with Applications, Journal of Cleaner Production. He has published three books—Structural Equation Modeling, Multi-Criteria Decision Making and Project Management with Springer. He is contributing as Associate Editor of the journals—i) International Journal of Productivity and Performance Management, Emerald and ii) OPSEARCH, Springer. He has executed number of large scale research and consultancy projects sponsored by DST, DRDO, NLC, ICSSR, and MHRD. He has extended training/consulting to various organizations such as L&T, DRDO, Food Corporation of India, Neyveli Lignite Corporation, Tata Steel, Tata Hitachi, Essar Steel, Ordnance Factory, Lakshmi Machine Works Ltd., Godrej, Qatar Chemical. His online course on "Six Sigma" offered by NPTEL has been attended by more than 35,000 University students and Industry professionals in last five years. He has delivered more than 100 invited sessions at various reputed platforms. He has been invited as a faculty expert by IIM Ahmedabad, IIM Indore, IIT Kanpur, IIT Madras, IIT Bombay, IIT Kharagpur, NITIE Mumbai, NIT Surat, NIT Trichy, IRMA, AMA, BCCI Kolkata, L&T-IPM.

Chapter 1
Project Complexities and Scope

Critical Questions

What is project and project management?
Why project management is important?
What is time–cost–performance relationship and what are its implications for a typical project?
What is the importance of S-curve in project management?
What is the role of project management, project leader, and project engineer?
What are the different forms of projects?

Project Management in Practice

Why is Project Management critical for India?—National Infrastructure Pipeline (NIP)

> Infrastructure is much more than cement and concrete. Infrastructure guarantees a better future. Infrastructure connects people
>
> *Shri Narendra Modi, Prime Minister of India*

The National Infrastructure Pipeline (NIP) for FY 2019–25 is a first-of-its-kind, whole-of-government exercise to provide world-class infrastructure to citizens and improving their quality of life. It aims to strengthen project preparation and attract investments into infrastructure. To draw up the NIP, a High-Level Task Force was constituted under the chairmanship of the Secretary, Department of Economic Affairs (DEA), Ministry of Finance, India.

J. J. Thakkar, *Project Management*, Management and Industrial Engineering,
https://doi.org/10.1007/978-981-15-3695-3_1

Infrastructure Vision and Goals:

Vision:

Infrastructure services that raise the quality of life and ease of living in India to global standards.

Mission:

- Develop a Five-Year Plan of infrastructure development for India in key sectors.
- Facilitate design, delivery, and maintenance of public infrastructure as per global standards.
- Facilitate generic and sectoral reforms in regulation and administration of public infrastructure services as per global best practices.
- Push India up in global rankings in public infrastructure.

Strategic Goals:

- Provide a positive and enabling environment for significant private investment in infrastructure at all three levels of government.
- Design, deliver, and maintain public infrastructure projects to meet efficiency, equity, and inclusiveness goals.
- Design, construct, and maintain public infrastructure to meet disaster resilience goals.
- Create a fast-track institutional, regulatory, and implementation framework for infrastructure.
- Benchmark infrastructure performance to global best practices and standards.
- Leverage technology to enhance service standards, efficiency, and safety.

Key Benefits of NIP:

1. *Economy:*

Well-planned NIP will enable more infrastructure projects, power business, create jobs, improve ease of living, and provide equitable access to infrastructure for all, thereby making growth more inclusive.

2. *Government:*

Well-developed infrastructure enhances the level of economic activity, creates additional fiscal space by improving the government's revenue base, and ensures the quality of expenditure focused on productive areas.

3. *Developers:*

Provides better-prepared projects, reduces aggressive bids/failure in project delivery, and ensures enhanced access to sources of finance as a result of increased investor confidence.

4. *Banks/financial institutions/investors:*

Builds investor confidence as identified projects are better prepared, exposures less likely to suffer stress given active project monitoring by the competent authority, thereby ensuring better returns.

Sources:

(a) *National Infrastructure Pipeline, Report of the Task Force, Department of Economic Affairs, Ministry of Finance, Government of India, Volume I (*https://dea.gov.in/sites/default/files/Report%20of%20the%20Task%20Force%20National%20Infrastructure%20Pipeline%20%28NIP%29%20-%20volume-i_1.pdf*) accessed on 25 June 2021.*
(b) *Department of Economic affairs, Invest India, Government of India (*https://indiainvestmentgrid.gov.in/national-infrastructure-pipeline*) accessed on 25 June 2021.*
(c) *The Indian Construction Industry: A Brief Summary by Kartikeya Saigal, Invest India, 17 November 2020 (*https://www.investindia.gov.in/team-india-blogs/indian-construction-industry-brief-summary*)*

1.1 Project and Project Management

What is a Project?

A project is an undertaking, typically a temporary endeavor to achieve predefined objectives. It is constituted from a set of resource-consuming and interconnected activities whose completion leads to the achievement of the project's objective or goal. Project Management Institute (PMI) adopts the notion that projects deliver exclusive, non-repetitive or unique products, services, or outcomes. The key constraints under which projects need to be executed include scope, cost, schedule, quality, resource, and risk.

Why are Projects Complex?

The projects tend to be complex. A project usually consists of multiple activities that have some precedence relations among themselves. These activities are to be performed by a diverse group of individuals with varying expertise and resources. The project consumes resources such as time, money, equipment, and raw materials each having its constraints.

PMBOK stands for Project Management Body of Knowledge. The sixth edition of "A Guide to the Project Management Body of Knowledge" published by Project Management Institute, Inc., 2017[1] contains 10 knowledge areas and 49 project

[1] Project Management Institute. (2017). A guide to the Project Management Body of Knowledge (PMBOK guide) (6th ed.). Project Management Institute.

management processes, distributed among five process groups: initiating, planning, executing, monitoring and controlling, and closing.

The presence of multiple decision points, imbalance in the power structure, and the influence of critical attributes of ecosystem such as social, environmental, and political add to the project's complexity and significantly impact project execution and completion.

The critical concern for today's project managers is how can they effectively manage six conflicting constraints of scope, cost, schedule, quality, resource, and risk) and execute the projects whose requirements are constantly fluctuating?

1.2 Classification of the Projects

It is essential to classify the projects and understand the relative importance of each. Table 1.1 provides a broader classification of various types of projects with suitable examples. This helps to understand the scope and complexity of the projects at different levels.

1.3 Difference Between Project and Project Management

There are some key differences between project and project management. A project is a temporary endeavor designed to deliver a unique product/service with a defined beginning and end, undertaken to meet individual goals with a prime objective to add value. Integrating three domains such as project management tools, techniques, people, and systems leads to project management. Project management is well-recognized as a strategic competency that enables organizations to:

- Tie project results to business goals
- Compete more effectively in their markets
- Sustain the organization
- Respond to the impact of business environment changes on projects.

The scope of project management activities includes planning, scheduling, and controlling.

- Planning: goal setting, defining the project, team organization
- Scheduling: relates people, money, and supplies to specific activities, identifies interrelationships among activities
- Controlling: monitoring resources, costs, quality, and budgets; based on feedback updates, plan and mobilize resources to meet time and cost demands.

Effective project management is influenced by the factors such as flexibility, autonomy, control, power, resource allocation, communication, and negotiation. Appropriate integration of these factors provides a management system to support

Table 1.1 Types of projects

Personal Project Examples: 1. Exam preparation 2. Writing a book 3. Organizing Picnics 5. Organizing Marriage functions	**Local Neighborhood Project Examples**: 1. University function 2. Cleanliness drive 3. Construction of toilets 4. Tree plantation drives 5. Jogging track development	**Organizational Project Examples**: 1. Construction of buildings or highways 2. Planning and launching of new product 3. Disposal of dead stock 4. Completing a financial audit 5. Refinery turnaround	**National Project Examples**: 1. Launching a new satellite 2. Poverty alleviation 3. Literacy campaign 4. COVID-19 lockdown 5. Preparation of annual budget	**Global Project Examples**: 1. Polio Eradication 2. Space exploration 3. Conducting world trade 4. COVID-19 vaccination 5. Wildlife preservation

Table 1.2 Project versus project management

Project	Project management
A project is a temporary endeavor designed to produce a unique product/service with a defined beginning and end to meet unique goals and objectives and add value	Project management is teamwork to achieve specific goals and meet specific success criteria at the specified time using initiation, planning, execution, controlling, and termination of the project
Different types of projects include: • Personal-level projects • Organizational projects • National projects • Global projects Examples: • Construction • Health care • Engineering • IT • Energy	Different methods of project management include: • Agile—for extreme flexibility and speed • Traditional • Critical path analysis • Adaptive • Program evaluation and review technique • Rational unified process • Scrum • Six sigma
A project fails due to the following reasons: • Improper alignment of goals with organizational goals • Unrealistic expectations • Lack of executive sponsorship • Lack of proper management skills and techniques	A project manager might fail due to the following reasons: • Disillusionment among people • Chaos • Search for the guilty • Punishment of the innocent • Promotion of non-participants

project needs more effectively and efficiently. The key differences between "project" and "project management" are narrated in Table 1.2.

1.4 Time–Cost–Performance (T-C-P) Relationship in Project Management

Time (T) is the available time to complete the project, cost (C) represents the funds or resources available, and performance (P) represents the quality that the project must accomplish to be a success. It is challenging to maintain the proper balance among these three components, typically recognized as project management trade-offs, as shown in Fig. 1.1. The conflicts arising in managing the exchanges in an uncertain business environment are characterized by longer lead time of the suppliers, delays in government/regulatory approvals (such as environmental clearances), land acquisition, mismatch in subcontractor's schedules, inflations in material, and labor prices, etc.

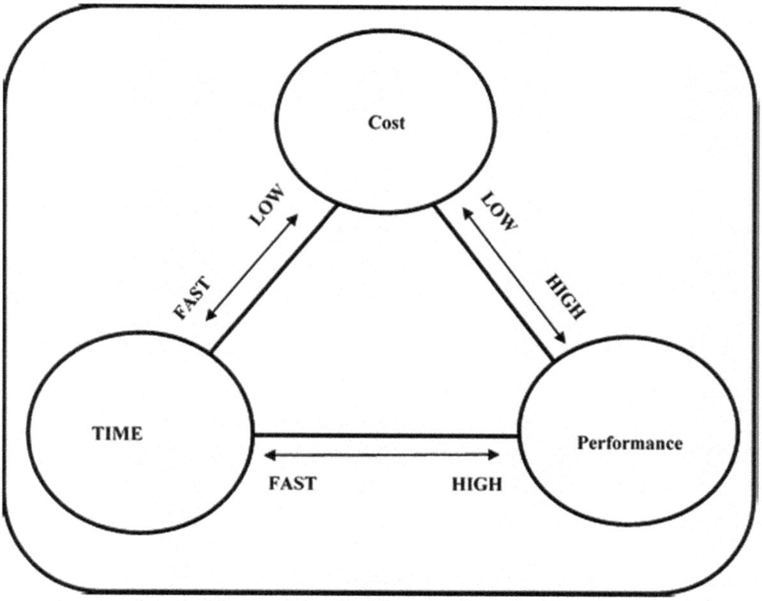

Fig. 1.1 Project management trade-offs

The T-C-P triangle in Fig. 1.1 illustrates that if one of these factors is fixed, the other two will vary inversely to each other. For example, if time is fixed, the quality of the product will be subject to the funds or resources available. Therefore, it is extremely demanding for the project managers to ensure the timely completion of projects within budget to deliver the needed scope and quality. The project complexity for a construction and a pharmaceutical R&D project is explained in Table 1.3.

Project Management in Practice

Project Complexities in Metro Rail Projects in India.

Metro railways are more complex and complicated engineering projects compared to the construction of roads and airports. Metro rail projects must be executed considering various environmental factors, rock formations, the presence of building foundations, other railway lines, statues, climate, socio-economic factors, etc. These projects are considered as a solution for most of the traffic- and environmental pollution-related problems which are affecting the major cities throughout the world. With metro rail construction activities being undertaken in a big way in India, key focus areas which add to the complexity of projects include:

- Land/Property Acquisition and Resettlement and Rehabilitation (R&R)-Related Issues

Table 1.3 Complexity of project in a typical construction and pharmaceutical R&D project

Project complexity criteria	Construction project	Pharmaceutical R&D project
Investment	Very high investment	High investment
Rate of return	High rate of return	Very high rate of return
Risk	Low risk	High risk
Payback	Faster payback	Slow payback
Likely profit	High profit	Very high profit
Similarity to existing business	Highly similar	No similarity
Expected life	Very long project life	Very long project life
Flexibility	Not flexible	Highly flexible
Environmental impact	High impact	Depends on the research domain
Competition	Very high competition	High competition
	T-C-P Implications:	
What are the key challenges?	The construction industry is exposed to rising material and labor costs, increased competition, and shrinking profit margins. The emphasis is on reducing the time and cost of the project. The norms for quality are well-established in the market	R&D projects aim to deliver new products or services for the organization to enhance its competitiveness. This expects R&D project managers to emphasize upon the quality and timely delivery of the project. Therefore, R&D cost is considered less important as it is perceived as a sunk cost

- Loss of trees/green cover
- Noise pollution and vibration issues
- Accidents during construction phase
- Traffic Issues during construction phase.

1.5 Project Life Cycle Management

A project can be divided into five phases: initiation, planning and scheduling, execution, monitoring and control, and termination. This helps the project manager and team to structure their efforts and resource utilization for the various milestones of the project management. Figure 1.2 illustrates the various phases of the project life cycle as per PMBOK (6th edition). It is important to note that the cost and staffing-level requirements increase during the first three phases of the project, namely "starting the project," "organizing and preparing," and "carrying out the work." Subsequently, this becomes steady and goes down in the final phase of the project life cycle—"closing the project." The cost of change increases as the project advances in its life

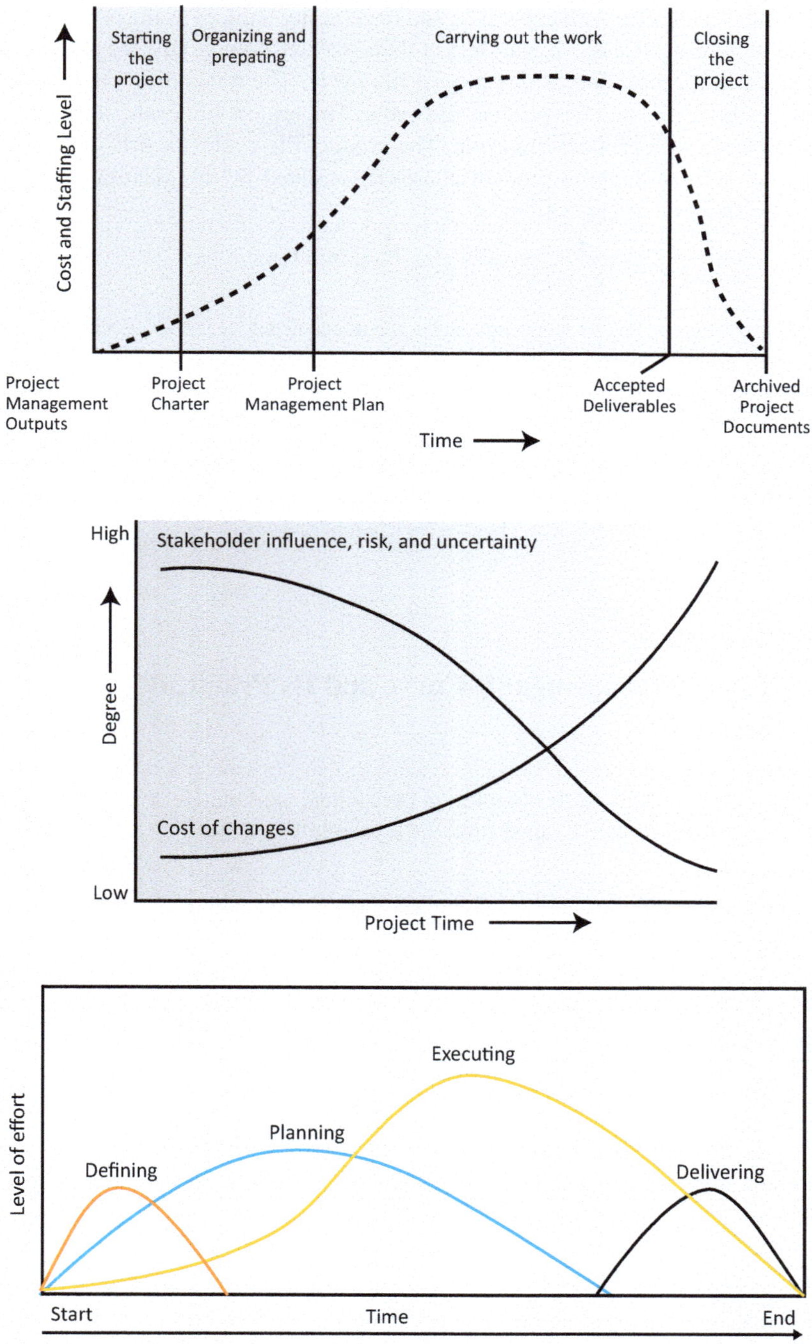

Fig. 1.2 Project life cycle (*Adapted from : PMBOK, 6th edition*)

cycle while stakeholder influence, risk, and uncertainty reduces as the project moves into a more advanced stage. Regardless of their size and complexity, the projects can be mapped to the structure of this project life cycle. There is a difference between the project life cycle and the product life cycle. The project life cycle occurs in one or more phases of a product life cycle. The product life cycle consists of generally sequential, non-overlapping product phases determined by the manufacturing and control needs of the organization.

The project life cycles can be broadly classified into three types:

Type 1	Predictive life cycles	Project scope, time and cost required to deliver that scope are determined as early as possible
Type 2	Iterative and incremental life cycle	• Project phases intentionally repeat one or more project activities as the team's understanding of the product increases • Iterations develop the product through a series of repeated cycles while increments successively add to the functionally of the product
Type 3	Adaptive life cycle	Intended to respond to high levels of change and ongoing stakeholder involvement. They are fixed in time and costs

1.6 Project Management S-Curve and Its Practical Relevance

S-curve displays the project's completion percentage, cumulative costs, labor hours, or other quantities plotted against time (as shown in Fig. 1.3).

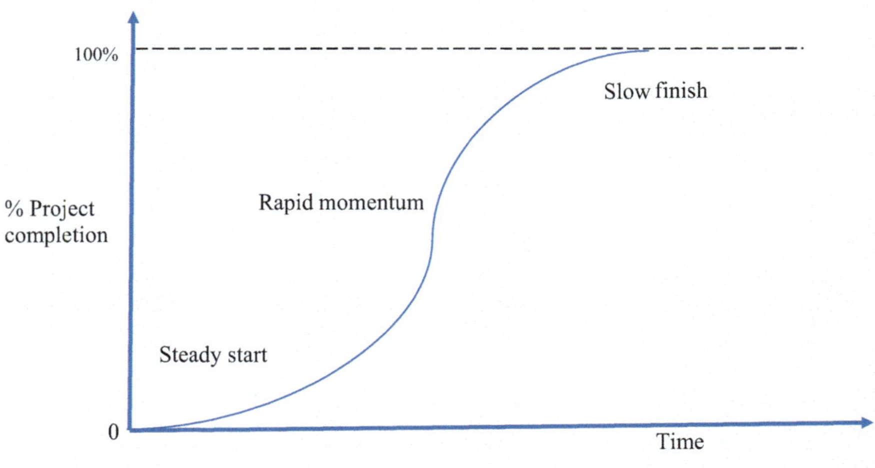

Fig. 1.3 S-curve

Table 1.4 Comparison of S-curve for projects: construction project, R&D, ERP development project, HIV control project

Project	Starting phase (slow)	Middle phase (quick momentum)	Finishing phase (slow)
Construction project	Planning, design, feasibility check	Construction	Inspection of structure, safety parameters checking
R&D	Idea selection, roadmap and timeline creation for the project, research and brainstorming a feasibility check	Developing the selected product or service	Evaluation of the project in terms of profit
ERP development	Finding the stakeholders, planning, design	Integration of the business process using software	Efficiency checking
HIV control	Strategy, planning, medicine collection	Awareness program (advertisements, etc.), medicine distribution	Evaluation of the success of the drive

The S-shape of the curve is flatter at the beginning and the end and steeper in the middle. The opening represents a steady, deliberate, slow but accelerating start, while the end shows a slowdown as the work is finishing. S-curves are great graphical project management tools for monitoring, planning, analyzing, controlling, and forecasting project status, progress, and performance. Table 1.4 presents a comparison of S-curve for various types of projects.

1.7 Role of Project Manager, Project Leader, and Project Engineer

Successful project management demands the effective contribution of three key entities—project managers, project leaders, and project engineers. The role of a project manager is highly demanding as it spreads over the creation of a plan to execution, validation, and evaluation of the project. An effective project manager is expected to achieve the stipulated project milestones within the allocated budget and specified quality standards. In addition, a project manager must ensure that the people assigned to the project receive motivation, direction, and information for executing various tasks. Finally, a project leader looks after the overall coordination among people, while a project engineer monitors and controls various engineering and technical activities. Table 1.5 narrates the key differences between the project manager, project leader, and project engineer.

Table 1.5 Difference between the project manager, project leader, and project engineer

Project manager	Project leader	Project engineer
A project manager's role is successful planning and ensuring that the team performs at their best	A team leader is responsible for successful coordination among people to complete the different tasks	A project engineer's responsibility includes schedule preparation, preplanning, and resource forecasting for engineering and other technical activities related to the project
Space missile development project		
Project manager	*Project leader*	*Project engineer*
• Establishes project objectives and specifications • Prepares project implementation plan (PIP) • Prepares WBS, master schedule budget, earned value plan, etc.	• Authorizes work and allocation of resources • Implements risk control, review of risks, and evaluation of impacts • Decides on corrective actions	• Carries out technical assessments and feasibility studies • Hardware/software design and development • Launch operations • On-orbit verification and tests Mission operations
Computer software development project		
Project manager	*Project leader*	*Project engineer*
• Defines the requirement of the project • Acts as a link between the client, his team, and the supervisor • Lays out a blueprint of the whole project, including project scope and parameters • Allots timeframes for each task • Budget allocation and cost estimates	• Team builder—making teamwork together requires a keen understanding of people • Delegation of tasks • Autonomy—provides the team with clear goals and lets them find the best way to achieve them • Motivates the team • Guides the team through the problem-solving process	• Reviews current systems • Presents ideas for system improvements • Works closely with the analyst's designer and staff • Produces detailed specifications and writes program codes • Tests the product in real controlled situations before going live • Maintains the system once the project has started
Equipment and system installation project		
Project manager	*Project leader*	*Project engineer*
• Identifies and reviews program, space, and scope with the end user for approval • Initiates purchase order • Sets up and leads a series of the team to review design drawings with clients • Creates a project budget and tracks project costs	• Ensures the project gets completed on time and within budget • Prioritizes and delegates the task to team members to maximize efficiency • Analyzes equipment and employee performance	• Oversees product blueprint design—CAD software • Designs and implements policies and procedures • Observes and refines design and production based on feedback

Table 1.6 Illustrative example of critical characteristics of project manager for a football tournament

Characteristic No.	What is essential for management of a typical "Football team"?
Characteristic 1	Appropriate team composition, without which a football team cannot execute strategy effectively
Characteristic 2	Ensuring availability of backup players in case of injuries
Characteristic 3	Prioritizing selection of team players to maximize the chances of winning
Characteristic 4	Planning of intelligent strategies to defeat opponents
Characteristic 5	Ability to bring out the best style of play in each player and deserve and command respect from the team

1.8 Key Characteristics for Becoming a Successful Project Manager

The most crucial role of a project manager is to create an environment where the team can work productively. To achieve this, a project manager must ensure effective communication with the team and stakeholders throughout the project, focus on solving problems, and develop contingency plans to accommodate failures. The key characteristics of a successful project manager include:

Characteristic 1	Adept in building team skills and in tactically executing a vision
Characteristic 2	Demonstrates an ability to encounter unexpected events and come up with solutions on the spot
Characteristic 3	Believes that collaboration, negotiation, and problem-solving are the keys to a successful outcome
Characteristic 4	Adept at problem-solving and plotting the steps needed to achieve an outcome
Characteristic 5	Utilizes each person's skills to their fullest extent

The relevance of these five characteristics is illustrated using the example of a coach to a football team which is participating in a tournament project in Table 1.6.

1.9 Project Management in Services Versus Manufacturing Sector

The deployment of systematic project management practices in the various sectors, such as agriculture, IT, automobile, defense, energy, infrastructure, and real estate, enhances organizational productivity and leads to savings by lowering project execution costs. It is a process of managing change irrespective of the nature of the industry (manufacturing or service) to accomplish goals and objectives while respecting real-life challenges and constraints. It enables organizations to introduce or improve

Table 1.7 Project management in service versus manufacturing sector

Criteria	Service sector	Manufacturing sector
Tangibility	The service sector provides intangible services which cannot be quantified UUExampleUU: banking, hospitality, advertising, and consultancy	The manufacturing sector provides tangible products which users can hold and see UUExampleUU: smartphones, soaps, notebooks
Standardization	Have more opportunities to offer customizations and client-centric services UUExampleUU: hairdressers customize the hairstyle for the facial features of the customer	Manufacturers have a standardized way of producing goods. Goods are delivered in mass quantities UUExampleUU: Production of soft drinks
Production environment	Plan the environment as to how it affects the customer experience UUExampleUU: Layout of furnishing, color, and sound to enhance customer experience	Focused more on manufacturing layout. It can be process-focused or product-focused UUExampleUU: Manufacturing of tube lights
Operations management	Schedule workers to handle customer demands. Must coach and train employees to provide proper customer care UUExampleUU: Initial training of recruits	Overseeing activities to produce goods from raw material, product to produce, and output quality UUExampleUU: Storage and flow of material across the plant

new or existing products and services. Project management principles are holistic and applied to the projects executed in both manufacturing and service industries. However, it is necessary to recognize the critical differences between the service sector and manufacturing sector, as listed in Table 1.7, to appropriately decide the project's scope, structure, and execution policy. Services are intangible and have dyadic relationships in executing the customer requirements.

1.10 Project Organizations: Key Characteristics

A project organization is a structure that facilitates the coordination and implementation of project activities. The structure helps to specify the relationships among members of the project management team and the relationships with the external environment. There are three different forms of project management organizations: functional, pure project, and matrix.

1. **Functional**: The company's structure is based on the division of workers according to their specialization. Once the project begins operation, the various components of the project are executed by the functional units. The key advantages of this structure are: (i) the use of personnel with greater flexibility; (ii) the ability of experts to share their knowledge and experience; (iii) the continuity

of the project in case any member leaves, etc. The disadvantages include (i) clients are not the focus; (ii) the manager may not have sufficient authority; and (iii) response is slow. For example, a mid-sized manufacturing company could be divided into functional areas, including administration, marketing and sales, and production.

2. **Pure project**: Refers to the creation of an independent project team within the company; the team's management is separated from the parent organization's other units. The team consists of its technical staff and management, and the enterprise allocates the necessary resources to the project team. The key advantages of this structure are: (1) the project manager is given complete authority for the execution of the project; (2) effective communication among project team members; (3) the project team has its own identity. The disadvantages include: (i) overstaffing or duplication of efforts; (ii) organizational inconsistency; and (iii) project team members lack business continuity and security. For example, a national hotel chain could have regional divisions to retain the necessary facilities and technical skills for executing various projects. Workers from multiple specialties work together, along with administrative specialists who support them.
3. **Matrix**: A hybrid structural form: it loads a project management structure on the functional structure. The main advantages of this structure are: (i) project is the focus, and the manager will coordinate and integrate the activities between different units; (ii) clients receive prompt responses; and (iii) project members can return to their respective functions on completion of the project. The disadvantages of this structure include: (i) sharing of resources and personnel may lead to conflict; (ii) delay in decision making; and (iii) project team members need to work under dual authority, which can induce a split in loyalty and may hamper their performance. For example, a software company aiming to develop a new application for use by school children might pull together a team that includes engineers and programmers from specific functional areas and artists and market specialists from other departments.

1.11 PMBOK Knowledge Areas

PMBOK stands for Project Management Body of Knowledge. It documents an entire collection of tools, techniques, methodologies, terminologies, and best practices for project managers. It is developed and maintained by the Project Management Institute (PMI). PMBOK outlines a framework and best practices that project managers can utilize to manage projects successfully. It is mainly focused on waterfall project management methodologies. PMBOK contains 10 knowledge areas, 5 process groups, and 49 processes that project managers need to be familiar with. A comprehensive review of the 10 PMBOK knowledge areas is given in Table 1.8.

Table 1.8 PMBOK knowledge areas

PMBOK knowledge area	Relevance	Processes	Which chapter of the book deals with this knowledge area?
Integration management	Integration management is where the project comes together. All the moving parts are combined and unified. This is where all the outputs of other processes are combined to form the project deliverables	1. Develop project charter 2. Develop project management plan 3. Direct and manage project work 4. Manage project knowledge 5. Monitor and control project work 6. Perform integrated change control 7. Close project or phase	Chap. 4 Chap. 10 Chap. 16
Scope management	Project scope includes all of the work required to complete the project. Scope management involves defining what will be included in the project scope and what will not be, and actively managing scope throughout the entire project to prevent scope creep	1. Plan scope management 2. Collect requirements 3. Define scope 4. Create WBS 5. Validate scope 6. Control scope	This Chapter Chap. 2 Chap. 3
Schedule management	The project schedule tells you when each of the project activities need to start and finish. Schedule management includes the creation of the project schedule, ensuring the project team follows the schedule, and managing changes to it	1. Plan schedule management 2. Define activities 3. Sequence activities 4. Estimate activity durations 5. Develop schedule 6. Control schedule	Chap. 4 Chap. 5 Chap. 6 Chap. 7
Cost management	Every project manager has a budget to adhere to. Cost management involves estimating project activity costs, creating the project budget, tracking project expenditures, and keeping overruns to a minimum	1. Plan cost management 2. Estimate costs 3. Determine budget 4. Control costs	Chap. 8

(continued)

Table 1.8 (continued)

PMBOK knowledge area	Relevance	Processes	Which chapter of the book deals with this knowledge area?
Quality management	Quality refers to how closely the project deliverables adheres to the stakeholders' requirements. Quality management includes the processes for creating, managing, and executing quality policies. The goal of this knowledge area is to ensure high quality on the project	1. Plan quality management 2. Manage quality 3. Control quality	Chap. 13
Resources management	Resource refers to the people and materials that you will need to complete the project. Resource management includes the processes to identify the types of resources needed, going out and acquiring the required resources, and managing them throughout the project life cycle	1. Plan resource management 2. Estimate activity resources 3. Acquire resources 4. Develop team 5. Manage team 6. Control resources	Chap. 9
Communications management	Project team members communicate with each other to ordinate on project tasks and discuss project strategies. Communications management includes the processes that ensure the planning, distribution, storage, and management of project information	1. Plan communications management 2. Manage communications 3. Monitor communications	Chap. 15
Risk management	Risk is an uncertain event that may or may not happen in the future. A risk can be positive or negative. Risk management includes the processes of identifying and analyzing new risks as well as planning and implementing risk responses	1. Plan risk management 2. Identify risks 3. Perform qualitative risk analysis 4. Perform quantitative risk analysis 5. Plan risk responses 6. Implement risk responses 7. Monitor risks	Chap. 11
Procurement management	When a project team cannot produce something in-house, they have to obtain a third-party vendor to do it for them. The processes required to obtain and manage the vendors are called procurement management	1. Plan procurement management 2. Conduct procurements 3. Control procurements	Chap. 12

(continued)

Table 1.8 (continued)

PMBOK knowledge area	Relevance	Processes	Which chapter of the book deals with this knowledge area?
Stakeholders management	A stakeholder is anyone who has an interest (or stake) in your project. Stakeholder management includes the processes required to identify your project stakeholders, analyze their power and impact, and develop appropriate strategies to manage them effectively	1. Identify stakeholders 2. Plan stakeholder engagement 3. Manage stakeholder engagement 4. Monitor stakeholder engagement	Chap. 14

Source (PMBOK 6th Edition)

1.12 Key PMBOK Terminologies

Terminology	Relevance
Project constraints	Scope, cost, schedule, quality, resource, risk
Program management	Group of related projects managed in a coordinated way to obtain the benefits and control not available from managing them individually
Portfolio management	Collection of projects or programs and other work grouped together to facilitate effective management of that work to meet strategic business objectives
Project management office	Is an organizational body or entity assigned various responsibilities related to the centralized and coordinated management of those projects under its domain
Project manager	Is a person assigned by the performing organization to achieve the project objectives
Project integration management	Includes the processes and activities needed to identify, define, combine, unify, and coordinate the various processes and project management activities within the project management process groups
Project charter	Defines the high-level project boundaries of the project. Project manager uses the project charter as the starting point for the initial planning throughout the initiation process group
Project management plan	The project management plan integrates and consolidates all of the subsidiary management plans and baselines from the planning processes
Schedule forecasts	The schedule forecasts are derived from progress against the schedule baseline and computed time estimate to complete (ETC). This is typically expressed in terms of schedule variance (SV) and schedule performance index (SPI)
Cost forecasts	The cost forecasts are derived from progress against the cost baseline and computed estimates to complete (ETC). This is typically expressed in terms of cost variance (CV) and cost performance index (CPI)
Project documents	Assumption log, basis of estimates, schedule and cost forecasts, issue log, lessons learned register, milestone list, quality reports, risk register. Risk report
Configuration control	Is focused on the specification of both the deliverables and the processes
Change control	Is focused on identifying, documenting and controlling changes to the project and the product baselines
Scope management	Includes processes required to ensure that the project will include all the work required and only the work required to complete the project successfully
Project scope statement	It details of the project's deliverables and the work required to create those deliverables. It provides a common understanding of the project scope among project stakeholders. It includes project exclusions, assumptions, and constraints
Requirements traceability matrix	Linking requirements to their origin and traces them throughout the project life cycle

(continued)

(continued)

Terminology	Relevance
Project scope management plan	Scope management plan specifies how to create the WBS from detailed project scope statement and how the WBS will be maintained and approved
Scope baseline	Scope baseline is a component of the project management plan and the approved version includes: project scope statement, WBS, WBS dictionary, work package, planning package
Work performance data	Includes the number of change requests received, the number of requests accepted or the number of deliverables completed
Organizational process assets	Organization repository that can influence control scope process
Work performance information	Includes correlated and contextualized information on how the project scope is performing compared to the scope baseline
Project schedule management	Focuses on the processes that are used to help ensure the timely completion of the project
Schedule management plan	Establishes the criteria and the activities for developing, monitoring and controlling schedule
Schedule network analysis	Technique that generates the project schedule
Critical path method	Sequence of activities which represents the longest path in the project which determines the shortest possible project duration
Leads and lags	Leads and lags are refinements applied during network analysis to develop a viable schedule
Schedule compression	Shorten the schedule duration without reducing the project scope
Agile release planning	High-level timeline of the release schedule based on product roadmap and product vision
Project cost management	Process involved in planning, estimation, budgeting, financing, funding, managing, and controlling costs so that the project can be completed within the approved budget
Cost management plan	Component of project management plan and describes how the project costs will be planned, structured, and controlled
Cost estimation	Is the process of developing an approximation of the monetary resources needed to complete the project activities
Three point estimating	Based on PERT using three estimates to define an approximate range for an activity's cost
Bottom up estimating	Detailed cost is summarized or rolled up to higher levels for subsequent reporting and tracking
Project management information system	Can include spreadsheets, simulation software, and statistical analysis tools to assist with cost estimating
Cost aggregation	Rolling up cost estimates to cost control accounts
Cost baseline	An authorized time-phased budget at completion (BAC)

(continued)

(continued)

Terminology	Relevance
Change requests	Analysis of project performance may result in a change request to the cost baseline or other components of project management plan
Project quality management	Describes the process involved to determine quality policies, objectives and responsibilities so that the project will satisfy needs for which it was undertaken
Benchmarking	Comparing actual or planned project practices to those of comparable projects to identify best practices, generate ideas for improvement, and provide a basis for measuring performance
Cost of quality	All costs incurred over the life of the product by investment in preventing nonconformance to requirements, appraising the product or service for conformance to requirements, and failing to meet requirements (rework)
Quality metrics	An operational definition that describes, in very specific terms, a project or product attribute and how the quality control process will measure it, e.g., on-time performance, defect frequency, etc.
Resource management	Includes the processes to identify, acquire, and manage the resources needed for the successful completion of the project
Resource calendars	Identifies the working days, shifts, start, and end of normal business hours, weekends, and public holidays when each specific resource is available. It also specify when and for how long identified team and physical resources will be available during the project
Communication management	Describes the process involved to ensure timely and appropriate generation, collection, distribution, storage, retrieval, and ultimate disposition of project information
5 Cs of communication management	1. Correct grammar and spelling 2. Concise expression and elimination of excess words 3. Clear purpose and expression directed to the needs of the reader 4. Coherent logical flow of ideas 5. Controlling flow of words and ideas
Risk management	Describes the process for conducting risk management planning, identification, analysis, response planning, response implementation, and monitoring risks of a project
Probability and impact matrix	Risks are prioritized according to their potential implications for having an effect on the project's objectives
Revised stakeholders' tolerances	Stakeholders' tolerances, as they apply to the specific project, may be revised in the plan risk management process
Risk reporting formats	Defines how the outcomes of the risk management processes will be documented, analyzed, and communicated
Risk categories	Grouping individual project risks to structure risk categories with risk breakdown structure (RBS)
Risk register	The primary outputs from identify risks, the individual risks, are the initial entries into the risk register. The risk register ultimately contains the outcomes of the other risk management processes as they are conducted, resulting in an increase in the level and type of information contained in the risk register over time

(continued)

(continued)

Terminology	Relevance
Procurement management	It includes the processes necessary to purchase or acquire products, services, or results needed from outside the project team. It also administer agreements such as contracts, purchase orders, memoranda of agreements (MOAs), or internal service level agreements (SLAs)
Requirements documentation	It may include: technical requirements that the seller is required to satisfy and requirements with contractual and legal implications that may include health, safety, security, performance, environmental, insurance, intellectual property rights, equal employment opportunity, licenses, permits, and other nontechnical requirements
Requirements traceability matrix	Links product requirements from their origin to the deliverables that satisfy them
Make-or-buy decision	Make-or-buy decisions document the conclusions reached regarding what project products, services, or results will be acquired from outside the project organization, or will be performed internally by the project team
Stakeholder management	Includes the processes required to identify the people, groups, or organizations that could impact or be impacted by the project, to analyze stakeholder expectations and their impact on the project, and to develop appropriate management strategies for effectively engaging stakeholders in project decisions and execution

Project Management in Practice

Dedicated Freight Corridors (DFCs).

Rail transport provides an important mode of conveyance for both goods and people. Transport of freight over rail (especially for bulk commodities such as coal and iron ore) can be both more energy- and cost-efficient for long-distance transport. In India, Dedicated Freight Corridors (DFCs) were planned to be built with the Golden Quadrilateral route to enhance the growth of the economy. Dedicated Freight Corridor Corporation of India Limited (DFCCIL) will decongest the already saturated road network and promote shifting of freight transport to more efficient rail transport. This shift is expected to significantly reduce greenhouse gas (GHG) emissions in the transport sector in India.

Key Project Management Challenges:

Time: The completion of the DFCCIL project sanctioned in 2006 was expected by 2011. The final location survey and detailed project report assessment were done in 2014, eight years after getting the approval. The target completion was first shifted to 2016–17, then to 2017–18. The high number of affected patches caused an additional time delay.

Cost: The conditions of the Japan International Cooperation Agency (JICA) loan for Western Dedicated Freight Corridor (WDFC) (constituting 80% of DFCs costs) required that 30% of the JICA funding be used to import equipment and goods from Japan, and all contracts for DFCL must have a Japanese firm.

Performance: With an inclination toward using renewable resources in the future, most of its traffic was expected for National Thermal Power Corporation Ltd. (NTPC). Double line compared to two single lines on the feeder, and the route lines were not as per DFCL standard. Hence, this route cannot be operated as a streamlined double line.

Source:

Dedicated Freight Corridor: Current Challenges by G Raghuram and Apoorva Verma, World Conference on Transport Research - WCTR 2019, Mumbai, 26–31 May 2019, Transportation Research Procedia (https://www.iimb.ac.in/sites/default/files/inline-files/Transportation%20Research%20Procedia_WCTR.pdf).

Summary

A project in a general sense is basically an undertaking or endeavor to achieve some objective or goal. It is constituted of a set of resource-consuming and interconnected activities whose completion leads to the achievement of the project's objective or goal.

Projects tend to be complex in nature. A project usually consists of multiple activities that have some precedence relations among themselves. These activities have to be done by a group of people by using the resources available to them.

Time (T) is the available time to complete the project, cost (C) represents the funds or resources available, and performance (P) represents the quality that the project must accomplish to be a success.

S-curve is a display of the percentage of completion of the project, cumulative costs, labor hours, or other quantities plotted against time.

The role of a project manager is challenging as it spans over the activities such as creation of a plan to execution, validation, and evaluation of the project.

Questions for Discussion

1. What is a project? Why are projects complex? Classify the projects and give at least five examples under each category.

2. What is the difference between project and project management? Provide your response in a tabular format.
3. Can we use project management for all the forms of organization and situation? Enlist and explain the key criteria with suitable examples to be used for the use of project management.
4. What is the T-C-P relationship in a project? How do you compare the complexity of a project in a typical construction and R&D project? Provide your comparison against a set of criteria in a tabular format. Also, comment on the T-C-P relationship in both construction and R&D projects.
5. Why is the role of a project manager challenging? What is the difference between the project manager, project leader, and project engineer? Explain the role of each one for at least three different kinds of projects such as space missile development project, computer software development project, and equipment and system installation project.
6. What is S-curve? Explain the nature and practical relevance of the S-curve? Compare and explain the S-curve for at least four projects: construction project, R&D, ERP development project, HIV removal project, and setting up a new plant project.
7. What are the key characteristics of becoming a successful project manager? Why each one of these is essential. Explain with a suitable project example—why a specific feature is vital for a particular project?
8. How is project management different in the service sector compared to the manufacturing industry? First, compare the onset of critical criteria in a tabular format. Then, explain with a suitable real-life example.
9. What are the different types of project organizations? Explain each one with a suitable example.
10. How do you compare the project complexities in three projects such as R&D, information technology (IT), and infrastructure? First, compare the projects on a set of criteria in a tabular format. Then, prioritize the challenges involved based on a group consensus.

Web-based Exercise

Study the details of the Hyderabad Metro Rail Project. List down the complexities and issues involved in handling this public–private partnership project (PPP).

Group Project

A group of five students should study the site of a construction project and interact with the project leader, project manager, project sponsor, contractors, and subcontractors. Based on their interaction, a group should highlight the T-C-P trade-offs involved in executing this project and how the project manager is trying to resolve the key conflicts in executing this real-life project.

Chapter 2
Project Organization

Critical Questions

What is the importance of organizational structure in project management?
What are the different types of organizational structure and what are their key characteristics?
What is the role of organizational structure in conflict management and negotiation?

Project Management in Practice

Some perspectives on development projects in Higher Technical Institutes *by* Professor S. G. Deshmukh, Professor, Indian Institute of Technology Delhi

Infrastructure development projects in publicly funded higher technical institutes (HTIs) in India typically involve projects such as the construction of lecture complexes, housing for faculty/staff, and sports and recreation facilities. These projects are characterized by (i) tight timelines, (ii) government funding, (iii) layered decision-making (organized into Building & Works Committee, Finance Committee, Board of Governance, etc.), and (iv) manpower shortages of the Institute at the site. Typically, in such projects, the project management agency is from the government (such as the Central Public Works Department).

The following points are relevant from a project management perspective:

(a) Construction projects are vital in HTIs. They act as facilitators for providing ambiance and an ecosystem for a conducive learning environment. The coordination between the architect and the construction agency is crucial in translating the institute's vision into action.

J. J. Thakkar, *Project Management*, Management and Industrial Engineering,
https://doi.org/10.1007/978-981-15-3695-3_2

(b) Communication and coordination between the various stakeholders are two critical success factors for executing these projects on time, within the budget and with quality.
(c) The role of contractors and subcontractors cannot be undermined. Often, factors such as local culture, conditions, and liaison with other local agencies play an essential role in getting the deliverables on time.
(d) A deep understanding of the overall picture is very much necessary for these projects. Understanding formal tools of project management (like CPM/PERT, project crashing, resource leveling) helps a lot in translating the bigger picture into actual deliverables. However, one must also understand that there is a difference between a theoretical concept and actual practice. The understanding becomes mature over a period and with exposure to harsh realities on the ground.
(e) There are various challenges at every stage of the project—be it in project initiation, project execution, or project closure. These challenges need to be countered by having a constant interaction between various stakeholders. This helps a lot in having a shared perception about the bigger picture.
(f) Understanding and appreciating "contract management" is very crucial for the construction project.
(g) In a publicly funded HTI, the role of audit is also very important. The internal and external audit provides an opportunity to carefully look at the financial resources required for the completion of the projects. Therefore, documentation is essential at every stage of the project.
(h) Managing "people issues" becomes a critical factor in the successful execution of the project. The project team needs to evolve its mechanisms to solve and resolve "people issues." Getting work done from people requires interpersonal and strong communication skills.

2.1 Organizational Structure: Issues and Characteristics

Organizational Structure

The organization hierarchy is defined through the organizational structure. It recognizes each function and where they report within the organization. This helps to determine how the organization works and achieve its intended objectives to allow for future development. In addition, it is a visual representation of an organization to describe the roles and responsibilities of the employee and the decision-making hierarchy across the organization.

What Is the Relevance and Importance of Structure in the Organization for Decision-Making?

An organizational structure is necessary to grow a business profitably. A well-designed organizational structure ensures an optimal level of resource allocation at various levels in the organization. In addition, it improves the flow of information and hence coordination across the organization. The key advantages of a well-designed organizational structure are listed below.

1. The company's success largely depends upon the flow of information. The organization structure helps the individuals and departments better coordinate their efforts.
2. It helps to define work with a specific goal and deadline and facilitates faster decision-making.
3. It ensures appropriate connections across multiple business locations.
4. It brings clarity on roles and responsibilities and hence enhances employee morale and performance.
5. It develops greater clarity on organizational activities, which results in improved operating efficiency and reduced employee conflict.
6. It minimizes the possibility of duplication of work and redundant operations in the system.

The foresightedness of the top management is crucial in developing a project-based organization. The development of a project-based organization demands clarity on vision—on how the organization will function and what it is expected to achieve. A team-based coordinated approach among top management and project managers is essential in developing the organizational structure. This brings threefold advantages to an organization and improves operational efficiencies. Firstly, this can help in the appropriate mobilization of the resources; secondly, it facilitates a positive change in the people's behavior and, lastly, improves stakeholder's participation in the project. A well-designed organizational structure exhibits the following key characteristics.

1. Easy

The organizational structure should be straightforward. A simple organizational structure can help achieve the objectives and outcomes of the projects more effectively in terms of time, cost, and performance by avoiding the scope of improper communication and lack of clarity on roles.

2. Accountability and continuity

The organizational structure should be dynamic and able to evolve with emerging business requirements. Therefore, it is necessary to have an adequately synchronized design that can establish a smooth flow of instructions and ensure knowledge retention and management in the long run.

3. Clear line authority

No matter how the structure is accepted, there should be clear lines of power running either from top to bottom or in the preferred direction. This helps people at all strategic, tactical, and operational levels develop a clear understanding of their roles, responsibilities, and accountability.

4. The use of the executive body

It suggests that although a project manager delegates work to the subordinates he is ultimately responsible for completing the project. In this sense, everyone in an organization has two responsibilities: responsibility for their work and their undersigned work. Thus, the manager is responsible for both the complete work assigned to him by senior executives and accountable for the delegated work to his subordinates.

5. Delegation of authority

The concept of total authorization will only apply if there is an adequate independence at various organizational levels. Power transfer means the authority of the manager to make certain decisions. A common problem encountered by managers is that when they fail to delegate the authority in the project meetings, they invite multiple issues in the long run, such as improper planning, delayed completion of the project, dissatisfied customer or client, etc. The key characteristics of organizational structure for a typical IT company and manufacturing organization are summarized in Table 2.1.

2.2 Impact of Organizational Structure on Efficiency

1. A good organizational structure makes it easier to achieve goals by establishing coordination among the activities and employees.
2. It eliminates overlap and repeats the action. A good organizational structure clearly defines roles and responsibilities.
3. It reduces the possibility of "runarounds." A runaround happens under a lack of clarity on roles and responsibilities. Therefore, the project manager should ensure the progress of activities without any runarounds.
4. It facilitates the allocation and promotion of employees. Since organizational charts identify the roles and responsibilities of the employee, it is easier to identify the suitable position for the employee with varied skills. Thus, it facilitates skill development and promotion of the employees to appropriate positions.
5. It can help to define the amount and value of work a person is executing. This can assist in determining wages and salaries.

Table 2.1 Key characteristics of organizational structure for a typical IT company and manufacturing organization

S. No.	Key characteristics	IT company	Manufacturing organization
1	*Easy (Simple and transparent in operations)*	• Organizational structure for an IT company should be simple and not affect the service quality • Tasks assigned to the project team members should be well-defined to avoid modifications and enlargements in the project's scope • Minimize the possibility of conflict between client and project team	• Minimize the duplication and wastages of resources • Improve coordination across functions and hence enhance operational efficiency • Improve the relationships among various entities of the supply chain
2	*Accountability and continuity (clarity on flow of ownership)*	• Frequent changes in organizational structure can hamper the performance of IT projects and relationships between client and project team • Facilitate the technological advancements through appropriate accountability built into the project structure	• Manufacturing organizations evolve in their scope and functions. Therefore, timely changes in the organizational structure should adequately support this • Accountability and power need to be clearly defined at various levels
3	*Clear line authority (transparency on approval process)*	Scope modifications are inevitable, and hence, a clear line of authority helps maintain a good relationship between client and project team	Helps minimize delays in various operational decisions and hence improves coordination and operational efficiency
4	*The use of the executive body (responsibility for governance and scope control)*	Accommodating client expectations in terms of scope change is a challenge	Accommodating timely delivery within the allocated budget is a challenge
5	*Delegation of authority (empowering the employee for operational modifications to expedite the project)*	It should include decisions on necessary technological modifications, intellectual developments, and building relationships with clients	It should facilitate various operational activities by allocating necessary budget and purchasing powers

6. It facilitates communication at all levels of the organization. It can ease communication among various stakeholders like engineers, contractors, workers, transporters, government agencies, etc.
7. It provides a foundation for effective resource planning. It sets the basis for estimating and modifying the resource requirements.
8. It promotes innovation. A well-designed structure promotes the employee's transparency and recognition, which fosters initiative-taking and innovation in the organization.

2.3 Relationship Between Organizational Structure and Project Management

There are three different organizational structures—functional, matrix, and experimental. The relationship between each of the organizational structures and project management is explained as below.

Functional Organizational Structure

It intends to group specific skills and knowledge. It puts the various departments in the vertical structure and defines the roles from the chief executive officer (CEO) to accounts and sales departments to customer service, structuring the organization into various divisions including engineering, human resources, finance, IT, planning, and policy. Each functional division operates independently under an executive manager. The three characteristics that define the scope and nature of available organizational structure are: (i) integration between sub-units handled by rules, procedures; (ii) management chain handles problems; and (iii) works well in a stable environment.

Project Organizational Structure

The project manager executes full power in determining the priorities, use of resources, and selecting team members for the project. All members of the project team are directly responsible to the project manager. Once the project is completed, the resources are utilized for another project.

Matrix Organizational Structure

In this structure, employees report to two or more managers. This may include their respective functional manager and a project manager. For example, an employee may have a primary manager in the purchasing department and the project manager to assign additional project responsibilities. The functional part helps an employee to maintain his expertise and sustained employment in the organization. The involvement in the project helps to utilize his unique skills and abilities for the requirements of the project. This type of structure ensures the most effective utilization of resources. Hence, it is widely adopted by the companies that have the set of routine functional and project activities to be carried out by the existing workforce.

Issues and Challenges in R&D and Infrastructure Projects

Issues and Challenges in R&D Projects

The problems in R&D projects are internal or external. Internal factors influencing R&D are characterized by resource mobilization, resource allocation, and prioritization. External challenges include requirements of regulatory agencies, such as government policies, regulators and enforcement, and other organizations.

The key challenge is to address the concerns of the stakeholders about the environment, the use of rare or protective resources, and desirability of the outcomes. For example, the development of vaccine for COVID-19 demands the pharmaceutical companies and research laboratories to develop a cost-effective and safe vaccine in a shorter period of time.

R&D managers have to face challenges in purchasing or developing new technologies, determining small- and long-term priorities, and attracting and retaining top talent.

Issues and Challenges in Infrastructure Projects

Delays in initial stages: Project time may be increased due to delays found in the project planning and design stage.

Communication breakdowns: Communication failures can lead to vague project objectives and targets. This increases lack of trust and mutual respect among contractors, subcontractors, transporters, and clients in an infrastructure project.

Unrealistic deadlines: For many managers, challenge is to find alternative methods for tasks and schedules to complete the project "on time" or to pass the deadline. It is primarily due to the unrealistic deadline.

Scope changes: Due to change in scope, a planned schedule gets disturbed which creates priority conflicts in a complex infrastructure project.

Resource competition: Large-scale infrastructure projects compete for resources (people, money, time), and an appropriate allocation of resources is extremely important for a project manager.

Absence or improper use of project management techniques: Many companies in India still do not have the proper concept of using tools and techniques for project management. Typically, this is very much evident in the infrastructural projects which encounter many uncertainties and delays because of external factors.

Failure to manage risk: The failure to identify and mitigate the risks in the design stage of the project severely affects the quality and timely completion of the project. For example, a bullet train project between Ahmadabad and Mumbai is delayed because of problems in land acquisition.

Insufficient team skills: For many projects, the team members are allocated on their availability and are not trained for their assigned tasks.

Customers and end users are not engaged during the project: An inadequate involvement of clients and customers in the design stage of an infrastructure project is one of the prime reasons for time and cost overruns. This results in frequent changes in the scope of the project during implementation phase and creates priority conflicts.

2.4 Types of Organizational Structures

Line structure

It is a type of structure that has a sequence of specific commands. Policies and directions in this structure come from top to bottom. The project manager performs duties based on position or authority in the hierarchy. This type of structure is suitable for small companies like private accounting firms or law firms (Fig. 2.1; Table 2.2).

Staff and Line Structure

It is a modification of the line organization. In this structure, functional specialists work with line managers to guide and advise them. This structure is widely used at present day by many larger enterprises. Staff consists of two categories: general and specialized teams (Fig. 2.2; Table 2.2).

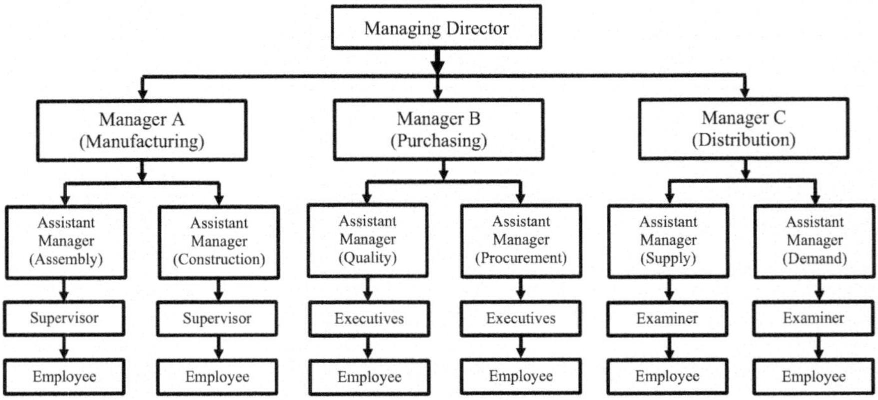

Fig. 2.1 Line structure

Table 2.2 Merits and demerits of organizational structures

Organizational structure	Merits	Demerits
Line structure	• It is the simplest type of structure • It facilitates prompt, quick, and efficient decision-making • There is clarity in authority and responsibility structures • People who provide quality work have good career prospects	• There is a possibility of being partial head bias • Lack of specialization is an ongoing issue • Branch head work can be very burdensome • Communication only takes place from top to bottom • Superior officers can be misused for their purpose
Staff and line structure	• Allows employees to work at a faster rate • It considers job responsibilities and provides specialized skills in a particular area • It helps organizers handle diligence at work • There is little or no resistance to institutional changes	• It induces confusion among employees • Employees do not have the functional knowledge to offer feedback-based suggestions • High-level hierarchy • Employees may have differing opinions, and it slows down the job • Slower decision-making
Functional structure	• A clear line of authority • Increased accountability for work • Higher speed and efficiency • No duplication of work • Each of the functions is of equal importance	• Communication has many obstacles • It is challenging to coordinate • Greater focus on people than the company • With the company's expansion, it is challenging to control its activities
Matrix structure	• Decision-making is decentralized • Coordination is vital in the project • Quick response to change • Convenient use of resources • Effective use of support systems	• High administration cost • It creates potential confusion over authority and responsibility • High prospects of conflict • Overemphasis on group decision-making • Excessive focus on internal relations

Functional Structure

It is an extension of line organization that clusters workers based on their specialization areas. The functional managers are responsible for leading the team and managing the associated operations. This concept originated with Fredrick W. Taylor. This type of structure is effectively used in customer service, engineering or production, accounts, and administration departments. (Fig. 2.3; Table 2.2).

Matrix Organizational Structure

It is a hybrid structure that combines the features of a project and functional organization. It helps to overcome the limitations of each organization. Here, both the functional and project managers share their respective authorities (Fig. 2.4; Table 2.2).

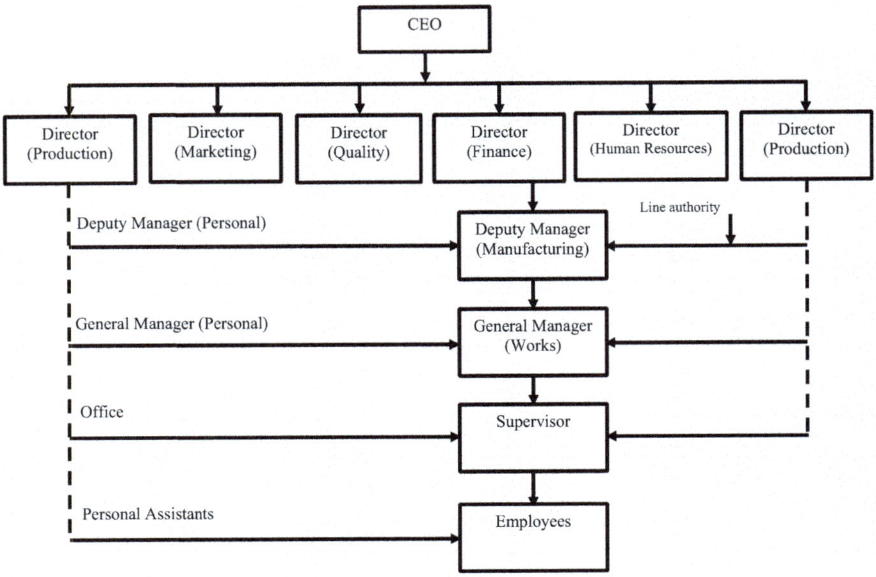

Fig. 2.2 Staff and line structure

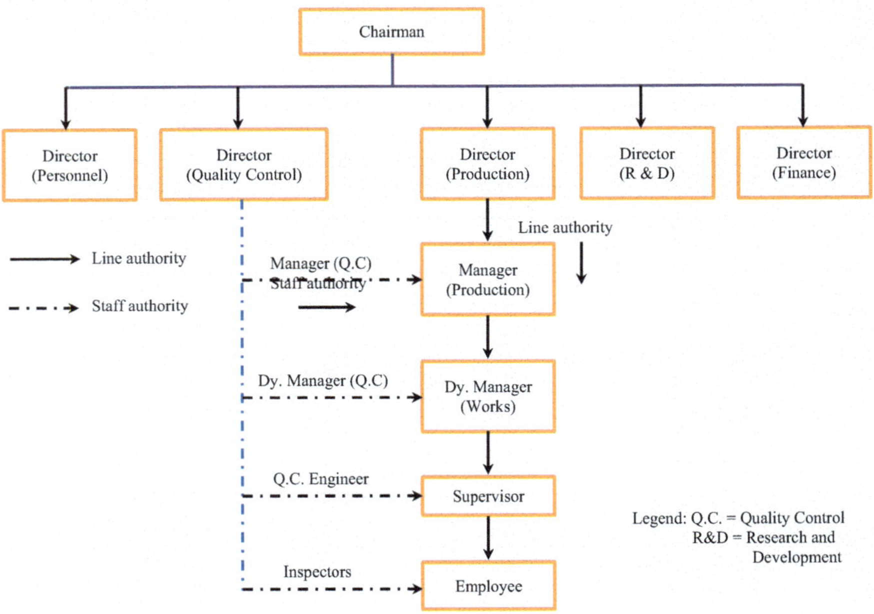

Fig. 2.3 Functional structure

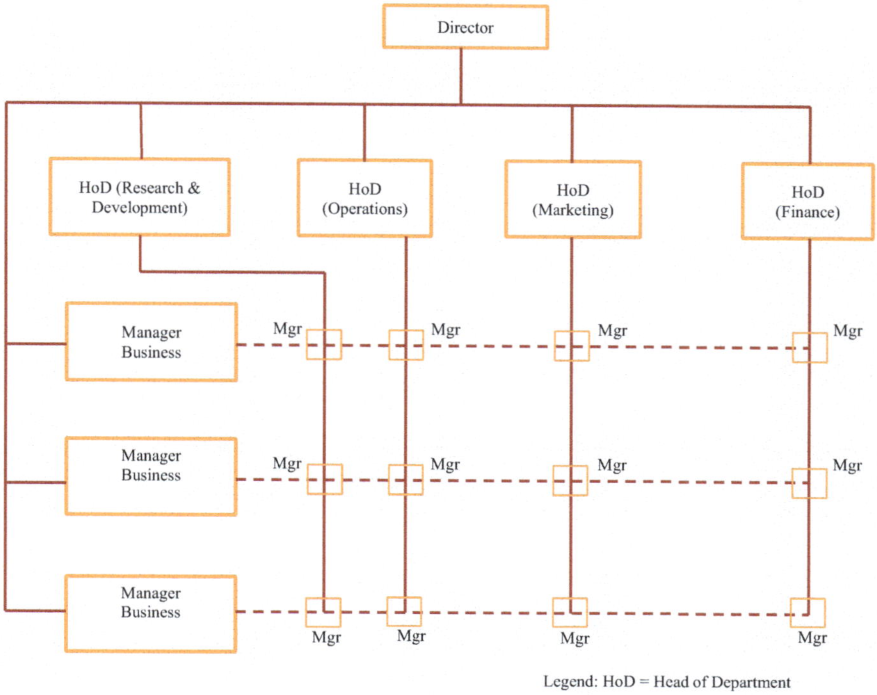

Fig. 2.4 Matrix structure

Project Management in Practice

Priority Conflict for a faculty in a matrix structure.

Internal Memo

To: Project Officer

From: Assistant Program Director

Dated: 14th March 2021

Dear Project Officer,

I am sorry to bring to your kind attention that I would like to be relieved from the assistant program director position of the post-graduate management program. Trust between a program director and assistant program director is the core for the smooth working of any program.

The Program Director does not trust my work and is not satisfied with my coordination as assistant program director and coordinator of flagship post-graduate programs with foreign Universities. This mistrust and mismatch in working style may not be suitable for the program and students.

Hence keeping in view and concerns of students as a priority, I request you to appoint a new assistant program director. I will support the activities until the new assistant program director and coordinator for the flagship post-graduate program is appointed and takes charge. Then, I can share all the information available with me.

Yours Sincerely,
Assistant Program Director.

Example 2.1: Organizational Structure for ERP Development

Figure 2.5 depicts the organizational structure for a typical software project like the development of ERP system, which is an example of line and staff structure.

Example 2.2: Organizational Structure for Organizing a Conference Project

Figure 2.6 is an example of the conference project's organizational structure, which is an example of functional structure.

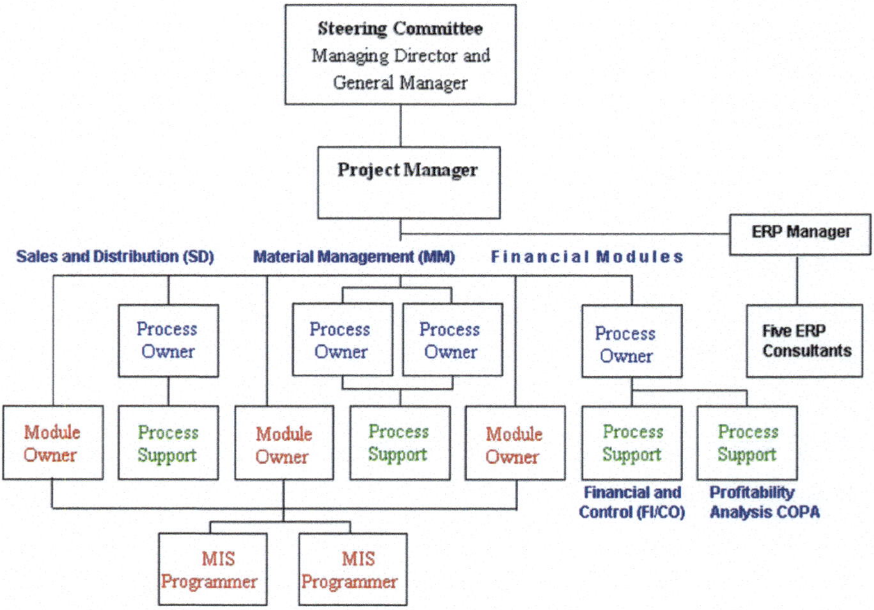

Fig. 2.5 Organizational structure for ERP development project

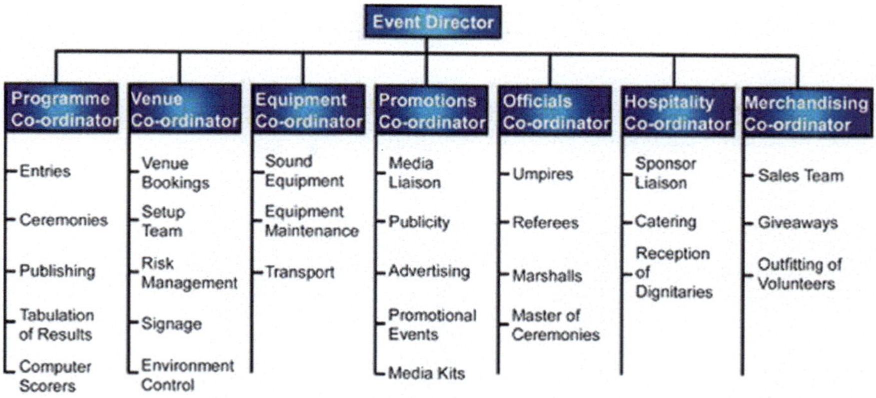

Fig. 2.6 Organizational structure for conference project

2.5 Organizational Structure and Conflict Resolution

The five modes for conflict resolution include confronting, compromising, smoothing, forcing, and avoiding.

Confronting can be of problem-solving, integration, collaborative, or win–win styles. It consists of face-to-face meetings and collaboration to reach an agreement that resolves the concerns of both parties. The particular situations in which confrontation can take place include:

- Both parties need to win.
- A need to reduce costs.
- A need to create simpler lines of authority.
- Skills are complementary.
- Inadequate time management/paucity of time.
- Lack of trust.
- Training is the ultimate goal.

Compromise is also described as a "Give and Take" style, with opposing parties bargaining for a settlement. Both parties give up something in order to come to a decision with mutual satisfaction. The particular situations in which comprise works can include:

- Both parties need to win.
- Inadequacy of time.
- Continued need for good relations between the affected parties.
- If we do not compromise, we will get nothing.
- Stakes are moderate.

The critical areas of a contract are stressed, and the areas of disagreement are minimized. Controversies do not always resolve conflicts. A party can sacrifice its concerns or objectives to satisfy the other party's concerns or objectives. This style is effective in the situations such as:

- Overcoming the goal reached.
- We need to create responsibility in later trading.
- Stakes are small.
- Liability is limited.
- Any solution is sufficient.
- Where harmony and goodwill are to be maintained.
- The likelihood of loss is otherwise high.
- To gain time.

Forcing is also known as competing, controlling, or dominating styles. It is about ignoring the needs and concerns of the other party; a party will be forced to go for everything to win its position. The intensity of the growth of the conflict and the tendency of a weak controversy are high. It is the victory of the party at the expense of other parties. The situations in which this style is helpful include:

- "Do or die" situations.
- Where the stakes are high.
- Important formulas are in stock.
- The relationship between the parties is not important.
- There is a need for a quick decision.

Avoiding is also described as a withdrawal style. This procedure is followed by postponing the problem or completely withdrawing from the situation. It offers only a temporary solution because the problem and the conflict continue again and again. This style can be used in the conditions such as:

- One party knows they cannot win.
- Stakes are small.
- Stakes are high, but the interested party is not prepared.
- In order to gain time.
- There is a need to maintain neutrality or reputation.
- The problem is anticipated to go away.
- A delay will allow gains.

Researchers have investigated the styles used by project managers and their impact on conflicts and stress levels in the project. It has been realized that people who use a particular style may create environments with a diverse range of contradictions. People who use the smoothing style create environments with low levels of work dispute, which reduces stress and pressure among project team members. By contrast, people who are more likely to use forcing or avoidance styles tend to create environments with more controversial situations which can increase conflict and stress.

Cognitive analysis is a widely used technique for dispute resolution. First, it identifies the primary source of contention for cognitive disagreements between the parties. Second, it facilitates the controversial parties reaching an acceptable solution

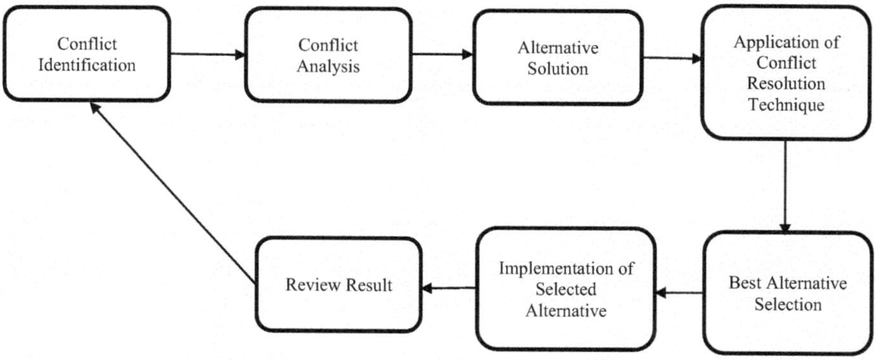

Fig. 2.7 Project conflict management process

to the conflict. The proposed systematic method for disputed solutions recognizes signals, distance variables, and judgments and then determines relationships between these variables. The six steps of this method are given below.

1. Identity of the conflict domain
2. Conflict case generation
3. Exercise
4. Results analysis
5. Conversation of judgment conflicts (cognitive opinion)
6. Negotiation between disputing parties.

A pictorial representation of the conflict management process is given in Fig. 2.7.

Use of the conflict analysis method has led to a more significant agreement between the parties in the conflict. Cognitive feedback identifies the reasons why parties need to speak and reach an agreement. Thus, the use of cognitive feedback can be effectively utilized for problem-solving.

Active listening techniques can be used to help resolve conflicts. Listening allows conflict to take its natural course by allowing the expression of different opinions. Awareness about a complete understanding of the controversial resolution and awareness of their consequences provides project managers with special tools to create an optimal working environment. Furthermore, an appropriate selection of organizational structure for the key characteristics of the project helps to set the mechanism for conflict resolution and hence sets a path for the smooth execution and control of the project.

2.6 Impact of Organizational Structure on Teamwork and Decision-Making

2.6.1 Impact of Organizational Structure on Teamwork

Team-based lateral organizational structures are one of the newest forms of business organization. It groups employees at the same level into teams for the performance of specific job functions. This structure typically consists of low overheads and requires low maintenance and hence is highly beneficial. The key advantages of this structure include:

Low-Overhead Management

Team-based lateral structures are similar to traditional lateral structures, carrying decisions and taking overhead management to cause better techniques or delayed implementation of new ideas. As there is no need to follow long chains to accept ideas or changes to the business model, team-based lateral structures can ensure necessary adjustments rapidly and allow for faster responses to changing market conditions. It is essential in today's digital economy.

Collective Effort

By spreading the responsibility of a group of people rather than having one person in the management or management of a business area, decisions can be reached by correction. Decisions made by a team are sometimes better than decisions made by a person and are more effective.

Remove Delay

Team-based lateral structures eliminate the traditional scalar chains, which would lead to delays and labor disillusionment with confusing communications channels. A team can comfortably express their concerns to the management without being seen as unsatisfactory or disheartened. It makes communication between employees and management more effective. It motivates people to be more likely to talk about workplace problems or inefficiencies.

Less Management

The removal of management layers can foster faster decision-making and reduce administrative costs. Employees can experience an increase in power and develop greater courage. Using a group-based structure, employees usually solve problems without consulting experts, which reduces the time required to complete the actions.

Better Relationships

When teams work, they are responsible for completing the schedule. If an employee is unable to complete the work, another team member will fulfill that responsibility. By creating a comprehensive role and responsibility matrix, accountability is clearly defined, and misunderstandings and conflicts can be minimized. It helps to develop matured and long-term professional relationships among project team members. Additionally, connected team members can use social media technology, including wikis, blogs, and forums, to provide information to other employees.

Increased Productivity

People in a team struggle and invest for common goals instead of focusing on individual motives. This creates an environment of increased cooperation, reduced fear, increased creativity, and joint problem-solving. It has been observed that people who work closely with problems usually have the best ideas to solve them. In a team-based approach, the leader encourages his colleagues to act collectively and creatively.

Balance

Successful teams require three key attributes: action-, people-, and thought-based members. Action-based team members and groups are encouraged to address the challenges and motivate them to perform their work. People-based roles guide the team, providing support and resources. Finally, thought-oriented team members provide ideas, evaluation options, and exceptional knowledge. This balanced group allows resolution of all aspects of complex problems effectively.

2.6.2 Impact of Organizational Structure on Decision-Making in the Project

Proper organizational structures help assign jobs to the project team members and improve decision-making in the organization. A well-defined organizational structure may help in the following ways for decision-making.

In Understanding the Decisions We Should Take

It may seem obvious, but the first step in determining the decision is to find the decision. In other words, we have to choose and define the type of decision we should take, how to change our work process, or improve our product or service.

Collecting Information

We should evaluate all the information and data we collect to make the right decision. For example, if we consider a marketing consultancy conducting a pay-per-click advertising campaign, we would like to have the keywords that most customers use when searching for results related to consulting.

Identifying Alternatives

Once we have analyzed the information, we must develop different options regarding the decision we should make. Therefore, it helps to know about alternatives and their evaluation.

Analyzing Pros and Cons

Analyzing each option's pros and cons helps us understand the trade-offs and eliminate the less effective alternatives. It facilitates the search for an option that can ensure the least possible chance of failure. In some cases, there is not much difference between the two options. In this situation, we have to seek the opinions of top management and rank them in descending order.

Choosing the Best Alternative

Once we rank our options, we have to choose what we think is a strong chance to achieve our goal. We can combine several options in some cases, but we want to take the clear-cut direction in most cases.

Making the Final Decision

A decision is just a choice until we take action. This means we have to understand the available resources in order to make the decision successful. We need to conduct several meetings with our managers and team leaders to explain our decision, and we should know how it affects their daily work or how it affects our clients and customers.

Evaluating the Effect of Decision

It is very complicated to analyze decisions after time has passed. Hence, it is necessary to investigate the effectiveness of our decision (choice) at this stage and develop confidence. The question we need to examine at this stage is: "Can we resolve our problem with this decision?" If the answer is "yes," then we can go ahead with our choice. If the answer is "No," we should recognize the mistake and repeat the decision-making process.

2.7 Selection of Organizational Structure for New Product Development Project

A matrix organizational structure is recommended for a new development project which demands extensive cross-functional integration.

The following are three core parts of a cross-functional team:

- **Dedicated core team:** The primary team consists of three entities dedicated full-time to cross-team operations.

Cross-functional (CF) leader: Provides attention on what is needed and is responsible for achieving and resolving critical issues and disputes.

Program manager (PM): Monitors the progress of each sub-functional component and dependencies. Each sub-functional team needs to collaborate with project managers. The program manager will also disclose the risks that will arise anytime and plan and implement cross-team operations to the leader.

Systems architect (SA): Recognizes all modules and their combination or integration in different functional areas. The system architect is probably the most experienced technician and helps other sub-functional technicians execute a project.

- **Functional Managers and Architectures**: Functional managers are responsible for planning, funding, and tracking individual parts. Unlike program managers, functional managers manage cross-functional work, get involved in many other projects, and officially report to their own divisional company.
- **Sub-functional Team Members**: Represents one subset of a large team under the functional manager.

Leadership

- How is the leader selected?

Cross-team leaders must possess qualities such as (i) high conceptual skills with the ability to quickly capture extensive technical concepts, extending across multiple domains; and (ii) high collaborative skills instead of the fact that they would have to drive others without direct authority.

- Providing team vision and mission

The leader's goal is to clarify clearly and provide clear criteria for success and project delivery period.

He/she should focus on the project's business value and how it will be meeting personal assessment in terms of gifts or growth.

Culture

Character conflicts: In a typical matrix company, functional managers, technical leaders, and members have dual roles and responsibilities: cross-functional and in their division. This role comes from the stirring that the conflict would be cross-functional work valid by the member's own company.

Power struggle: This is the "character clash." Some members can understand that a cross-functional object is an opportunity and forum and seeks to control its decision-making. On the other hand, cross-functional members using their position in the organizational structure try to exercise their influence in decision-making, which creates a power struggle.

Team dynamics: Honesty and transparency in all information-sharing help establish trust. Furthermore, the beliefs and voice of each participant should be respected. This promotes active problem-solving and creativity. Finally, members must have confidence in their leadership team to avoid unexpected obstacles in achieving goals.

Decision-Making

Centered versus group decisions: Cross-functional leaders and architects must guarantee the balance between implementing their decisions (centralized decisions) and group reporting decisions (group decisions). Upcoming decisions (centered) should focus on strategic thinking, product definition, structure, and ecosystem. Make decisions up below (group) every part of the particulars and its reliability should be focused. Group decision-making reduces the chances of overlooking inadequacies.

Project Management in Practice

Developing an Organizational Structure of Student Council.

Consider the Following Conversation in a Meeting Discussing the Development of an Organization Structure for a Student Council

Faculty In-Charge, Student Council (FI-SC)

I am sharing the draft document of the Student Council organization structure. This includes positions, roles and responsibilities, functionaries, election procedure, ethics, and budget approval process.

Principal Academic Advisor (PAA)

Financial powers should be as follows:

President and Treasurer for up to Rs. 2000. (Trust the students).

President and Registrar up to Rs. 20,000.

Above that, President, Registrar, and Faculty-in-charge.

(Also, how does it work for electronic transfers?).

The Audit Board should be on the recommendation of the President and Faculty in charge.

We can't get into all the details right now. We should add:

Provisions in the document can be modified with a proposal from the Executive Council and the Faculty-in-charge. To be approved by VC.

Has the Registrar seen the document? And is okay with it? Then it can be sent to VC.

The elections should be organized at the earliest now.

FI-SC

The document is revised as follows: Initially, the term of all approved new clubs is for one year. However, based on performance evaluation after one year, it could be extended up to 5 years. Clubs such as Gender Equity and Sensitization, PlaceCom are permanent.

PAA

The President should be a student right from the start, and not a faculty member. In a sense, the very meaning of a Student Cell and the spirit of students doing extra-curricular and co-curricular activities seems not to be in place if we do not trust them.

I also worry about interview-based selection among the top four. I have not come across this process anywhere else. (Of course, my experience is in a few institutions only).

I know we said there may not be enough 'maturity' yet, but then I don't know when there will be enough maturity unless we give them responsibility and self-determination.

There are a few other suggestions I have, but after finalizing the above.

An essential item I wanted to see was approving new clubs (unless I am missing something). New clubs should go through the process, and so we need not mention them here. Do we want a life term for a club, after which it will be reviewed for continuation? We should mention that existing clubs are being accepted as is and presented in an annexure, with when it will be reviewed.

Regarding the President appointing Returning Officer, please make it Faculty-in-Charge.

Regarding the President appointing the Head of Clubs/activities, I think the first step should be elections in the appropriate constituency. If it is not through elections, then appointment, but in consultation with Faculty-in-charge.

Please keep a proviso that this document can be modified based on the Executive Council and Faculty-in-Charge recommendation, approved by IAC (Internal Academic Committee).

Why does the President chair club meetings? The President has to be arms-length from club activities. And certainly can't be spending all the time on club activities.

The Head of the club/activity is the right person.

Am I missing something?

FI-SC

The Executive Council (composed of five members, one each from all programs) will look after the club affairs, including meetings. The Executive Council head is President, and hence he is solely responsible for chairing the club meetings and the approval of meeting minutes. The President shall chair the club's meeting with the Executive Council.

Programme Director, Post Graduate (PD-PG)

Let us make all our communication and documents gender unbiased!!

PAA

I think we are making this too top-heavy. Not suitable for students. President will have no time for academics. And should be arm's length.

Clubs/Activities have to be managed by the Head, who will have the passion for the concerned domain.

One EC member can be a standing nominee to the Club/Activity meetings. I guess what Ananth mentioned as Rep (for representative).

If there are concerns, let us discuss them.

PAA

I don't understand!

Why should a programme elected representative Chair some club or activity meeting? None of the program representatives may know enough about music or sports, or cultural activities…

(In the premier institutions, the Student cell itself is the union of the students who become Sports, Cultural, Academic, Placement, Business-festival heads. There is no separate Executive Council. But in addition, they do have independent members).

What is your concern about the Head of that club chairing the meetings?

Please organize a discussion on this.

FI-SC

Why the Member of EC Chair the Meeting?

The Club Heads will only chair the clubs meeting internally, whenever they want (in case of proposing/scheduling an event, finalizing the internal working module, assigning tasks, etc.).

But when it comes to the performance evaluation, rules/norms modification, conducting elections, or deciding/planning something for any external event, where it is a matter of Institute reputation, they need to be first asses by some committee, i.e., Executive Council. If everything goes internally, the assessment procedure may go wrong. Therefore, a nominated person of the Executive Council shall chair the meeting every two months to assess Clubs' operational/any other related activities.

EC Importance in Clubs Election?

A highly deserving candidate may not make it to the highest post for any club due to conflict issues (friendship, favoritism towards any program-specific club), despite his interest in the position. For example, Business Club, maybe engineering background student is much more capable than an existing lot of core members from BBA/MBA or vice-versa in technology-related clubs). Hence, we need Executive Council to look after such issues during elections.

Why Is Student Cell not Doing This Alone (Above Point)?

They will be loaded heavily and cannot manage everything alone. Brainstorming can be another factor here.

Composition of Executive Council

The Executive Council, composed of one each representative from the different program (not CR of that course), will be selected based on their interest in extra-curricular activities other than the academics, such as sports, administration, singing, dancing, technological interest, cooking, blogging, designing, etc.

Why We Need Executive Council?

As I said, for any matter, there is a chance of biased decision even within the Student Cell (President, Vice President, General Secretary, Treasurer, Secretary). To avoid such biases, we need the Executive Council to have a diverse group of students who can put together unbiased opinions for the particular problem, which can be used to think critically and decide.

When I was at IIT Kanpur, I saw these very closely, and the Executive Council is essential in those terms.

Whether it is okay, otherwise, I can put Club Heads will chair the meeting.

Programme Director, Under Graduate (PD-UG)

What are the individual and organizational attributes we wish to evolve?

1. Leadership
2. Transparency
3. Flexibility
4. Responsiveness
5. A culture of commitment and ownership
6. Empowerment to decide scope, measures, and policies of each club.

A flat structure empowering club heads (individuals with requisite skills) for various decisions can better satisfy the above six attributes. A bureaucratic structure demoralizes the people and creates the scope for opportunism.

Registrar

FI-SC proposes to have an EC member or office bearer of the Student Cell to be the Chair as a neutral person in inter-club meetings to balance out the possible upper hand by any club and address the possible unrest. Though it appears a little bureaucratic, he has come across such issues earlier and is trying to address such possibilities. Whether the Cell should have an EC at all is a bigger question. But the idea to have a Rep of the higher body in inter-club meetings is not misplaced. As the institute has more students and the activities increase, there must be a good hierarchy and clear role definition to avoid concentration of power and ease functioning.

PAA

You are distinguishing the meetings for activity planning and activity evaluation.

Does the document identify these two types of meetings?

Given this, activity evaluation meetings may not have to be every two months. You had started with six months. May be once in three months? Or say every term?

Overall, the principle should encourage and facilitate students for synergistic extra and co-curricular activities without compromising academics and overbearing supervisory structure.

In your explanations, you seem to have given thought and brought in experience from other institutions.

Having considered the above inputs, do go ahead.

FI-AC

I agree. These terms (planning and activity evaluation) are there. A meeting schedule is suggested once every three months.

This document is ready to be sent to the VC for approval. I will wait for two days and then start with the election process.

Summary

The organization hierarchy is defined through the organizational structure. It recognizes each function and where it reports within the organization.

The structure helps to determine how the organization works and to help the organization to achieve its objectives to allow for future development. The structure is depicted using a structural chart.

A structure is necessary to grow a business in profitable way. The structure of the organization helps to identify the talent to add to the senior management company.

Key characteristics of organizational structure are: easy, accountability and continuity, delegation of authority, the use of the executive body, clear authority line.

Team-based lateral organization structure is one of the newest forms of business organization. The lateral structure based on a team is an organizational table that groups the same level of employee groups as teams for maintaining specific job functions.

Questions for Discussion

1. What do you mean by organizational structure? What is its relevance and importance in the organization for decision-making? How does it improve the efficiency of the organization? What are the generic key characteristics of a good organizational structure? Compare the key characteristics of organizational structure for a typical IT company and manufacturing organization on a set of critical criteria in the tabular format.
2. How does an organization's structure help the organization to meet with its environment? Explain in detail with a suitable example.
3. What is the relationship between organizational structure and project management? How can an appropriate selection of organizational structure help to have smooth and effective execution of the project? How authority, delegation, power, and ownership issues can be resolved in the project through an appropriate organizational structure? Illustrate the issues for a case of R&D and Infrastructure projects.
4. What are the different types of organizational structures (develop pictorial presentations)? How do you define them? Explain the relative merit and demerit of each organizational structure in a tabular format. Then, select an appropriate organizational structure for select projects such as a missile development project, ERP development project, and organizing conference project. Justify your choices critically.

5. How can an appropriate organizational structure help resolve conflicts in the projects and set a basis for reasonable negotiations?
6. What is the impact of organizational structure on teamwork and decision-making in the project? Explain this with a suitable example for an infrastructure project executed by the government.
7. There is a new product development project which demands extensive cross-functional integration. What is an appropriate organizational structure you would like to recommend? Critically justify your recommendation to the top management.
8. What do you mean by organizational structure? What is its relevance and importance in the organization for decision-making? How it improves the efficiency of the organization? What are the generic key characteristics of a good organizational structure? Compare the key characteristics of organizational structure for a typical IT company and manufacturing organization on a set of critical criteria in the tabular format.
9. How an organization's structure helps the organization to meet with its environment? Explain in detail with a suitable example.
10. What is the relationship between organizational structure and project management? How can an appropriate selection of organizational structure help to have smooth and effective execution of the project? How authority, delegation, power, and ownership issues can be resolved in the project through an appropriate organizational structure? Illustrate the issues for a case of R&D and infrastructure projects.
11. What are the different types of organizational structures (develop pictorial presentations)? How do you define them? Explain the relative merit and demerit of each organizational structure in a tabular format. Then, select an appropriate organizational structure for select projects such as a missile development project, ERP development project, and organizing conference project. Justify your choices critically.
12. How can an appropriate organizational structure help resolve conflicts in the projects and set a basis for reasonable negotiations?
13. What is the impact of organizational structure on teamwork and decision-making in the project? Explain this with a suitable example for an Infrastructure project executed by government.
14. There is a new product development project which demands extensive cross-functional integration. What is an appropriate organizational structure you would like to recommend? Critically justify your recommendation and convince the top management of your recommendation.
15. Develop a case study (50 lines) indicating the impact of organizational structure for a typical government R&D organization. Feel free to define various roles, stakeholders, processes, etc.

Web-Based Exercise

Collect the details of at least three businesses of a diversified company. List down the peculiarity of each of the businesses. Then, justify an appropriate selection of organizational structure for each of the firms. What are the advantages and limitations of selecting a particular organizational structure? What implications will it have on executing special projects like new product or service development in a specific business segment?

Group Project

Undertake a detailed brainstorming session in a group of three to five persons on the relevance of organizational structure for the following projects:

1. New product development
2. ERP implementation
3. Organizing picnic
4. Missile launch
5. Tree plantation for green city.

Enlist all your responses on key characteristics of each project and critically justify the selection of organizational structure for smooth and hassle-free execution of the above projects.

Chapter 3
Project Selection

Critical Questions

What is the importance of project selection?
What are the key considerations in project appraisal and selection?
How do we prioritize projects using multi-criteria decision-making approach?
How do we formulate the project selection problem as a typical Operations Research (OR) problem?

Project Management in Practice

Chenab River Bridge

The Chenab Bridge is a 467-m long arch main span located about 320 m above waterway steel and a substantial curve among Bakkal and Kauri in Jammu and Kashmir, India. The scaffold will be braced with 63 mm thick unique impact evidence steel and substantial steel that can withstand blasts and seismic activity. An online warning and a monitoring system on the bridge will also provide added security. The Chenab Bridge was scheduled to be ready in December 2009. In September 2008, however, it was announced that the project was interrupted despite the completion of the substructure of the approach viaduct. The Chenab Bridge will be the biggest, longest-spanning, and highest railway arch bridge ever built in the world. Its design offered significant challenges, which will be even more demanding before the gap between the main arch halves is filled.

J. J. Thakkar, *Project Management*, Management and Industrial Engineering,
https://doi.org/10.1007/978-981-15-3695-3_3

Project management challenges

Time:

Problems were encountered in constructing foundations having complex conglomerate soil strata due to difficulties in well sinking because of the high water level. The altitude of 5000 m and elevation of 2300 m from the river were all factors that contributed to the project's delay. Made of blast-proof steel designed to withstand both earthquakes and explosives, DRDO's study and research in addition to other uncertainties also contributed to delays. Launched in 2003 and the project was expected to be completed in December 2021. In 2009, an extended argument about slopes broke out between two privately owned businesses that had been charged to chip away at discrete areas of the line.

Cost:

JKPCC and R&B authorities failed to finalize tenders even after the lapse of one and half years.

No running or advance payment for materials or machinery was allowed. Delaying the work and halting it made the cost of the bridge high as the new total cost was INR 12,000 crores.

Performance:

After detailed investigations of the behavior of this river and based on a thorough examination, it was found that the bridge should have a lifespan of 120 years, and it is just 60 km to the international border, so it should be strong enough to withstand an explosion. Under DRDO's advice, the bridge has been so designed that even if one of the 17 piers supporting it is blown up, it will not collapse. It indicates the challenges in making the bridge strong. The bridge is located in a windy zone, so it is also a challenge for the Indian railway.

Sources:

(a) *Conceptual Design of the Chenab Bridge in India by P. Pulkkinen, S. Hopf and A. Jutila, Procedia Engineering, Vol. 40, 2012, pp. 189–194,* https://doi.org/10.1016/j.proeng.2012.07.078 (https://www.sciencedirect.com/science/article/pii/S1877705812024642)

(b) *Chenab Bridge: A bridge of hope and challenges, webindia123, IANS indianarrative/in* (https://news.webindia123.com/news/Articles/Asia/20200729/3600656.html), *accessed on 12th June 2021.*

3.1 Importance of Project Selection

Project selection is a process to evaluate project ideas against well-defined criteria to understand the relative priorities of the projects for an organization. It helps the management understand the feasibility and achievability of various projects. It is a multi-criteria decision-making problem.

An organization must examine the feasibility of projects concerning the allocated budget and resources. The feasibility of projects can be evaluated based on the inputs received from various stakeholders such as the funding agencies, community, local and national governments, and NGOs. Therefore, the need for an appropriate mechanism for project selection rises. It helps evaluate the rationale for selecting a specific project and helps to establish a priority order for the projects under consideration.

Suppose a company or organization does not have the necessary experience in conducting eradications. In that case, it is recommended to focus on a lesser number of projects, preferably one project, until the employees in the organization have acquired the required skills and experience. It must then build up and grow its capacity for undertaking multiple projects simultaneously. Ideally, the company/organization must initially take up more straightforward projects and then move on to more challenging and rewarding ones. The expertise gained from more straightforward projects can help in solving issues/answering questions while carrying out more complex tasks.

In a mix of straightforward and complex projects, managers may face difficulties in finding ways to start them. This is where project selection comes into play. The project selection stage assists by providing a procedure to weigh up the projects' importance and select the best appropriate projects.

The project selection stage is a step-by-step, objective method for prioritizing proposed projects. It can help explain to stakeholders the reasons behind selecting a specific project. A typical process of project selection is represented in Fig. 3.1, which comprises of five steps. It begins with the generation of various ideas on developing a list of potential projects for selection. The proposed potential projects will then be subject to a cost–benefit analysis using multiple financial and non-financial criteria associated. As it can be considered a multi-criteria decision-making problem, the objective and subjective inputs from experience, the company's annual reports, and an intuitive understanding of the experts can prioritize the projects. The prioritized list of potential projects is presented to the top management for necessary discussion and approval. If there is any discrepancy in the prioritized proposed projects, it will go through the analysis process with the same set of experts or a different team. A detailed process of project selection is illustrated in Fig. 3.2.

The significant advantages of implementing project selection are:

- A well-documented and transparent record of the reason behind the choice of a specific project can be generated.
- Order of preference for projects based on the considerations of importance and the attainability of a project can be generated.

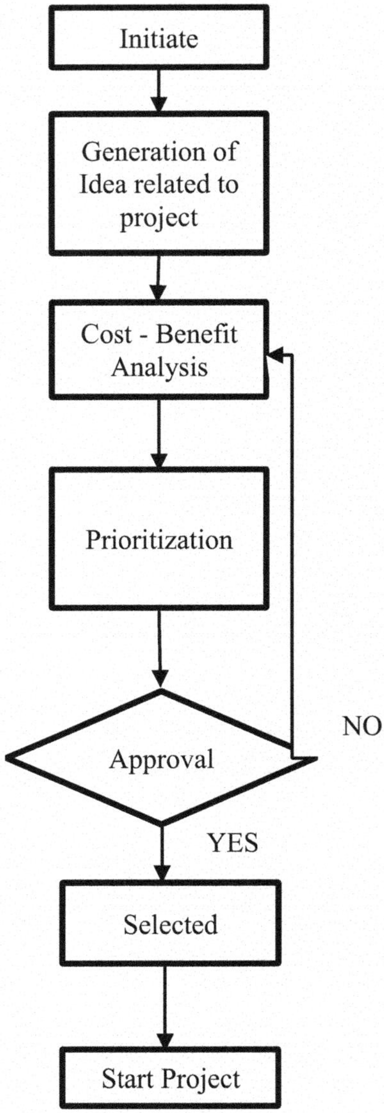

Fig. 3.1 Typical project selection process

What Are the Consequences of Arbitrary Selection of Projects?

When a project or a set of projects are selected arbitrarily, the chances of their failure rise exponentially compared to those chosen by proper project selection procedure. It compromises the project execution quality and can affect the organization in various aspects, as discussed below.

1. Lost opportunity cost
2. Debt problem due to irregular cash flows

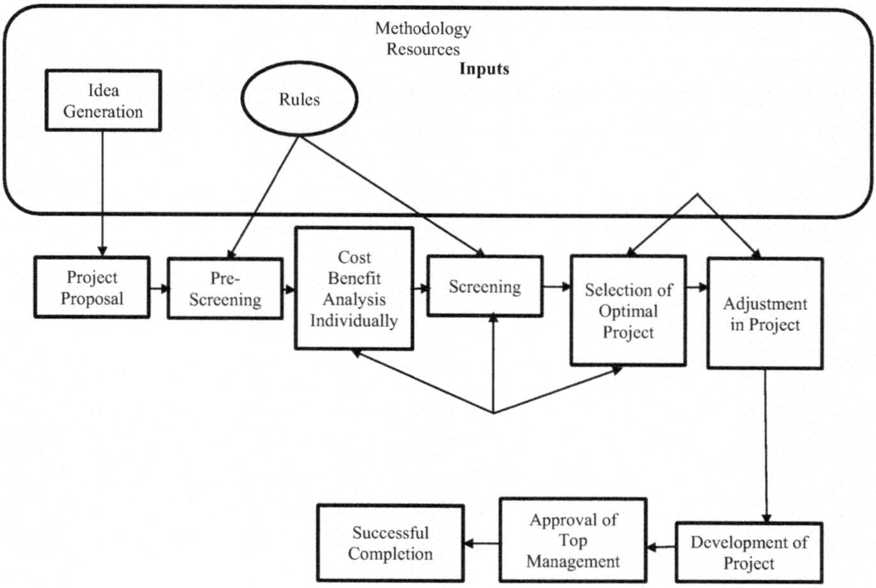

Fig. 3.2 A detailed process of project selection

3. Deterioration of the firm's value
4. Lost finance and business
5. Low team morale.

Lost opportunity cost	When an organization arbitrarily decides to take up a project, resources are allocated to respective components of the same. If a proper project selection had been carried out, the investment of the same amount of resources could have brought back a higher return. Thus, there lies an opportunity cost in randomly selecting the projects
Debt issues due to insufficient cash flows	The selection of a wrong project can bring in many associated issues. One such issue is of cash flows: irregular inflows and unforeseen outflows can cause the organization to run into debt. This may be prevented by project appraisal in the preliminary phase
Deterioration of the firm's value	Failed projects in a firm's portfolio bring down its financial value as they directly affect profits, degrade the firm's value from a client's perspective, and create hurdles in competing for future projects
Lost business	Future business is affected by failed projects in a portfolio. Credibility for getting loans and finances is also affected, which makes future projects more vulnerable to problems and failures, bringing down the organization's value
Low team morale	When results are not fruitful even after putting in the best possible efforts to a wrongly selected project, the project team gets demotivated. A lowered team morale further affects other projects which were being implemented in the right direction

3.2 Project Appraisal

Project appraisal is the evaluation of the overall ability of a feasible project to succeed. After the preparation of a detailed project report (DPR), project appraisal is done to comprehensively evaluate the feasibility, sustainability, and relevance of a project before deciding whether to undertake it or not, i.e., determine how viable a project is.

The primary aim of project appraisal is to consider and compare feasible projects and select the best one that meets the objectives. All feasibility studies serve as the groundwork for project appraisal. The aspects covered in feasibility studies are re-examined during project appraisal.

The project appraisal can be explained as a detailed examination of several aspects of a given project before resources are committed. A typical project appraisal cycle is illustrated in Fig. 3.3. It begins with a market appraisal to investigate the utility for an individual customer, society, or the nation. The selected potential projects are then evaluated for various dimensions such as technical, social, economic, ecological, financial, and commercial aspects.

Phases in Project Appraisal

The process of project appraisal starts from the beginning of the project. Starting a project appraisal process early allows the organization to be better positioned to decide on how the capital can be invested into the project. It also helps prevent overspending on the project, and allows dismissal of a project that is not economically feasible.

There are five phases in the process of project appraisal:

1. Initial evaluation
2. Defining problems and long-list
3. Consulting and short-list
4. Developing alternatives
5. Comparing and selecting the project.

Types of Appraisals

Various kinds of project appraisals are mentioned below.

1. Technical Appraisal
2. Financial Appraisal
3. Market Appraisal
4. Socio-Economic Appraisal
5. Environmental Appraisal.

Technical Appraisal

A technical appraisal is the technical review of the proposed plant and equipment to the prescribed norms. It is essential as it ensures the technical feasibility of the

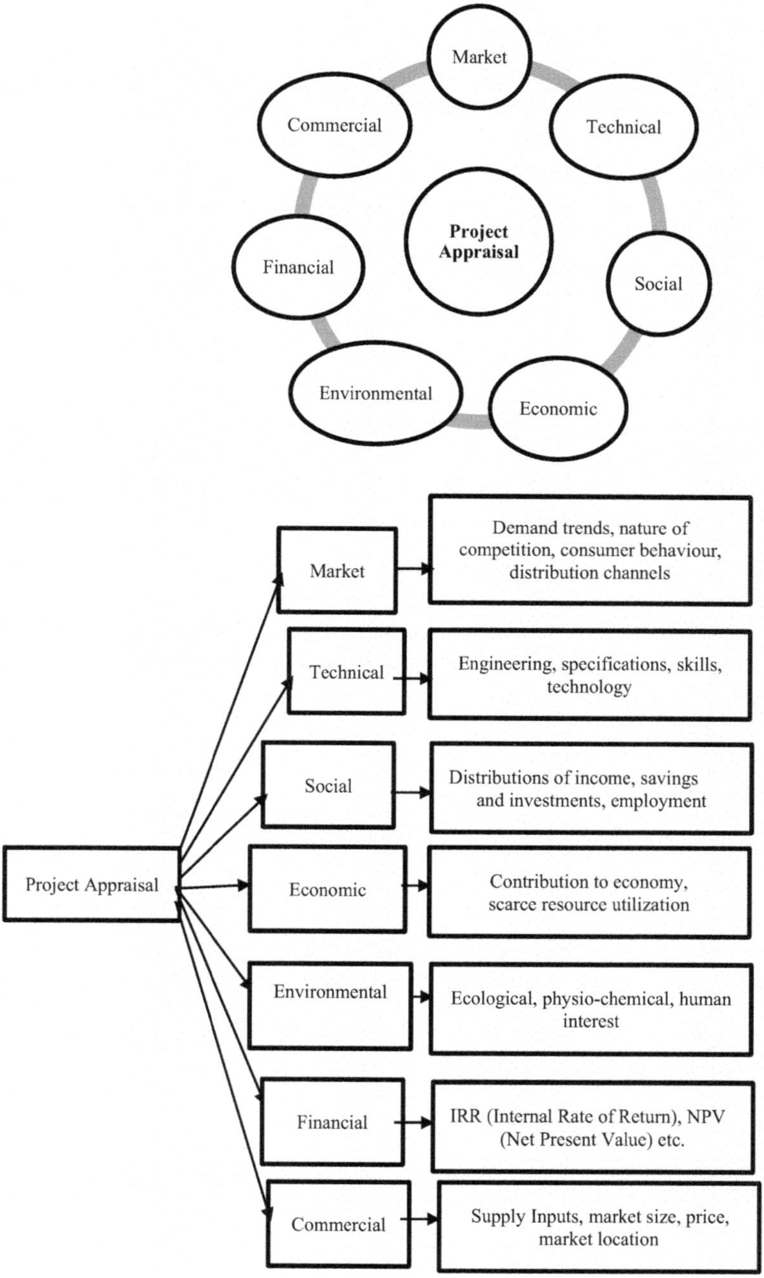

Fig. 3.3 Typical project appraisal process

project and contributes to optimal project formulations. The following aspects of the project should also be taken into consideration.

- Availability of inputs at reasonable cost
- Consistency and soundness of engineering design
- Economics of scale in production
- Availability of appropriate technology and alternative ways of production
- Advantageous location of the project
- Maintenance and repairs
- Provision for expansion
- Balancing of equipment.

Financial Appraisal

Finance is one of the essential prerequisites for sustainable organizational operations. It can facilitate an entrepreneur to procure labor services, machines, raw materials and combine them to manufacture the required products. The financial viability of the project can be evaluated using the following aspects.

- Exhaustive and realistic cost estimation
- Sound capital structure: Fund Source
- Provision for working capital requirement
- Generation of sufficient cash flows to cover debt service liability
- Generation of adequate profit
- Safety margin
- Break-even point
- Payback period.

Market Appraisal

Before the production starts, an entrepreneur needs to realize the potential market and value. Identifying the market potential can help in adopting strategies for the promotion and distribution of the product. In addition, it can help analyze the competitors involved. The key factors that should be taken into consideration are:

- Past and current demand trends
- Past and present supply position
- Production possibilities and constraints
- Imports and exports
- Nature of competition
- Cost structure
- Elasticity of demand.

Socio-Economic Appraisal

The viability of a project also needs to be judged from the viewpoint of society or a nation. Socio economic appraisal is known as social cost–benefit analysis.

- Direct economic benefits and costs in terms of shadow prices
- Impact of the project on distribution of income in society

- Impact on level of savings and investments in society
- Impact on fulfillment of national goals
  (a) Self-sufficiency
  (b) Employment
  (c) Social order

Environmental Appraisal

In recent years, environmental concerns have garnered a great deal of significance. An environmental appraisal should be done particularly for significant projects like power plants, irrigation schemes, and industries which have substantial environmental implications such as in chemical and leather processing. The key factors considered for environmental appraisal are:

- Ecological: fisheries, tree plantation, wetland/wetland habitats, and forests.
- Physio-Chemical: flood control and drainage erosion, drainage, congestion/waterlogging, obstruction to waste water flow, soil fertility, early flooding.
- Human Interest: areas of settlements, agricultural lands, navigation/boat communication, irrigation facilities, landscape, land values.

Generic Checklist for the Project Appraisal

- Availability of inputs at reasonable cost
- Consistency and soundness of engineering design
- Economics of scale in production
- Appropriate technology and alternative ways of production
- Advantageous location of the project
- Maintenance and repairs
- Provision for expansion
- Balancing of equipment
- Exhaustive and realistic cost estimation
- Sound capital structure: Fund Source
- Provision for working capital requirement
- Generation of sufficient cash flows to cover debt service liability
- Generation of adequate profit
- Safety margin
- Break-even point
- Payback period
- Past and current demand trends
- Past and present supply position
- Production possibilities and constraints
- Imports and exports
- Nature of competition
- Cost structure
- Elasticity of demand
- Direct economic benefits and costs in terms of shadow prices
- Impact of the project on distribution of income in society

- Impact on level of savings and investments in society
- Impact on fulfillment of national goals

  (a) Self-sufficiency
  (b) Employment
  (c) Social order.

3.3 Project Selection: Illustrative Example

A detailed brainstorming is conducted to identify the set of project options for controlling global warming. First, the project options are evaluated for various key criteria, as discussed in the previous section. Subsequently, each of the project options is ranked to identify the top five options with most potential for controlling global warming. The various project options for controlling global warming are listed below.

1. Identification of sources of greenhouse gases (P_1)
2. Innovation labs for sustainable future (P_2)
3. Hi-tech refrigeration (P_3)
4. Generation and distribution of clean energy (P_4)
5. Forest plantation and protection (P_5)
6. Recycling (P_6)
7. Population control (P_7)
8. Developing silvopasture (P_8)
9. Cleaning up polluted air (P_9)
10. Public awareness drives (P_{10})

Evaluation of The Project Options Using the Checklist for Project Appraisal

There are many criteria for evaluating different projects, out of which a few have been selected, keeping in mind the performance expected from projects for controlling global warming. The evaluation is done on a scale of 1–10, as presented in Table 3.1.

The scale of 1–10 identifies "1" as the least preferred option, and "10" as a highly preferred project for a given benefit criterion ("higher is better"). However, for cost criteria ("lower is better"), the reverse interpretation is applicable. After evaluating all the projects on different criteria of project appraisal, projects *P*4, *P*5, *P*6, *P*7, *P*10 are ranked as the most potential top five options for controlling global warming. The rank-wise list of projects is as follows:

1. Population control (97) (Score: 97)
2. Generation and distribution of clean energy (*P*4) (Score: 95)
3. Forest plantation and protection (*P*5) (Score: 92)
4. Recycling (*P*6) (Score: 91)
5. Public awareness drives (*P*10) (Score: 81).

Table 3.1 Evaluation of global warming projects

Criteria/projects	P_1	P_2	P_3	P_4	P_5	P_6	P_7	P_8	P_9	P_{10}
Availability of inputs at reasonable cost	10	6	6	7	10	7	8	5	6	9
Economics of scale in production	1	5	7	10	8	7	8	7	7	10
Appropriate technology and alternative ways of production	8	8	6	9	9	7	6	8	6	10
Maintenance and repairs	10	6	5	5	9	8	9	8	6	8
Exhaustive and realistic cost estimation	8	6	7	6	8	7	8	7	5	8
Sound capital structure: fund source	7	5	6	7	6	5	7	6	7	9
Generation of sufficient cash flows	9	6	8	9	6	7	6	6	5	6
Cost structure	10	7	7	6	8	7	9	8	6	8
Direct economic benefits and costs in terms of shadow prices	2	4	5	8	3	5	6	6	4	5
Impact of project on distribution of income in society	1	4	1	5	7	6	6	5	1	5
Impact on level of savings and investments in society	1	3	6	7	5	7	5	2	1	1
Impact on fulfillment of national goals:										
• Self-sufficiency	1	4	1	5	3	7	8	4	1	1
• Employment	3	2	2	4	4	5	5	2	2	3
• Social order	3	3	1	7	6	6	6	5	1	2
Total	74	69	68	95	92	91	97	79	58	81

3.4 Multi-criteria Decision Making (MCDM) for Project Selection

MCDM is a widely used approach for evaluating project options, which accommodates the trade-off between both tangible and intangible criteria for project selection. This section demonstrates an application of MCDM techniques such as Technique for Order Preference using Similarity to Ideal Solution (TOPSIS), Elimination et Choice Translating Reality (ELECTRE), and Analytical Hierarchy Process (AHP) for ranking the top five project options for global warming identified in the previous section.

The top five projects identified for controlling global warming in the previous section are renamed in sequence as $P1$–$P5$ as follows:

1. Population control (P_1)
2. Generation and distribution of clean energy (P_2)
3. Forest plantation and protection (P_3)
4. Recycling (P_4)
5. Public awareness drives (P_5).

Table 3.2 Decision matrix for the project identified for controlling global warming

	C_1	C_2	C_3	C_4	C_5
P_1	7	5	9	8	5
P_2	10	9	6	3	7
P_3	7	8	7	5	6
P_4	8	9	6	6	6
P_5	9	8	6	5	5

An application of MCDM techniques for the project selection is demonstrated by considering a few of the most important criteria for evaluating identified projects $P1$–$P5$. This includes:

1. Availability of inputs at a reasonable cost (C_1)
2. Lower Maintenance and repairs (C_2)
3. Generation of sufficient cash flows (C_3)
4. Direct economic benefits and costs in terms of shadow prices (C_4)
5. Impact of the project on distribution of income in society (C_5).

The evaluation scale was set such that all the criteria are evaluated on the benefit scale. The resultant decision matrix for the projects is given in Table 3.2.

3.4.1 Technique for Order Preference Using Similarity to Ideal Solution (TOPSIS)

TOPSIS is a multi-criteria decision-making tool developed by Yoon and Ching-Lai Hwang and has been appraised by various levels of decision-makers. It takes a compensatory aggregation approach for identifying the best alternative among the identified set of alternatives. The method is based on the concept that the best alternative should have the least geometric distance from a positive ideal solution and similarly farthest from any negative ideal solution. The alternatives are initially graded based on their similarity with an ideal solution, which is a solution that is best in all aspects and therefore practically may not exist. The alternative with higher similarity to the best solution is rated higher than alternatives with a lower similarity value. The TOPSIS approach involves identifying the distance of each alternative from the ideal solution and choosing the best alternative based on this distance.

Step 1: Normalize the decision matrix

Since different criteria will have different scales for measurement, it becomes essential to standardize the scale for each factor. Various linear scale transformation approaches exist. The normalization method, which is widely used, is shown by the equation below.

Table 3.3 Normalized decision matrix for TOPSIS

	C_1	C_2	C_3	C_4	C_5
P_1	0.378	0.270	0.583	0.634	0.382
P_2	0.544	0.486	0.389	0.238	0.535
P_3	0.378	0.432	0.454	0.397	0.459
P_4	0.435	0.486	0.389	0.476	0.459
P_5	0.490	0.432	0.389	0.397	0.382

$$r_{ij} = x_{ij}^2 / \sqrt{\sum_1^n x_{ij}^2}$$

The normalized decision matrix for the decision matrix given in Table 3.2 is obtained as shown in Table 3.3.

Step 2: Determine criteria weightages

The weights are assigned to the various criteria according to their importance.

$$w_{ij} \quad j = 1, 2, \ldots n$$

The weight for each criterion is assumed to obtain the weighted decision matrix. The assumed weights for the projects $P1$, $P2$, $P3$, $P4$, and $P5$ are 0.3, 0.2, 0.3, 0.1, and 0.1, respectively.

Step 3: Obtain weighted normalized decision matrix

The weighted normalized decision matrix calculates the product of the normalized decision matrix and the associated weights. For example, the weighted normalized value v_{ij} can be computed as below:

$$v_{ij} = w_{ij} \times r_{ij}$$

The weighted decision matrix is obtained as given in Table 3.4.

Step 4: Determine the positive and negative ideal solution

The positive ideal solution and negative ideal solution are determined as follows.

$$A* = \{(\max v_{ij} | j \in J), (\min v_{ij} | j \in J')\}$$
$$A- = \{(\min v_{ij} | j \in J), (\max v_{ij} | j \in J')\}$$

$J = 1, 2, 3 \ldots, n$ where J is associated with the benefit criteria.
$J' = 1, 2, 3 \ldots, n$ where J' is associated with the cost criteria.
The weighted decision matrix is used to compute ideal ($A*$) and negative ideal ($A-$) solutions.

Table 3.4 Weighted decision matrix for TOPSIS

	C_1 ($W_1 = 0.3$)	C_2 ($W_2 = 0.2$)	C_3 ($W_3 = 0.3$)	C_4 ($W_4 = 0.1$)	C_5 ($W_5 = 0.1$)
P_1	*0.1134*	*0.0540*	**0.1749**	**0.0634**	*0.0382*
P_2	**0.1632**	**0.0972**	*0.1167*	*0.0238*	**0.0535**
P_3	0.1134	0.0864	0.1362	0.0397	0.0459
P_4	0.1305	0.0972	0.1167	0.0476	0.0459
P_5	0.1470	0.0864	0.1167	0.0397	0.0382

the numbers in "bold" and "italic" indicate ideal (A*) and negative ideal (A-) solutions. For benefit criteria - "higher is better" and cost criteria—"lower is better" are considered as ideal (A*) solutions. All C1 to C5 are benefit criteria and hence numbers in bold indicate "ideal (A*) solution". The lowest value in each column (C1 to C5) indicated as "italic" represent "negative ideal (A-) solution"

$$A* = (\mathbf{0.1632, 0.0972, 0.1749, 0.0634, 0.0535})$$
$$A- = (0.1134, 0.0540, 0.1167, 0.0238, 0.0382)$$

Step 5: Calculate separation measure

The separation of each alternative from the positive ideal one is given by:

$$S_i^* = \sqrt{\sum \left(v_{ij} - v_j^*\right)^2} \quad j = 1 \text{ where } i = 1, 2 \ldots, m$$

Similarly, the separation of each alternative from the negative ideal one is given by:

$$S_i^- = \sqrt{\sum \left(v_{ij} - v_j^-\right)^2} \quad j = 1 \text{ where } i = 1, 2 \ldots, m$$

The separation measures are calculated using the values of ideal and negative ideal solutions.

Ideal solution	Negative ideal solution
$S_1^* = 0.0677$	$S_1^- = 0.0704$
$S_2^* = 0.0704$	$S_2^- = 0.0677$
$S_3^* = 0.0700$	$S_3^- = 0.0417$
$S_4^* = 0.0690$	$S_4^- = 0.0528$
$S_5^* = 0.0675$	$S_5^- = 0.0493$

Step 6: Calculate the relative closeness (RC) to the ideal solution

The relative closeness of Ai to $A*$ is defined as:

$$C_i^* = S_i^- / (S_i^* + S_i^-), 0 \leq C_i^* \leq 1 \text{ where } i = 1, 2 \ldots, m$$

The larger the C_i^* value, the better the performance of the alternatives.

The relative closeness values are calculated using the values of separation measures for ideal and negative ideal solutions. This provides relative closeness to the ideal solution.

$$\text{RC}_1^* = 0.510$$
$$\text{RC}_2^* = 0.490$$
$$\text{RC}_3^* = 0.373$$
$$\text{RC}_4^* = 0.433$$
$$\text{RC}_5^* = 0.422$$

Step 7

Arrange the options in the decreasing order of the C_i^* values to rank the alternatives.

These relative closeness values indicate the ranks (priorities) of the projects as P_1, P_2, P_4, P_5, and P_3.

3.4.2 Elimination Et Choice Translating Reality (ELECTRE)

Elimination Et Choix Traduisant la Reality (ELECTRE) is a decision-making method proposed by Bernard Roy and his team at the SEMA consultancy company. The approach is classified as the "outranking method" of decision-making. It consists of two significant steps: first, outranking relations that provide the basis for pair-wise comparison of various alternative course of actions available, followed by the second exploitation phase that elaborates the multiple recommendations based on the first step. The ELECTRE method employs the concept of ranking various alternatives based on a set of attributes or criteria. A solution or alternative is considered to exhibit dominance over other alternatives if it performs better than other alternatives in some set of criteria and at least equal to other alternatives in the remaining set of criteria. The ranking is done based on pair-wise comparisons, and they are ranked relative to each other. Outranking the relationship between two solutions suggests that even when one alternative does not dominate the other quantitatively, the decision-maker is likely to consider one alternative as better than the other.

Step 1 Normalizing the Decision Matrix

Since different criteria will have different scales for measurement, it becomes essential to standardize the scale for each factor. Various linear scale transformation approaches exist. The normalization method, which is widely used, is shown by the equation below.

Table 3.5 Normalized decision matrix for ELECTRE

	C_1	C_2	C_3	C_4	C_5
P_1	0.378	0.270	0.583	0.634	0.382
P_2	0.544	0.486	0.389	0.238	0.535
P_3	0.378	0.432	0.454	0.397	0.459
P_4	0.435	0.486	0.389	0.476	0.459
P_5	0.490	0.432	0.389	0.397	0.382

$$r_{ij} = x_{ij}^2 / \sqrt{\sum_1^n x_{ij}^2}$$

The normalized matrix X of the dimensionless values is defined as follows:

$$X = \{x11x12 \cdots x1Nx21x22 \cdots x2NxM1xM2 \cdots xMN\}$$

where M is the number of alternatives and N is the number of criteria, and x_{ij} is the new and dimensionless preference measure of the ith alternative in terms of the jth criterion.

The normalized decision matrix is given in Table 3.5.

Step 2 Weighting the Normalized Decision Matrix

The values from the previous step are weighted by multiplying the matrix X with the weights associated with each factor/criteria.

The weighted matrix Y is given by $Y = XW$, where

$$\begin{aligned} Y &= \{y11y12 \cdots y1Ny21y22 \cdots y2NyM1yM2 \cdots yMN\} \\ &= \{w1y11w2y12 \cdots wNy1Nw1y21w2y22 \\ &\quad \cdots wNy2Nw1yM1w2yM2 \cdots wNyMN\} \end{aligned}$$

and $W = \{y11y12 \cdots y1Ny21y22 \cdots y2NyM1yM2 \cdots yMN\}$ Also, $\sum_{i=1}^{M} wi = 1$.

The weighted decision matrix obtained is given in Table 3.6.

Step 3 Determine the Concordance and Discordance Sets

The concordance set *Ckl* of two alternatives *Ak* and *AR*, where $M.k, l \geq 1$, is defined as the set of all criteria for which *Ak* is preferred to *Al*. That is, the following is true.

Ckl = {*j, such that: ykj* ≥ *ylj*}, for $j = 1, 2, 3, \ldots, N$. The complementary subset is called the discordance set, and it is described as follows:

Dkl = {*j, such that: ykj* < *ylj*}, for $j = 1, 2, 3 \ldots, N$.

The concordance and discordance sets are calculated in Table 3.7.

Step 4 Construct the Concordance and Discordance Matrices

Concordance index represents the relative value of all the elements in concordance matrix C. The concordance index *ckl* is the sum of the weights associated with the criteria contained in the concordance. That is, the following is true:

$$Ckl = \sum_{j \in Ckl} Wj \quad \text{for } j' \, 1, 2, 3, \ldots, N.$$

Table 3.6 Weighted decision matrix for ELECTRE

	C_1 ($W_1 = 0.3$)	C_2 ($W_2 = 0.2$)	C_3 ($W_3 = 0.3$)	C_4 ($W_4 = 0.1$)	C_5 ($W_5 = 0.1$)
P_1	0.1134	0.0540	0.1749	0.0634	0.0382
P_2	0.1632	0.0972	0.1167	0.0238	0.0535
P_3	0.1134	0.0864	0.1362	0.0397	0.0459
P_4	0.1305	0.0972	0.1167	0.0476	0.0459
P_5	0.1470	0.0864	0.1167	0.0397	0.0382

Table 3.7 Concordance and discordance sets

Concordance set	Discordance set
$C_{12} = \{3, 4\}$	$D_{12} = \{1, 2, 5\}$
$C_{13} = \{1, 3, 4\}$	$D_{13} = \{2, 5\}$
$C_{14} = \{3, 4\}$	$D_{14} = \{1, 2, 5\}$
$C_{15} = \{3, 4, 5\}$	$D_{15} = \{1, 2\}$
$C_{21} = \{1, 2, 5\}$	$D_{21} = \{3, 4\}$
$C_{23} = \{1, 2, 5\}$	$D_{23} = \{3, 4\}$
$C_{24} = \{1, 2, 3, 5\}$	$D_{24} = \{4\}$
$C_{25} = \{1, 2, 3, 5\}$	$D_{25} = \{4\}$
$C_{31} = \{2, 5\}$	$D_{31} = \{1, 3, 4\}$
$C_{32} = \{3, 4\}$	$D_{32} = \{1, 2, 5\}$
$C_{34} = \{3, 5\}$	$D_{34} = \{1, 2, 4\}$
$C_{35} = \{2, 3, 4, 5\}$	$D_{35} = \{1\}$
$C_{41} = \{1, 2, 5\}$	$D_{41} = \{3, 4\}$
$C_{42} = \{4\}$	$D_{42} = \{1, 2, 3, 5\}$
$C_{43} = \{1, 2, 4\}$	$D_{43} = \{3, 5\}$
$C_{45} = \{2, 3, 4, 5\}$	$D_{45} = \{1\}$
$C_{51} = \{1, 2\}$	$D_{51} = \{3, 4, 5\}$
$C_{52} = \{4\}$	$D_{52} = \{1, 2, 3, 5\}$
$C_{53} = \{1\}$	$D_{53} = \{2, 3, 4, 5\}$
$C_{54} = \{1\}$	$D_{54} = \{2, 3, 4, 5\}$

The concordance index indicates the relative importance of alternative Ak to alternative Al. Apparently, $0 \prec ckl \succ 1$. Therefore, the concordance matrix C is defined as follows:

$$C = \{-C12 \cdots C1MC21 - \cdots C2MCM1CM2 \cdots -\}$$

It should be noted here that the entries of matrix C are not defined when $k = l$.

The discordance matrix D expresses the degree that a certain alternative Ak is worse than a competing alternative Al. The elements $dk\ l$ of the discordance matrix are defined as follows:

$$Dkl = \frac{\text{Max } |ykj - ylj|j \in Dkl}{\text{Max}|ykj - ylj|}$$

The discordance matrix is defined as follows:

$$D = \{-d12 \cdots d1Md21 - \cdots d2MdM1dM2 \cdots -\}$$

As before, the entries of matrix D are not defined when $k = l$.

It should also be noted here that the previous two $M \times M$ matrices are not symmetric.

The concordance discordance matrices are derived as below.

Concordance Matrix

$$[-0.40.70.40.50.6 - 0.60.90.90.30.4 - 0.40.70.60.10.6 - 0.70.50.10.30.3-]$$

Discordance Matrix

$$[-0.8560.8370.7420.5771 - 0.2170.7280.98111 - 0.8771111 - 1110.5800.655-]$$

Step 5 Determine the Concordance and Discordance Dominance Matrices

The concordance dominance matrix is constructed utilizing a threshold value for the concordance index. For example, Ak will only have a chance to dominate Al if its corresponding concordance index ckl exceeds at least ascertain threshold value c. That is, the following is true:

$$ckl > c$$

The threshold value c can be determined as the average concordance index. That is, the following relation is true:

$$\underline{c} = \frac{1}{M(M-1)} \sum_{k=1 \text{ and } k \neq l}^{M} \sum_{l=1 \text{ and } l \neq k}^{M} ckl$$

Based on the threshold value, the concordance dominance matrix F is determined as follows:

$$fkl = 1, \quad \text{if}\, ckl \geq c,$$
$$fkl = 0, \quad \text{if}\, ckl < c.$$

Similarly, the discordance dominance matrix G is defined by using a threshold value d, where d is defined as follows:

$$\underline{d} = \frac{1}{M(M-1)} \sum_{k=1 \text{ and } k \neq l}^{M} \sum_{l=1 \text{ and } l \neq k}^{M} dkl$$

$$\text{and: } gkl = 1, \text{if}\, dkl \geq d,$$
$$gkl = 0, \text{if}\, dkl < d.$$

Concordance Dominance Matrix

Finding average concordance index c,

$$c = \frac{\begin{array}{c}0.4 + 0.7 + 0.4 + 0.5 + 0.6 + 0.6 + 0.9 + 0.9 + 0.3 + 0.4 + 0.4 \\ +0.7 + 0.6 + 0.1 + 0.6 + 0.7 + 0.5 + 0.1 + 0.3 + 0.3\end{array}}{20}$$

$$c = 0.5$$

All entries in the concordance matrix lesser than 0.5 are treated as 0, and those greater than or equal to 0.5 are treated as 1.

The resultant concordance dominance matrix is as follows.

$$[-01011 - 11100 - 01101 - 11000-]$$

Discordance Dominance Matrix

Finding average discordance index d,

$$d = \frac{\begin{array}{c}0.856 + 0.837 + 0.742 + 0.577 + 1 + 0.217 + 0.728 + 0.981 \\ +1 + 1 + 0.877 + 1 + 1 + 1 + 1 + 1 + 1 + 1 + 0.580 + 0.655\end{array}}{20}$$

$$d = 0.852$$

All entries in the discordance matrix lesser than 0.5 are treated as 0, and those greater than or equal to 0.5 are treated as 1.

The resultant discordance dominance matrix is as follows.

$$[-01110 - 11000 - 00000 - 00011-]$$

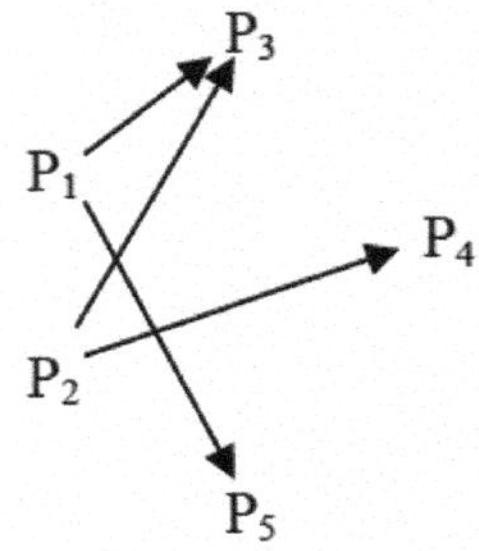

Fig. 3.4 Outranking relationships among projects

Aggregate Dominance Matrix

$$[-01010-11000-00000-00000-]$$

It is evident from Fig. 3.4 that projects 1 and 2 dominate projects 3, 4, and 5. However, we cannot tell the preference relation between projects 1 and 2. Hence, projects 3, 4, and 5 can be eliminated by ELECTRE.

3.4.3 Analytical Hierarchy Process (AHP)

Developed by Thomas L. Saaty, the AHP technique accommodates the linear hierarchical structure of the problems, including criteria, sub-criteria, and alternatives. It can include inputs in both forms like actual measurement of criteria such as cost, profit; and subjective assessment by experts that capture their satisfaction, opinions, beliefs, and priorities related to any decision. AHP tries to identify the relative significance of each element of a decision by comparing each pair of alternatives at every hierarchy level. The technique thus examines multiple alternatives and provides the decision-maker with relative priorities of the various considered alternatives. A decision-maker has eased in converting its subjective opinion into an objective while implementing AHP for a real-life problem. The procedural steps of AHP are illustrated in Fig. 3.5.

AHP involves a large number of pair-wise comparisons. Therefore, it is essential to check the consistency at each stage and ensure the reliability of subjective inputs collected from the experts. The procedure used for checking consistency in AHP is explained below.

Consistency Checking Procedure for AHP

Step 1 Multiply each column of the pair-wise comparison matrix by the corresponding weight $C_1, C_2, C_3 \ldots C_m$.

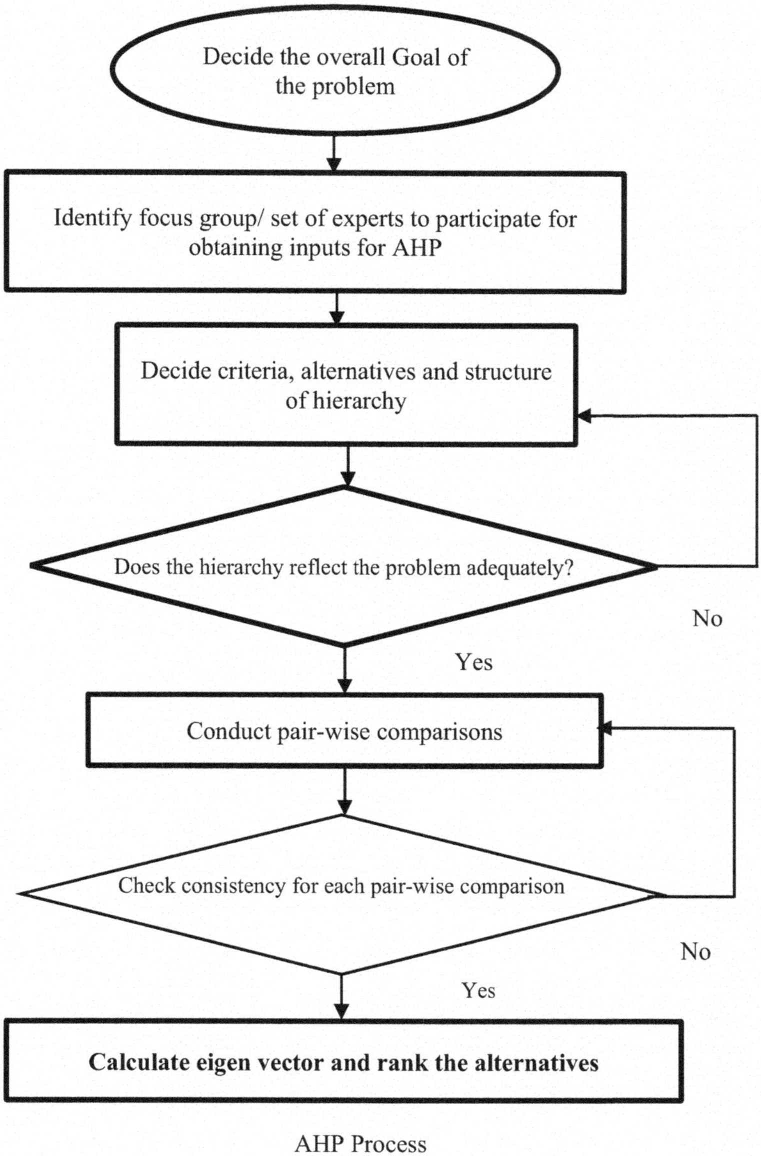

Fig. 3.5 Flow diagram for procedural steps of AHP

$$A.C = \begin{bmatrix} a_{11} & a_{12} & \ldots & a_{1m} \\ a_{21} & a_{22} & \ldots & a_{2m} \\ \ldots & \ldots & \ldots & \ldots \\ a_{m1} & a_{m2} & & a_{mm} \end{bmatrix} \begin{bmatrix} c_1 \\ c_2 \\ \ldots \\ c_m \end{bmatrix} = \begin{bmatrix} x_1 \\ x_2 \\ \ldots \\ x_m \end{bmatrix}$$

Step 2 Calculate the eigenvector λ by the following equation

$$\begin{bmatrix} \lambda_1 \\ \lambda_2 \\ \lambda_3 \\ \ldots \\ \ldots \\ \lambda_n \end{bmatrix} = \begin{bmatrix} x_1/c_1 \\ x_2/c_2 \\ x_3/c_3 \\ \ldots \\ \ldots \\ x_n/c_n \end{bmatrix}$$

Step 3 Compute the average of the values, denote it by λ_{avg}, which is the maximum Eigenvalue of the pair-wise comparison matrix.

$$\lambda_{\text{avg}} = \left| \frac{\sum \lambda_i}{n} \right|$$

Step 4 Calculate Consistency Index (CI)

$$\left| \frac{\lambda_{\text{avg}} - n}{n - 1} \right|$$

Step 5 Calculate Consistency Ratio (CI/RI)

If this ratio (CI/RI) is substantial (Saaty suggests > 0.10), then we are not consistent enough, and the best thing to do is to go back and revise the comparisons.

The value of RI can be directly taken from the following standard table.

Saaty's table for CR calculation

Order of matrix (n)	1	2	3	4	5	6	7	8	9	10	11	12	13	14	15
Random index (RI)	0.00	0.00	0.58	0.90	1.12	1.24	1.32	1.41	1.45	1.49	1.51	1.48	1.56	1.57	1.59

For demonstrating an application of AHP for project selection, the preferences between various criteria are assumed as follows:

- C_1 is equally to moderately preferred over C_2.
- C_1 is very strongly preferred over C_3.
- C_1 is equally preferred over C_4.
- C_1 is strongly preferred over C_5.
- C_2 is moderate to strongly preferred over C_3.
- C_2 is moderately preferred over C_4.
- C_2 is very strongly to extremely preferred over C_5.
- C_3 strongly to very strongly preferred over C_4.
- C_3 is moderately preferred over C_5.
- C_4 is extremely preferred over C_5.

Criteria	C_1	C_2	C_3	C_4	C_5
C_1	1	1/2	1/7	1	1/5
C_2	2	1	1/4	1/3	1/8
C_3	7	4	1	1/6	1/3
C_4	1	3	6	1	1/9
C_5	5	8	3	9	1

Similarly, assuming project-wise preference for each criterion, the following matrices are obtained.

Criteria 1

Project	P_1	P_2	P_3	P_4	P_5
P_1	1	1/6	1/2	1/4	1/7
P_2	6	1	1/3	1/6	1
P_3	2	3	1	8	5
P_4	4	6	1/8	1	6
P_5	7	1	1/5	1/6	1

Criteria 2

Project	P_1	P_2	P_3	P_4	P_5
P_1	1	1/3	2	1/7	1/6
P_2	3	1	1/5	1/2	3
P_3	1/2	5	1	4	2
P_4	7	2	1/4	1	1/7
P_5	6	1/3	1/2	7	1

Criteria 3

Project	P_1	P_2	P_3	P_4	P_5
P_1	1	3	8	1/6	5
P_2	1/3	1	4	1/2	1/6
P_3	1/8	1/4	1	3	8
P_4	6	2	1/3	1	1/4
P_5	1/5	6	1/8	4	1

Criteria 4

Project	P_1	P_2	P_3	P_4	P_5
P_1	1	4	6	1/5	1/2
P_2	1/4	1	1/3	1/8	1/6
P_3	1/6	3	1	1/5	1/8
P_4	5	8	5	1	7
P_5	2	6	8	1/7	1

Criteria 5

Project	P_1	P_2	P_3	P_4	P_5
P_1	1	7	1/5	1/6	1/4
P_2	1/7	1	1/6	1/5	3
P_3	5	6	1	4	1
P_4	6	5	1/4	1	3
P_5	4	1/3	1	1/3	1

Normalized matrices are as follows:

Criteria 1

Project	P_1	P_2	P_3	P_4	P_5	Average
P_1	0.05	0.0149	0.2315	0.0261	0.0109	0.0667
P_2	0.3	0.0896	0.0299	0.0174	0.0761	0.1026
P_3	0.1	0.2688	0.4630	0.8350	0.3805	0.4095
P_4	0.2	0.5376	0.0120	0.1044	0.4566	0.2621
P_5	0.35	0.0896	0.0926	0.0174	0.0761	0.1251

Criteria 2

Project	P_1	P_2	P_3	P_4	P_5	Average
P_1	0.0571	0.0385	0.5063	0.0113	0.0264	0.1279
P_2	0.1714	0.1155	0.0506	0.0396	0.4754	0.1705
P_3	0.0285	0.5774	0.2532	0.3165	0.3170	0.2985
P_4	0.4	0.2309	0.0633	0.0791	0.0226	0.1592
P_5	0.3429	0.0385	0.1266	0.5538	0.1585	0.2441

Criteria 3

Project	P_1	P_2	P_3	P_4	P_5	Average
P_1	0.1307	0.2449	0.5944	0.0192	0.3467	0.2672
P_2	0.0436	0.0816	0.2971	0.0577	0.0116	0.0983
P_3	0.0163	0.0204	0.0743	0.3460	0.5548	0.2024
P_4	0.7843	0.1633	0.0248	0.1153	0.0173	0.2210
P_5	0.0261	0.4868	0.0092	0.4316	0.0693	0.2046

Criteria 4

Project	P_1	P_2	P_3	P_4	P_5	Average
P_1	0.1188	0.1667	0.2951	0.1198	0.0569	0.1515
P_2	0.0297	0.4170	0.0164	0.0749	0.0190	0.1114
P_3	0.0198	0.1250	0.0492	0.1198	0.0142	0.0656
P_4	0.5938	0.3333	0.2459	0.5988	0.7964	0.5136
P_5	0.2375	0.2500	0.3935	0.0855	0.1138	0.2161

Criteria 5

Project	P_1	P_2	P_3	P_4	P_5	Average
P_1	0.0620	0.3621	0.0763	0.0292	0.0303	0.1120
P_2	0.0089	0.0517	0.0636	0.0351	0.3636	0.1046
P_3	0.3098	0.3104	0.3816	0.7018	0.1212	0.3650
P_4	0.3717	0.2586	0.0954	0.1754	0.3636	0.2529
P_5	0.2478	0.0172	0.3816	0.0585	0.1212	0.1653

After finding the average for each criterion, the below matrix is obtained.

Project/criteria	C_1	C_2	C_3	C_4	C_5
P_1	0.0667	0.1279	0.2672	0.1515	0.1120
P_2	0.1026	0.1705	0.0983	0.1114	0.1046
P_3	0.4095	0.2985	0.2024	0.0656	0.3650
P_4	0.2621	0.1592	0.2210	0.5136	0.2529
P_5	0.1251	0.2441	0.2046	0.2161	0.1653

Normalizing the criteria preference matrix.

Criteria	C_1	C_2	C_3	C_4	C_5	Average
C_1	0.0625	0.0303	0.0137	0.0870	0.1130	0.0613
C_2	0.1250	0.0606	0.0240	0.0290	0.0706	0.0618
C_3	0.4375	0.2424	0.0962	0.0145	0.1883	0.1958
C_4	0.0625	0.1818	0.5775	0.0870	0.0628	0.1943
C_5	0.3125	0.4848	0.2887	0.7826	0.5650	0.4867

The row average obtained from the criteria matrix is the performance vector for the criteria.

Criteria	Performance vector
C_1	0.0613
C_2	0.0618
C_3	0.1958
C_4	0.1943
C_5	0.4867

Project Priority Calculation

	C_1	C_2	C_3	C_4	C_5		Performance vector
P_1	0.0667	0.1279	0.2672	0.1515	0.1120		0.0613
P_2	0.1026	0.1705	0.0983	0.1114	0.1046		0.0618
P_3	0.4095	0.2985	0.2024	0.0656	0.3650	X	0.1958
P_4	0.2621	0.1592	0.2210	0.5136	0.2529		0.1943
P_5	0.1251	0.2441	0.2046	0.2161	0.1653		0.4867

$$
\begin{aligned}
\text{Score } P_1 &= 0.0667(0.0613) + 0.1279(0.0618) + 0.2372(0.1958) \\
&\quad + 0.1515(0.1943) + 0.1120(0.4867) \\
\text{Score } P_1 &= 0.1424 \\
\text{Score } P_2 &= 0.1026(0.0613) + 0.1705(0.0618) + 0.0983(0.1958) \\
&\quad + 0.1114(0.1943) + 0.1046(0.4867) \\
\text{Score } P_2 &= 0.1086 \\
\text{Score } P_3 &= 0.4095(0.0613) + 0.2985(0.0618) + 0.2024(0.1958) \\
&\quad + 0.0656(0.1943) + 0.3650(0.4867) \\
\text{Score } P_3 &= 0.2736 \\
\text{Score } P_4 &= 0.2621(0.0613) + 0.1592(0.0618) + 0.2210(0.1958) \\
&\quad + 0.5136(0.1943) + 0.2529(0.4867) \\
\text{Score } P_4 &= 0.2921 \\
\text{Score } P_5 &= 0.1251(0.0613) + 0.2441(0.0618) + 0.2046(0.1958) \\
&\quad + 0.2161(0.1943) + 0.1653(0.4867) \\
\text{Score } P_5 &= 0.1853
\end{aligned}
$$

Based on the scores calculated for the projects *P*1, *P*2, *P*3, *P*4, and *P*5, the ranks (priorities) are derived below.

Project	Score	Rank
P_1	0.1424	4
P_2	0.1086	5
P_3	0.2736	2
P_4	0.2921	1
P_5	0.1853	3

Projects *P*4 and *P*3 are the top-ranking projects as per AHP analysis. However, the results of TOPSIS and ELECTRE indicate *P*1 and *P*2 as high-priority projects.

The AHP results may differ because of extensive pair-wise comparisons and a lack of consistency in the subjective opinions of the experts. Hence, we can consider projects 1, 2, and 4 as the three top-ranking projects based on TOPSIS and ELECTRE.

3.5 Financial Appraisal of Projects

Importance of Financial Appraisal

Financial appraisal involves analyzing information in the financial statements to make it easily comparable with data of similar other projects. In addition, it deals with the development of the economic profile of a project. The goal is to check if the project is lucrative enough to gather funds for the various activities and provide sustainable cash flows. Therefore, it is essential to perform a financial appraisal of the projects to check their long-term sustainability.

Financial analysis has emerged as an essential practice because of the significant interest of different parties in an organization's financial results. Looking at the last few years, the capital ownership of many public entities has turned broad-based. Many stakeholders are associated with the economic consequences, including creditors, suppliers, financial institutes like banks, industry finance corporations, investors, employees, tax authorities, and governments. Therefore, different stakeholders look at the financial statements according to their different perspectives. Many methods have been developed to analyze financial information for making inferences regarding an organization's financial health, profitability, and efficiency.

Criteria for Financial Appraisal

1. Net Present Value (N.P.V.)
2. Modified Net Present Value (Modified N.P.V.)
3. Benefit–Cost Ratio (B.C.R.)/Profitability Index
4. Internal Rate of Return (I.R.R.)
5. Modified Internal Rate of Return (M.I.R.R.)
6. Payback Period
7. Break-even point
8. Discounted Payback Period
9. Accounting Rate of Return
10. Debt Service Coverage Ratio (D.S.C.R.).

Net Present Value (N.P.V.)

The net present value is the difference between the present value of the cash inflow and the current value of cash outflows. It denotes the net profit over and above the compensation for time and risk. The decision associated with N.P.V. is,

- If N.P.V. is positive, accept the project
- If N.P.V. is negative, reject the project

The formula for finding out the N.P.V. of a project is,

$$\text{N.P.V.} = \sum_{t=1}^{n} \frac{C_t}{(1+r)^t}$$

where C_t is cash flow at the end of year "t", "n" is the life of the project and "r" is the discount rate.

Modified Net Present Value (Modified N.P.V.)

The N.P.V. method assumes that intermediate cash flows are reinvested at a rate of return equal to the cost of capital. When this assumption is invalid, reinvestment rates applicable to the intermediate cash flow need to be defined for calculating the modified N.P.V.

To find the modified N.P.V., the first terminal value of the project's cash inflows has to be calculated using the following equation,

$$\text{T.V.} = \sum_{t=1}^{n} \text{CF}_t (1+r')^{n-t}$$

where T.V. is the future value of the cash inflows, CF_t is the cash inflow at the end of year "t", and "r" is the reinvestment rate.

After calculating the terminal value, the modified N.P.V. is calculated using the following equation,

$$\text{N.P.V.}^* = \frac{\text{T.V.}}{(1+r)^n} - I$$

where N.P.V.* is the modified N.P.V., "r" is the cost of capital, and "I" is the investment outlay.

Benefit–Cost Ratio (B.C.R.)/Profitability Index

The benefit–cost ratio (B.C.R.), also known as the profitability index, is the ratio of the present value of benefits to the initial investment cost.

$$\text{B.C.R.} = \frac{\text{P.V.B.}}{I}$$

Internal Rate of Return (I.R.R.)

The internal rate of return is the discount rate at which the net present value of a project becomes zero. It is the value of r in the following equation,

$$\text{Investment} = \sum_{t=1}^{n} \frac{C_t}{(1+r)^t}$$

where C_t is cash flow at the end of year "t", "n" is the life of the project and "r" is the internal rate of return.

Modified Internal Rate of Return (M.I.R.R.)

The internal rate of return is not suitable when cash flows are unconventional or in a comparison between two projects to find out which one is better. It cannot differentiate between lending and borrowing.

To find the modified internal rate of return, calculate the present value of the costs (P.V.C.) of the project, using the cost of capital "r" as the discount rate.

$$\text{P.V.C.} = \sum_{t=0}^{n} \frac{\text{Cash Outflow}_t}{(1+r)^t}$$

Now the terminal value of the expected cash inflows has to be calculated.

$$\text{T.V.} = \sum_{t=0}^{n} \text{Cash Inflow}_t (1+r)^{n-t}$$

After finding out the present value of costs and terminal value of expected cash flows, the modified internal rate of return can be obtained from the following equation,

$$\text{P.V.C.} = \frac{\text{T.V.}}{(1+\text{M.I.R.R.})^n}$$

Payback Period

The payback period can be defined as the length of time required to recover the initial cash outlay on the project. It has a simple concept and does not consider the time value of money. It is a measure of the project's capital recovery, not profitability.

Break-Even Point

It is determined by break-even analysis by finding the point where the generated revenue and costs incurred in generating the revenue become equal. It calculates the margin of safety; it is the amount that revenues exceed the break-even point. It is the amount that revenues can fall while still staying above the break-even point.

Accounting Rate of Return

The accounting rate of return is the measure of profitability that relates income to investment. It is also known as the average rate of return on investment in a project.

There are many ways in which the accounting rate of return can be calculated since various methods of measuring investment and income are used.

(a) $\frac{\text{Average income after tax}}{\text{Initial investment}}$

(b) $\frac{\text{Average income after tax}}{\text{Average investment}}$

(c) $\frac{\text{Average income after tax but before interest}}{\text{Initial investment}}$

(d) $\frac{\text{Average income after tax but before interest}}{\text{Average investment}}$

(e) $\frac{\text{Average income before interest and taxes}}{\text{Initial investment}}$

(f) $\frac{\text{Average income before interest and taxes}}{\text{Average investment}}$

(g) $\frac{\text{Total income after tax but before depreciation} - \text{Initial investment}}{(\text{Initial investment}/2) * \text{years}}$

Debt Service Coverage Ratio (D.S.C.R.)

The Debt Service Coverage Ratio is a measure of the cash flow available to pay current debt obligations.

$$\text{D.S.C.R} = \frac{\text{Total cash accrual}}{\text{Debt service requirements}}$$

where total cash accrual includes,

- Profit after tax
- Depreciation
- Interest on a term loan

and debt service requirements include,

- Interest on a term loan
- Repayment of term loan.

Financial Appraisal of Top Three Projects

After applying various MCDM tools (TOPSIS, ELECTRE, and AHP) in the previous section, the top three projects derived for controlling global warming are as follows.

1. Population control (P_1)
2. Generation and distribution of clean energy (P_2)
3. Recycling (P_3).

The necessary financial data in Table 3.8 are assumed for conducting the financial appraisal of these three projects.

Table 3.8 Financial data for projects (figures indicated in Rs. crore, where 1 cr. = 10^7)

Year	Project 1	Project 2	Project 3
0	(₹ 2000 cr.)	(₹ 4000 cr.)	(₹ 3000 cr.)
1	(₹ 500 cr.)	₹ 2500 cr	₹ 1500 cr
2	₹ 750 cr	₹ 1500 cr	(₹ 200 cr.)
3	₹ 1000 cr	(₹ 500 cr.)	₹ 1500 cr
4	(₹ 500 cr.)	₹ 750 cr	(₹ 200 cr.)
5	₹ 1000 cr	₹ 1000 cr	₹ 1500 cr

Let us assume the cost of capital to be 10 and 20% at two different instances. Now we can compare these projects on different financial appraisal criteria.

Net Present Value (N.P.V.)

NPV for Project 1,

when the cost of capital is 10%,

$$\text{NPV}_{P1@10} = -2000 - \frac{500}{1.1^1} + \frac{750}{1.1^2} + \frac{1000}{1.1^3} - \frac{500}{1.1^4} + \frac{1000}{1.1^5}$$
$$\text{NPV}_{P1@10} = -803.98\,\text{cr.}$$

NPV for Project 1,

when the cost of capital is 20%,

$$\text{NPV}_{P1@20} = -2000 - \frac{500}{1.2^1} + \frac{750}{1.2^2} + \frac{1000}{1.2^3} - \frac{500}{1.2^4} + \frac{1000}{1.2^5}$$
$$\text{NPV}_{P1@20} = -1156.38\,\text{cr.}$$

NPV for Project 2,

when the cost of capital is 10%,

$$\text{NPV}_{P2@10} = -4000 + \frac{2500}{1.1^1} + \frac{1500}{1.1^2} - \frac{500}{1.1^3} + \frac{750}{1.1^4} + \frac{1000}{1.1^5}$$
$$\text{NPV}_{P2@10} = 269.92\,\text{cr.}$$

when the cost of capital is 20%,

$$\text{NPV}_{P2@20} = -4000 + \frac{2500}{1.2^1} + \frac{1500}{1.2^2} - \frac{500}{1.2^3} + \frac{750}{1.2^4} + \frac{1000}{1.2^5}$$
$$\text{NPV}_{P2@20} = -400.78\,\text{cr.}$$

NPV for Project 3,

when the cost of capital is 10%,

$$NPV_{P3@10} = -3000 + \frac{1500}{1.1^1} - \frac{200}{1.1^2} + \frac{1500}{1.1^3} - \frac{200}{1.1^4} + \frac{1500}{1.1^5}$$
$$NPV_{P3@10} = 120.10\,\text{cr.}$$

when the cost of capital is 20%,

$$NPV_{P3@20} = -3000 + \frac{1500}{1.2^1} - \frac{200}{1.2^2} + \frac{1500}{1.2^3} - \frac{200}{1.2^4} + \frac{1500}{1.2^5}$$
$$NPV_{P3@20} = -514.47\,\text{cr.}$$

	NPV @ 10%	NPV @ 20%
Project 1	–803.98 cr.	–1156.38 cr.
Project 2	269.92 cr.	–400.78 cr.
Project 3	120.10 cr.	–514.47 cr.

Modified Net Present Value (Modified N.P.V.)

For calculation of modified N.P.V., let us assume a reinvestment rate of 15% on cash inflows.

The terminal value of cash inflows of Project 1,

$$TV_{P1@15} = 750(1.15)^3 + 1000(1.15)^2 + 1000(1.15)^0$$
$$TV_{P1@15} = 3463.16\,\text{cr.}$$

Modified NPV for Project 1,

when the cost of capital is 10%,

$$NPV^*_{P1@10} = \frac{3463.16}{1.1^5} - 2000 - \frac{500}{1.1^1} - \frac{500}{1.1^4}$$
$$NPV^*_{P1@10} = 150.35\,\text{cr.}$$

when the cost of capital is 20%,

$$NPV^*_{P1@20} = \frac{3463.16}{1.2^5} - 2000 - \frac{500}{1.2^1} - \frac{500}{1.2^4}$$

$$\text{NPV}^*_{P1@20} = -1266.03\,\text{cr.}$$

The terminal value of cash inflows of Project 2,

$$\text{TV}_{P2@15} = 2500(1.15)^4 + 1500(1.15)^3 + 750(1.15)^1 + 1000(1.15)^0$$
$$\text{TV}_{P2@15} = 8516.39\,\text{cr.}$$

Modified NPV for Project 2,

when the cost of capital is 10%,

$$\text{NPV}^*_{P2@10} = \frac{8516.39}{1.1^5} - 4000 - \frac{500}{1.1^3}$$
$$\text{NPV}^*_{P2@10} = 912.35\,\text{cr.}$$

when the cost of capital is 20%,

$$\text{NPV}^*_{P2@20} = \frac{8516.39}{1.2^5} - 4000 - \frac{500}{1.2^3}$$
$$\text{NPV}^*_{P2@20} = -866.81\,\text{cr.}$$

The terminal value of cash inflows of Project 3,

$$\text{TV}_{P3@15} = 1500(1.15)^4 + 1500(1.15)^2 + 1500(1.15)^0$$
$$\text{TV}_{P3@15} = 6107.26\,\text{cr.}$$

Modified NPV for Project 3,

when the cost of capital is 10%,

$$\text{NPV}^*_{P3@10} = \frac{6107.26}{1.1^5} - 3000 - \frac{200}{1.1^2} - \frac{200}{1.1^4}$$
$$\text{NPV}^*_{P3@10} = 490.24\,\text{cr.}$$

Modified NPV for Project 3,

when the cost of capital is 20%,

$$\text{NPV}^*_{P3@20} = \frac{6107.26}{1.2^5} - 3000 - \frac{200}{1.2^2} - \frac{200}{1.2^4}$$
$$\text{NPV}^*_{P3@20} = -780.97\,\text{cr.}$$

	Modified NPV @ 10%	Modified NPV @ 20%
Project 1	150.35 cr.	–1266.03 cr.
Project 2	912.35 cr.	–866.81 cr.
Project 3	490.24 cr.	–780.97 cr.

Benefit Cost Ratio (B.C.R.)/Profitability Index

PVB for Project 1,

when the cost of capital is 10%,

$$\text{PVB}_{P1@10} = \frac{750}{1.1^2} + \frac{1000}{1.1^3} + \frac{1000}{1.1^5}$$
$$\text{PVB}_{P1@10} = 1992.07\,\text{cr.}$$

when the cost of capital is 20%,

$$\text{PVB}_{P1@20} = \frac{750}{1.2^2} + \frac{1000}{1.2^3} + \frac{1000}{1.2^5}$$
$$\text{PVB}_{P1@20} = 1501.41\,\text{cr.}$$

Investment (cost) for Project 1,

when the cost of capital is 10%,

$$I_{P1@10} = 2000 + \frac{500}{1.1^1} + \frac{500}{1.1^4}$$
$$I_{P1@10} = 2796.05\,\text{cr.}$$

Investment (cost) for Project 1,

when the cost of capital is 20%,

$$I_{P1@20} = 2000 + \frac{500}{1.2^1} + \frac{500}{1.2^4}$$
$$I_{P1@20} = 2657.79\,\text{cr.}$$

Benefit–cost ratio for Project 1,

when the cost of capital is 10%,

$$\text{BCR}_{P1@10} = \frac{1992.07}{2796.05}$$
$$\text{BCR}_{P1@10} = 0.71$$

when the cost of capital is 20%,

$$\text{BCR}_{P1@20} = \frac{1501.41}{2657.79}$$
$$\text{BCR}_{P1@20} = 0.56$$

PVB for Project 2,

when the cost of capital is 10%,

$$\text{PVB}_{P2@10} = \frac{2500}{1.1^1} + \frac{1500}{1.1^2} + \frac{750}{1.1^4} + \frac{1000}{1.1^5}$$
$$\text{PVB}_{P2@10} = 4645.58\,\text{cr.}$$

when the cost of capital is 20%,

$$\text{PVB}_{P2@20} = \frac{2500}{1.2^1} + \frac{1500}{1.2^2} + \frac{750}{1.2^4} + \frac{1000}{1.2^5}$$
$$\text{PVB}_{P2@20} = 3888.57\,\text{cr.}$$

Investment (cost) for Project 2,

when the cost of capital is 10%,

$$I_{P2@10} = 4000 + \frac{500}{1.1^3}$$
$$I_{P2@10} = 4375.66\,\text{cr.}$$

Investment (cost) for Project 2,

when the cost of capital is 20%,

$$I_{P2@20} = 4000 + \frac{500}{1.2^3}$$
$$I_{P2@20} = 4289.35\,\text{cr.}$$

Benefit–cost ratio for Project 2,

when the cost of capital is 10%,

$$\mathrm{BCR}_{P2@10} = \frac{4645.58}{4375.66}$$
$$\mathrm{BCR}_{P2@10} = 1.06$$

when the cost of capital is 20%,

$$\mathrm{BCR}_{P2@20} = \frac{3888.57}{4289.35}$$
$$\mathrm{BCR}_{P2@20} = 0.91$$

PVB for Project 3,

when the cost of capital is 10%,

$$\mathrm{PVB}_{P3@10} = \frac{1500}{1.1^1} + \frac{1500}{1.1^3} + \frac{1500}{1.1^5}$$
$$\mathrm{PVB}_{P3@10} = 3421.99\ \text{cr.}$$

when the cost of capital is 20%,

$$\mathrm{PVB}_{P3@20} = \frac{1500}{1.2^1} + \frac{1500}{1.2^3} + \frac{1500}{1.2^5}$$
$$\mathrm{PVB}_{P3@20} = 2720.87\ \text{cr.}$$

Investment (cost) for Project 3,

when the cost of capital is 10%,

$$I_{P3@10} = 3000 + \frac{200}{1.1^2} + \frac{200}{1.1^4}$$
$$I_{P3@10} = 3301.89\ \text{cr.}$$

Investment (cost) for Project 3,

when the cost of capital is 20%,

$$I_{P3@20} = 3000 + \frac{200}{1.2^2} + \frac{200}{1.2^4}$$
$$I_{P3@20} = 3235.34\ \text{cr.}$$

Benefit–cost ratio for Project 3,

when the cost of capital is 10%,

$$\mathrm{BCR}_{P3@10} = \frac{3421.99}{3301.89}$$
$$\mathrm{BCR}_{P3@10} = 1.04$$

when the cost of capital is 20%,

$$\mathrm{BCR}_{P3@20} = \frac{2720.87}{3235.34}$$
$$\mathrm{BCR}_{P3@20} = 0.84$$

	BCR @ 10%	BCR @ 20%
Project 1	0.71	0.56
Project 2	1.06	0.91
Project 3	1.04	0.84

Internal Rate of Return (I.R.R.)

Determining IRR for Project 1,

$$2000 = -\frac{500}{(1+r_{P1})^1} + \frac{750}{(1+r_{P1})^2} - \frac{1000}{(1+r_{P1})^3} - \frac{500}{(1+r_{P1})^4} + \frac{1000}{(1+r_{P1})^5}$$
$$r_{P1} = -31.02\%$$

Determining IRR for Project 2,

$$4000 = \frac{2500}{(1+r_{P2})^1} + \frac{1500}{(1+r_{P2})^2} - \frac{500}{(1+r_{P2})^3} + \frac{750}{(1+r_{P2})^4} + \frac{1000}{(1+r_{P2})^5}$$
$$r_{P2} = 13.61\%$$

Determining IRR for Project 3,

$$3000 = \frac{1500}{(1+r_{P3})^1} - \frac{200}{(1+r_{P3})^2} + \frac{1500}{(1+r_{P3})^3} - \frac{200}{(1+r_{P3})^4} + \frac{1500}{(1+r_{P3})^5}$$
$$r_{P3} = 11.60\%$$

	Internal Rate of Return
Project 1	–31.02 %
Project 2	13.61 %
Project 3	11.60 %

Modified Internal Rate of Return (M.I.R.R.)

Present value of costs of Project 1,

when the cost of capital is 10%,

$$\text{PVC}_{P1@10} = 2000 + \frac{500}{1.1^1} + \frac{500}{1.1^4}$$
$$\text{PVC}_{P1@10} = 2796.05\,\text{cr.}$$

when the cost of capital is 20%,

$$\text{PVC}_{P1@20} = 2000 + \frac{500}{1.2^1} + \frac{500}{1.2^4}$$
$$\text{PVC}_{P1@20} = 2657.79\,\text{cr.}$$

The terminal value of cash inflows of Project 1,

when the cost of capital is 10%,

$$\text{TV}_{P1@10} = 750(1.1)^3 + 1000(1.1)^2 + 1000(1.1)^0$$
$$\text{TV}_{P1@10} = 3208.25\,\text{cr.}$$

when the cost of capital is 20%,

$$\text{TV}_{P1@20} = 750(1.2)^3 + 1000(1.2)^2 + 1000(1.2)^0$$
$$\text{TV}_{P1@20} = 3736\,\text{cr.}$$

Equating PVC,

when the cost of capital is 10%,

$$\text{PVC}_{P1@10} = \frac{\text{TV}_{P1@10}}{(1 + \text{MIRR}_{P1@10})^n}$$
$$2796.05 = \frac{3208.25}{(1 + \text{MIRR}_{P1@10})^5}$$

$$\text{MIRR}_{P1@10} = 02.79\%$$

Equating PVC,

when the cost of capital is 20%,

$$\text{PVC}_{P1@20} = \frac{\text{TV}_{P1@20}}{(1 + \text{MIRR}_{P1@20})^n}$$
$$2657.79 = \frac{3736}{(1 + \text{MIRR}_{P1@20})^5}$$
$$\text{MIRR}_{P1@20} = 07.05\%$$

Present value of costs of Project 2,

when the cost of capital is 10%,

$$\text{PVC}_{P2@10} = 4000 + \frac{500}{1.1^3}$$
$$\text{PVC}_{P2@10} = 4375.66\,cr.$$

when the cost of capital is 20%,

$$\text{PVC}_{P2@20} = 4000 + \frac{500}{1.2^3}$$
$$\text{PVC}_{P2@20} = 4289.35\,\text{cr.}$$

The terminal value of cash inflows of Project 2,

when the cost of capital is 10%,

$$\text{TV}_{P2@10} = 2500(1.1)^4 + 1500(1.1)^3 + 750(1.1)^1 + 1000(1.1)^0$$
$$\text{TV}_{P2@10} = 3208.25\,\text{cr.}$$

when the cost of capital is 20%,

$$\text{TV}_{P2@20} = 2500(1.2)^4 + 1500(1.2)^3 + 750(1.2)^1 + 1000(1.2)^0$$
$$\text{TV}_{P2@20} = 3736\,\text{cr.}$$

Equating PVC,

when the cost of capital is 10%,

$$PVC_{P2@10} = \frac{TV_{P2@10}}{(1 + MIRR_{P2@10})^n}$$
$$4375.66 = \frac{3208.25}{(1 + MIRR_{P2@10})^5}$$
$$MIRR_{P2@10} = -06.02\%$$

Equating PVC,

when the cost of capital is 20%,

$$PVC_{P2@20} = \frac{TV_{P2@20}}{(1 + MIRR_{P2@20})^n}$$
$$4289.35 = \frac{3736}{(1 + MIRR_{P2@20})^5}$$
$$MIRR_{P2@20} = -02.72\%$$

Present value of costs of Project 3,

when the cost of capital is 10%,

$$PVC_{P3@10} = 3000 + \frac{200}{1.1^2} + \frac{200}{1.1^4}$$
$$PVC_{P3@10} = 3301.89\,\text{cr.}$$

when the cost of capital is 20%,

$$PVC_{P3@20} = 3000 + \frac{200}{1.2^2} + \frac{200}{1.2^4}$$
$$PVC_{P3@20} = 3235.34\,\text{cr.}$$

The terminal value of cash inflows of Project 3,

when the cost of capital is 10%,

$$TV_{P3@10} = 1500(1.1)^4 + 1500(1.1)^2 + 1500(1.1)^0$$
$$TV_{P3@10} = 5511.15\,\text{cr.}$$

when the cost of capital is 20%,

$$TV_{P3@20} = 1500(1.2)^4 + 1500(1.2)^2 + 1500(1.2)^0$$
$$TV_{P3@20} = 6770.40\,\text{cr.}$$

Equating PVC,

when the cost of capital is 10%,

$$\text{PVC}_{P3@10} = \frac{\text{TV}_{P3@10}}{(1+\text{MIRR}_{P3@10})^n}$$
$$3301.89 = \frac{5511.15}{(1+\text{MIRR}_{P3@10})^5}$$
$$\text{MIRR}_{P3@10} = 10.79\%$$

Equating PVC,

when the cost of capital is 20%,

$$\text{PVC}_{P3@20} = \frac{\text{TV}_{P3@20}}{(1+\text{MIRR}_{P3@20})^n}$$
$$3235.34 = \frac{6770.40}{(1+\text{MIRR}_{P3@20})^5}$$
$$\text{MIRR}_{P3@20} = 15.91\%$$

	MIRR @ 10% (%)	**MIRR @ 20% (%)**
Project 1	02.79	07.05
Project 2	–06.02	–02.72
Project 3	10.79	15.91

The summary of the financial appraisal is presented in Table 3.9.

Problem 3.1 In an engineering company, Mr. Rakesh, the CEO, is evaluating two mutually exclusive projects (Project ALPHA and Project BETA). The cash flows of these projects are given in Table 3.10.

Table 3.9 Summary of financial appraisal

Project	Cost of capital (%)	N.P.V.	Modified N.P.V.	B.C.R.	IRR (%)	MIRR (%)
Project 1	10	–803.98 cr.	150.35 cr.	0.71	–31.02	02.79
	20	–1156.38 cr.	–1266.03 cr.	0.56		07.05
Project 2	10	269.92 cr.	912.35 cr.	1.06	13.61	–06.02
	20	–400.78 cr.	–866.81 cr.	0.91		–02.72
Project 3	10	120.10 cr.	490.24 cr.	1.04	11.60	10.79
	20	–514.47 cr.	–780.97 cr.	0.84		15.91

Table 3.10 Cash flows for "ALPHA" and "BETA" projects

Year	Project ALPHA	Project BETA
0	Rs. (130,000)	(210,000)
1	(114,000)	46,000
2	(50,000)	50,000
3	(30,000)	45,000
4	190,000	85,000
5	360,000	22,000
6	(70,000)	(90,000)

* Numbers in bracket indicates investment/expenditure

Your Tasks

Task 1	Calculate NPV for projects ALPHA and BETA if the cost of capital is 10 and 25%
Task 2	What is the IRR of each project?
Task 3	Which project would you choose if the cost of capital is 10 and 20%?
Task 4	What is each project's MIRR if the cost of capital is 14%?

Solution

Task 1

NPV for Project ALPHA,

when the cost of capital is 10%,

$$\mathrm{NPV}_{A@10} = -130{,}000 - \frac{114{,}000}{1.1^1} - \frac{50{,}000}{1.1^2} - \frac{30{,}000}{1.1^3} + \frac{190{,}000}{1.1^4} + \frac{360{,}000}{1.1^5} - \frac{70{,}000}{1.1^6}$$

$$\mathrm{NPV}_{A@10} = 16{,}292.94$$

when the cost of capital is 25%,

$$\mathrm{NPV}_{A@25} = -130{,}000 - \frac{114{,}000}{1.25^1} - \frac{50{,}000}{1.25^2} - \frac{30{,}000}{1.25^3} + \frac{190{,}000}{1.25^4} + \frac{360{,}000}{1.25^5} - \frac{70{,}000}{1.25^6}$$

$$\mathrm{NPV}_{A@25} = -91{,}121.28$$

NPV for Project BETA,

when the cost of capital is 10%,

$$\mathrm{NPV}_{B@10} = -210{,}000 + \frac{46{,}000}{1.1^1} + \frac{50{,}000}{1.1^2} + \frac{45{,}000}{1.1^3} + \frac{85{,}000}{1.1^4} + \frac{22{,}000}{1.1^5} - \frac{90{,}000}{1.1^6}$$

$$\mathrm{NPV}_{B@10} = -72{,}136.58$$

when the cost of capital is 25%,

$$\mathrm{NPV}_{B@25} = -210{,}000 + \frac{46{,}000}{1.25^1} + \frac{50{,}000}{1.25^2} + \frac{45{,}000}{1.25^3} + \frac{85{,}000}{1.25^4} + \frac{22{,}000}{1.25^5} - \frac{90{,}000}{1.25^6}$$

$$\mathrm{NPV}_{B@25} = -99{,}728$$

Task 2

Determining IRR for Project ALPHA,

$$130{,}000 = -\frac{114{,}000}{(1+r_A)^1} - \frac{50{,}000}{(1+r_A)^2} - \frac{30{,}000}{(1+r_A)^3} + \frac{190{,}000}{(1+r_A)^4} + \frac{360.000}{(1+r_A)^5} - \frac{70{,}000}{(1+r_A)^6}$$

$$r_A = 11.64\%$$

Determining IRR for Project BETA,

$$\begin{aligned}210{,}000 &= \frac{46{,}000}{(1+r_B)^1} + \frac{50{,}000}{(1+r_B)^2} + \frac{45{,}000}{(1+r_B)^3} \\ &+ \frac{85{,}000}{(1+r_B)^4} + \frac{22{,}000}{(1+r_B)^5} - \frac{90{,}000}{(1+r_B)^6} \\ r_B &= -1.74866 - 0.37927i\end{aligned}$$

The above equation obtains no real solution for RB, so it is infeasible.

Task 3

NPV for Project ALPHA,

when the cost of capital is 10%,

$$\text{NPV}_{A@10} = 16{,}292.94$$

when the cost of capital is 20%,

$$\begin{aligned}\text{NPV}_{A@20} &= -130{,}000 - \frac{114{,}000}{1.2^1} - \frac{50{,}000}{1.2^2} - \frac{30{,}000}{1.2^3} \\ &+ \frac{190{,}000}{1.2^4} + \frac{360{,}000}{1.2^5} - \frac{70{,}000}{1.2^6} \\ \text{NPV}_{A@20} &= -64{,}222.18\end{aligned}$$

NPV for Project BETA,

when the cost of capital is 10%,

$$\text{NPV}_{B@10} = -72{,}136.58$$

when the cost of capital is 20%,

$$\begin{aligned}\text{NPV}_{B@20} &= -210{,}000 + \frac{46{,}000}{1.2^1} + \frac{50{,}000}{1.2^2} + \frac{45{,}000}{1.2^3} \\ &+ \frac{85{,}000}{1.2^4} + \frac{22{,}000}{1.2^5} - \frac{90{,}000}{1.2^6} \\ \text{NPV}_{B@20} &= -91{,}210.78\end{aligned}$$

Project ALPHA at the cost of capital 10% should be chosen because it has the maximum positive NPV.

Task 4

Present value of costs of Project ALPHA,

$$PVC_{A@14} = 130{,}000 + \frac{114{,}000}{1.14^1} + \frac{50{,}000}{1.14^2} + \frac{30{,}000}{1.14^3} + \frac{70{,}000}{1.14^6}$$
$$PVC_{A@14} = 320{,}613$$

The terminal value of cash inflows of Project ALPHA,

$$TV_{A@14} = 190{,}000(1.14)^2 + 360{,}000(1.14)^1$$
$$TV_{A@14} = 657{,}324$$

Equating PVC,

$$PVC_{A@14} = \frac{TV_A}{(1 + MIRR_A)^n}$$
$$320{,}613 = \frac{657{,}324}{(1 + MIRR_A)^6}$$
$$MIRR_A = 12.71\%$$

Present value of costs of Project BETA,

$$PVC_{B@14} = 210{,}000 + \frac{90{,}000}{1.14^6}$$
$$PVC_{B@14} = 251{,}001.79$$

The terminal value of cash inflows of Project BETA,

$$TV_{B@14} = 46{,}000(1.14)^5 + 50{,}000(1.14)^4 + 45{,}000(1.14)^3 + 85{,}000(1.14)^2 + 22{,}000(1.14)^1$$
$$TV_{B@14} = 375{,}232.55$$

Equating PVC,

$$PVC_{B@14} = \frac{TV_B}{(1 + MIRR_B)^n}$$
$$251{,}001.79 = \frac{375{,}232.55}{(1 + MIRR_B)^6}$$
$$MIRR_B = 6.93\%$$

Table 3.11 Financial information of projects

Project	Initial outlay	Contribution to net income			Contribution to sales growth (percentage)			NPV
		Year 1	Year 2	Year 3	Year 1	Year 2	Year 3	
1	14	1.4	1.2	1.6	1.0	1.7	1.8	3
2	12	1.6	1.2	1.5	1.2	1.0	1.2	6
3	15	0.8	1.2	2.0	0.5	1.2	3.0	7
4	16	1.5	1.8	1.9	1.8	2.0	2.2	8
5	13	0.8	1.2	1.5	0.6	1.4	1.8	4
6	25	1.2	2.5	4.5	1.0	3.0	3.5	10
7	22	1.8	2.0	2.5	2.3	3.0	3.5	5

Problem 3.2 Formulate the project selection problem as a goal programming problem for the information given in Table 3.11.

An infrastructure company is considering seven projects with financial information, as given in Table 3.11.

The capital budget constraint is 100. The goals sought by management are as follows:

(i) maximization of NPV, (ii) contribution to net income of 8.0, 10.0, and 12.0 in years 1, 2, and 3, and (iii) contribution to sales growth of 7, 9, and 11% in years 1, 2, and 3.

The priorities assigned to various goals are as follows:

Priority 1: Net income; Priority 2: Sales growth; Priority 3: NPV

At priority level 1, the relative weights attached to the net income of years 1, 2, and 3 are 3, 2, and 1, respectively. At priority level 2, the relative weights attached to the contribution to sales growth of years 1, 2, and 3 are 4, 2, and 1, respectively.

Formulate the Above Problem as a Goal Programming Problem

Solution

Let projects 1, 2, 3, 4, 5, 6, and 7 be represented by decision variables X_1, X_2, X_3, X_4, X_5, X_6, and X_7, respectively, where X_i = level of acceptance of project i for implementation during the plan period.

Priority level # 1: (P_1)

Minimize capital budget overrun,

(Capital budget = 100)

$$14X_1 + 12X_2 + 15X_3 + 16X_4 + 13X_5 + 25X_6 + 22X_7 - d_1^+ + d_1^- = 100$$

Objective: Min d_1^+

Priority level # 2: (P_2)

Maximize net income,

for year 1,

$$1.4X_1 + 1.6X_2 + 0.8X_3 + 1.5X_4 + 0.8X_5 + 1.2X_6 + 1.8X_7 - d_2^+ + d_2^- = 8$$

for year 2,

$$1.2X_1 + 1.2X_2 + 1.2X_3 + 1.8X_4 + 1.2X_5 + 2.5X_6 + 2.0X_7 - d_3^+ + d_3^- = 10$$

for year 3,

$$1.6X_1 + 1.5X_2 + 2.0X_3 + 1.9X_4 + 1.5X_5 + 4.5X_6 + 2.5X_7 - d_4^+ + d_4^- = 12$$

Objective

$$\text{Min}\, 3\left(-d_2^+ + d_2^-\right) + 2\left(-d_3^+ + d_3^-\right) + 1\left(-d_4^+ + d_4^-\right)$$

Priority level # 3: (P_3)

Maximize sales growth,

for year 1,

$$1.0X_1 + 1.2X_2 + 0.5X_3 + 1.8X_4 + 0.6X_5 + 1.0X_6 + 2.3X_7 - d_5^+ + d_5^- = 7$$

for year 2,

$$1.7X_1 + 1.0X_2 + 1.2X_3 + 2.0X_4 + 1.4X_5 + 3.0X_6 + 3.0X_7 - d_6^+ + d_6^- = 9$$

for year 3,

$$1.8X_1 + 1.2X_2 + 3.0X_3 + 2.2X_4 + 1.8X_5 + 3.5X_6 + 3.5X_7 - d_7^+ + d_7^- = 11$$

Objective

$$\text{Min}\, 4\left(-d_5^+ + d_5^-\right) + 2\left(-d_6^+ + d_6^-\right) + 1\left(-d_7^+ + d_7^-\right)$$

Priority level # 4: (P_4)

Maximize NPV,

$$3X_1 + 6X_2 + 7X_3 + 8X_4 + 4X_5 + 10X_6 + 5X_7 - d_8^+ + d_8^- = 20$$

Objective

$$\text{Min}\, d_8^-$$

Goal achievement function,

$$\begin{aligned}\text{Min } Z = {} & P_1 d_1^+ + P_2\{3(-d_2^+ + d_2^-) + 2(-d_3^+ + d_3^-) + 1(-d_4^+ + d_4^-)\} \\ & + P_3\{4(-d_5^+ + d_5^-) + 2(-d_6^+ + d_6^-) + 1(-d_7^+ + d_7^-)\} + P_4 d_8^-\end{aligned}$$

where

P_1, P_2, P_3, and P_4 are pre-emptive priority structures, and

d_1^- is the negative deviational variable of the ith goal. It represents the level by which the target level is under-achieved, and d_1^+ is the positive deviational variable of the ith goal. It represents the level by which the target level is overachieved.

Non-Negativity Constraints

$$X_1, X_2, X_3, X_4, X_5, X_6, X_7 \geq 0$$

$$\begin{aligned}& d_i^+, d_i^- \geq 0 \\ & i = 1, 2, 3, 4, 5, 6, 7, 8\end{aligned}$$

Complementary Constraints

$$\begin{aligned}& d_i^+ x d_i^- = 0, \\ & i = 1, 2, 3, 4, 5, 6, 7, 8\end{aligned}$$

Problem 3.3 The CEO of an engineering company, Mr. Rao, is evaluating six investment opportunities. Formulate the project selection problem as a goal programming problem for the data given in Table 3.12.

The budget available is limited to Rs. 180,000 in year 1 and Rs. 270,000 in year 2. Any amount not spent in year one can be transferred to year 2. The amount so transferred will earn a post-tax return of 8%.

There are two additional constraints: working capital constraints and managerial constraints. The requirements and constraints applicable in this project are given in Table 3.13.

Table 3.12 Financial data of an engineering company

Project (j)	Net present value (NPV_j) (Rs.)	Cash outflow in period t (CFj_1) (Rs.)	Cash outflow in period 2 (CFj_2) (Rs.)
1	15,000	20,000	6000
2	25,000	10,000	15,000
3	30,000	6000	42,000
4	45,000	37,000	28,000
5	70,000	80,000	15,000
6	120,000	60,000	115,000

Table 3.13 Requirements and constraints

Project (j)	Power requirement (W_j)	Managerial requirement (M_j)
1	8000	18
2	7000	22
3	5000	35
4	12,000	25
5	14,000	45
6	34,000	60
	$\sum XjWj \leq 68{,}000$	$\sum XjMj \leq 135$

Solution

Let projects 1, 2, 3, 4, 5, and 6 represent decision variables $X_1, X_2, X_3, X_4, X_5,$ and $X_6,$ respectively, where X_i = level of acceptance of project i for implementation of the plan period.

Priority level # 1: (P_1)

Minimize budget overrun for period 1,

(Budget = 180,000)

$$20{,}000X_1 + 10{,}000X_2 + 6000X_3 + 37{,}000X_4 + 80{,}000X_5 + 60{,}000X_6 - d_1^+ + d_1^- = 180{,}000$$

Objective

$$\text{Min}\, d_1^+$$

Priority level # 2: (P_2)

Minimize budget overrun for period 2,

(Budget = 270,000)

$$6000X_1 + 15{,}000X_2 + 42{,}000X_3 + 28{,}000X_4 + 15{,}000X_5 + 115{,}000X_6 - d_2^+ + d_2^- = 270{,}000 + 1.08d_1^+$$

Objective

$$\text{Min } d_2^+$$

Resource Constraints

Power requirement,

$$8000X_1 + 7000X_2 + 5000X_3 + 12{,}000X_4 + 14{,}000X_5 + 34{,}000X_6 \leq 68{,}000$$

Managerial requirement,

$$18X_1 + 22X_2 + 35X_3 + 25X_4 + 45X_5 + 60X_6 \leq 135$$

Goal achievement function,

$$\text{Min } Z = P_1 d_1^+ + P_2 d_2^+$$

where

P_1 and P_2 are pre-emptive priority structures, and

d_1^- is the negative deviational variable of the ith goal. It represents the level by which the target level is under-achieved, and d_1^+ is the positive deviational variable of the ith goal. It represents the level by which the target level is overachieved.

Non-Negativity Constraints

$$X_1, X_2, X_3, X_4, X_5, X_6 \geq 0$$

$$d_i^+, d_i^- \geq 0$$
$$i = 1, 2$$

Complementary Constraints

$$d_i^+ x d_i^- = 0,$$
$$i = 1, 2$$

3.6 Capital Budgeting

Capital is the company's total investment, and budgeting is the art of building budgets.

The budget serves as a standard for comparison, a baseline from which to measure the difference between the actual and planned uses of resources. It is essential for a project manager to carefully monitor the resource usage while deploying the resources to various work packages. The critical features of budgeting are:

- It is a plan for allocating scarce resources to the various projects of an organization.
- It helps recognize constraints.
- It makes managers sensible to their requirements and helps them develop a realistic schedule for various expenditures.
- It implies a commitment and support of top management.
- A higher budget and relatively lower cost extend higher managerial support.
- It acts as a control mechanism for prohibiting the expenses exceeding budget and comparing them against the allocated budget.

Capital budgeting is a formal procedure for evaluating future investments or expenditures of large amounts. It includes deciding to invest the current funds to add, disposition, modify, or replace fixed assets. In addition, it includes purchasing fixed assets like building and land, equipment, rebuilding and replacing existing equipment, and R&D. The capital spent for these kinds of projects is called capital expenditures. Capital budgeting is a tool for maximizing an organization's future profits since most companies cannot manage too many large projects simultaneously. Capital budgeting analyzes the number of years it takes for a project's cash flow to pay back the initial cash investment, an assessment of risk, and several other factors.

Value of the Firm

Managers often use a simple rule to make capital budgeting decisions: Invest in all the projects with a positive net present value and reject negative net current worth. Capital budgeting theory says that by adhering to this rule, the company's investment decisions will maximize shareholder's wealth. Since N.P.V. measures a project's contribution to shareholder wealth, it can be called a primary capital budgeting tool.

The general assumption of capital budgeting theory is that the prime goal of an organization's shareholders is to maximize the organization's value. Additionally, it is assumed that the firm has access to perfect financial markets, which allows it to finance all value-enriching projects. When these assumptions are fulfilled, an organization can separate finance and investment decisions and invest in all positive N.P.V. projects.

The N.P.V. theory focuses on how investments are to be chosen and gives the decision-makers some means by which they can assess different options when the investment period extends over time. The main objective is that the investments need to be self-sustaining; i.e., they should generate enough profits to repay the loan. If additional funds need to be developed for the repayment, it is considered a poor investment.

Role of Capital Budgeting in Quality of Investment Decision

The role of capital budgeting is significant because it creates measurability and accountability. Any organization that wants to devote its resources to any project without understanding its returns and risks would be held as irresponsible by the owners and shareholders. Moreover, if an organization does not measure the efficiency of its investment decisions, the organization will likely face survival challenges in a competitive market.

The prime goal for any organization is to make profits. The capital budgeting process is a quantifiable method for companies to estimate any investment's long-term financial and economic profitability.

Capital budgeting is also essential to a company as it creates a stepwise process that facilitates it to:

1. Formulate and develop long-term strategic goals
2. Seek out new investment projects
3. Forecast and estimate future cash flows
4. Facilitate the transfer of information
5. Control and monitor expenditure
6. Create decision rules.

Key Weaknesses

Capital budgeting is primarily based on estimations that are never exact. If they were precise, no organization would be making wrong decisions regarding project selection. Capital budgeting has its limitations, which are discussed below.

1. Long-term implementation
2. Insufficient investment
3. Cash flows
4. Time horizon.

Long Term Implementation

The implementation of capital budgeting is a long-term affair, so there are no short-term gains. A wrong selection at the preliminary phase can critically affect the organization's future survival.

Insufficient investment

Insufficient investment acts as a hurdle when an organization is trying to bring up its capital and budget.

Cash Flows

The outflows for a project are easily estimated as compared to the inflows. The revenue a project is going to generate can only be estimated with a high level of uncertainty. Overestimating revenues and underestimating costs can burden the organization with a considerable amount of capital, whereas overestimating costs and underestimating revenues can make it lose a probably profitable project.

Time Horizon

Estimating long-term cash flows becomes difficult as we go farther into time. Predictions are based on current knowledge. When the time horizon is expanded, the chances that a disruptive change in the working environment can affect the project's progress also rise. It can be new competition, technological advancement, changes in the legal or regulatory framework, or anything unprecedented that can hinder a project's success.

Project Management in Practice

Project Selection in Companies Providing Infrastructure Finance and Leasing Services

When a significant financing company defaults on some of its debt obligations, it can trigger broader concerns about risk in the rest of the country's financial sector.

NEIFLS Ltd. is a hypothetical core investment company which invests in infrastructure-related projects, forming an ecosystem of expertise across infrastructure, finance, and social and environmental services, and is involved in infrastructure-related project sponsorship, development and advisory, investment banking, corporate advisory, asset management, and advisory services in environmental and social management, spread across sectors like surface transportation, urban infrastructure, energy (thermal and renewable), education, maritime and ports. Given the scale of its involvement, when the company began defaulting on payments, the government intervened to provide the liquidity needed to ensure no further more defaults occurred and that the infrastructure projects were implemented smoothly. In such a case, the government has to come up with a proper project selection procedure for selecting the most viable projects and screening out projects with lesser potential.

Impact of a Good Project Selection

The current default in debt obligations had risen a lot due to the poor selection of projects. Such projects have occupied a significant investment, and due to incompletion, they were failing to generate cash flows, creating a debt liability mismatch. The organization was sitting on a debt of about Rs 91,000 crore. Nearly Rs 60,000 crore of debt is at the project level, including road, power, and water projects. A significant reason behind the troubles of NEIF&LS was complications in land acquisitions due to a change in law which made many of its projects unviable. All this could have been avoided if a proper project selection procedure had been incorporated in the beginning phase.

The records of project failures of the organization show that the project alternatives were passed through a project selection procedure. Most of these large-scale projects were stuck in the land acquisition phase, which means the social and ecological appraisal lacked earnestness. This had forced the government appointed management to practice utmost care in selecting the projects with special significance to the social and ecological aspects.

Merits and Demerits of the Methodology

The organization existed in a bad financial phase and may have gone bankrupt without government intervention. Despite conscious efforts to evaluate land acquisition issues, there was no surety that such problems would not arise in future projects as land-related issues can arise due to various unexpected reasons. On adopting a highly conservative approach, NEIF&LS might have to miss out on a potential financially profitable project due to too much emphasis on minor land acquisition issues during the selection process.

Final Project Selection

The final project selection is a cumulative result of all individual project appraisal techniques. Project alternatives go through economic, financial, technical, managerial, ecological, and social analyses.

NEIF&LS was aware that the purpose of the critical analysis and selection process is not to discourage projects but to ensure that only the best are pursued. Organizations rely on projects for survival and growth; ideally, they can choose projects from many proposals, RFPs, and initiation requests. Hence, the selection process can be likened to a funnel into which project ideas, recommendations, and concepts can flow and a filter that precludes all but the best from proceeding. Therefore, NEIF&LS adopted a funnel approach for the project selection. The funnel mouth is wide enough to take in lots of ideas, but the filter fine sufficient to screen out bad proposals yet provide a constant flow of high-quality projects (Exhibit 3.1).

Views of Stakeholders on a Prioritized List of Projects

The major stakeholders in NEIF&LS were banks and other financial institutions. Their view was to invest in projects which can generate more revenues at a quicker rate. Therefore, the investors were more interested in projects such as highways and

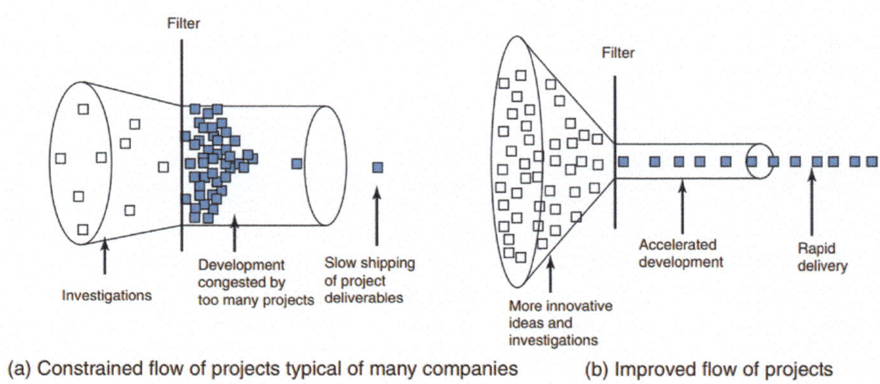

Exhibit 3.1 Project filtration process

expressways. Still, the decision-makers were worried about such projects' social and ecological feasibility due to difficulties faced during land acquisition in the past.

Conflicts and Negotiations Underlying Final Project Selection

The significant conflicts underlying final project selection were caused by differences in point of view and individual priorities. Old investors wanted the security of their investment, whereas the new investors wanted a high return as soon as possible. The projects which were stuck due to lack of funds demand investment as eagerly as new projects. Therefore, it was necessary to manage the conflicting priorities of various stakeholders.

MD, NEIF&LS was highly concerned about the survival of the organization. However, the resources were shrinking as it was difficult to find interested investors after its default. This has put a challenge for the top management to allocate the resources judiciously to various projects. The prevailing situation forced the MD and his team to select the less risky project and creates lesser returns, precisely when NEIF&LS encountered a debt of around 90,000 crores.

Questions for Class Discussion

1. What is the impact and implications of good project selection for the case organization?
2. What is the system approach the case organization should adopt in project selection? What are the fundamental limitations of this approach? What would you recommend as additional measures to improve the project selection process at NEIF&LS?

Summary

Project selection is a process to assess each project idea and select the project with highest priority so that the objectives of the parent organization will be achieved.

It is the process of evaluating proposed projects or group of projects on numerous factors that need to be considered before an organization decides to proceed with it.

When a project or a set of projects are selected arbitrarily, the chances of failing of the projects rise exponentially as compared to projects chosen by proper project selection procedure.

The main aim of project appraisal is to consider and compare the possible feasible project and select the best one that meets the objectives.

MCDM is a widely used approach for evaluating project options. It accommodates the trade-off between both tangible and intangible criteria for project selection.

Financial appraisal is analysis of information in the financial statements of a project and classifying and processing the numbers and data to make it easy to be compared with data of similar other projects.

Capital budgeting is a formal procedure followed by an organization for evaluating future investments or expenditures of large amount.

Question for Discussion

1. Why is project selection a critical issue? If you select a project or set of projects arbitrarily, what could be its consequences for the organization?
2. What is project appraisal? What are the phases in project appraisal? First, detail the appraisal points for each of the appraisals. Then, develop a generic checklist for the project appraisal.
3. Conduct a detailed brainstorming and identify a set of project options for controlling HIV infection in India. Next, evaluate each of the project options using the checklist developed for project appraisal in Question 2. Finally, rank each of the project options and identify the most potential top five options for controlling HIV infection in India.
4. Demonstrate an application of MCDM techniques—AHP, ANP, TOPSIS, ELECTRE, and PROMETHEE for ranking the top five project options for controlling HIV infection in India. Compare the results and declare the top three projects.
5. What is the importance of financial appraisal in today's competitive world? First, explain (in detail) at least ten financial appraisal criteria (NPV, IRR, Payback Period, etc.) with detailed explanation and equations. Then, consider the top three projects ranked in Question 4. Assume suitable data for conducting financial appraisal and calculate IRR, NPV, Payback Period, and MIRR for each top-rated three projects (in Question 4) for 10 and 20% interest rates. Finally, summarize all your calculations in a tabular format.
6. Mr. Roxwell, the VP of a tube manufacturing company, is evaluating three mutually exclusive projects: Project 1: Furnace for HDG (Hot Deep Galvanizing) Process, Project 2: Annealing Unit for Heat treatment, Project 3: Purchase of Boiler. The cash flows of these projects are given as below:

Year	Project 1: furnace for HDG (hot deep galvanizing) process	Project 2: annealing unit for heat treatment	Project 3: purchase of boiler
0	Rs. (155,000)	(210,000)	(300,000)
1	(125,000)	70,000	56,000
2	(75,000)	40,000	40,000
3	(20,000)	35,000	30,000
4	210,000	90,000	65,000
5	370,000	32,000	20,000

(continued)

(continued)

Year	Project 1: furnace for HDG (hot deep galvanizing) process	Project 2: annealing unit for heat treatment	Project 3: purchase of boiler
6	(90,000)	(40,000)	(100,000)

*Numbers in bracket indicates investment/ expenditure

Your Tasks

Task 1	Calculate NPV for ALL THE THREE projects if the cost of capital is 10 and 25%
Task 2	What is the IRR of each project?
Task 3	Which project would you choose if the cost of capital is 15 and 25%?
Task 4	What is each project's MIRR if the cost of capital is 12%?

7. Formulate the project selection problem as a goal programming problem for the data given below.

A multi-speciality hospital is considering seven projects with the following characteristics:

Project	Initial outlay	Contribution to net income			Contribution to sales growth (percentage)			NPV
		Year 1	Year 2	Year 3	Year 1	Year 2	Year 3	
1	15	1.5	1.7	1.6	1.2	2.7	1.0	4
2	13	1.7	1.3	1.6	1.2	1.0	1.9	8
3	14	0.9	1.2	2.3	0.5	1.4	3.0	6
4	17	1.5	1.8	2.0	1.8	2.0	2.2	10
5	14	0.8	1.2	1.6	0.6	1.4	1.8	4
6	28	1.2	2.8	4.7	1.0	3.2	3.8	12
7	20	1.8	2.2	2.6	2.3	2.8	3.4	6

The capital budget constraint is 100. The goals sought by management are as follows:

(i) maximization of NPV, (ii) contribution to net income of 9.0, 11.0, and 14.0 in years 1, 2, and 3, and (iii) contribution to sales growth of 6, 10, and 14% in years 1, 2, and 3.

The priorities assigned to various goals are as follows:

Priority 1: Net income; Priority 2: Sales growth; Priority 3: NPV

At priority level 1, the relative weights attached to the net income of years 1, 2, and 3 are 3, 2, and 1, respectively. At priority level 2, the relative weights attached to the contribution to sales growth of years 1, 2, and 3 are 4, 2, and 1, respectively.

Formulate the Above Problem as a Goal Programming Problem

8. Formulate the project selection problem as a goal programming problem for the data given below.

The CEO of a bearing manufacturing company, Dr. Kiran, is evaluating six investment opportunities as follows:

Project (*j*)	Net present value (NPV_j) (Rs.)	Cash outflow in period *t* (CFj_1) (Rs.)	Cash outflow in period 2 (CFj_2) (Rs.)
P1	35,000	40,000	8000
P2	45,000	10,000	15,000
P3	30,000	6000	32,000
P4	45,000	30,000	30,000
P5	80,000	85,000	25,000
P6	130,000	70,000	120,000

The budget available is limited to Rs. 250,000 in year 1 and Rs. 380,000 in year 2. Any amount not spent in year one can be transferred to year 2. The amount so transferred will earn a post-tax return of 8%.

There are two additional constraints: working capital constraints and managerial constraint. The requirements and constraints applicable in this project are:

Project (*j*)	Power requirement (W_j)	Managerial requirement (M_j)
*P*1	10,000	20
*P*2	12,000	34
*P*3	6000	28
*P*4	14,000	32
*P*5	15,000	55
*P*6	40,000	80
	$\sum XjWj \leq 88{,}000$	$\sum XjMj \leq 155$

9. What is capital budgeting? How it affects the value of the firm? What role it has to play in the quality of investment decision? First, indicate and explain the critical weaknesses in capital budgeting in detail. Then, evaluate a typical hypothetical infrastructure project and its viability against each of the essential weaknesses.

Web-Based Exercise

Collect necessary information on Tech Innovation Project "Akash Tablet." Investigate the reasons behind an unsuccessful initiative—"Akash Tablet" from the perspective of project selection. What are the precautions and key issues government should have addressed right at the project selection stage?

Group Project

Undertake a detailed brainstorming exercise in a group of four. Assume that you want to start a new company in the manufacturing or service sector. Anticipate the critical issues in the selection of a project called "New Start-up." Undertake a detailed qualitative and quantitative analysis using techniques such as SWOT, MCDM, and financial appraisal to check the worthiness of your project. Assume necessary financial data for analysis. Present your case to your instructor for evaluation.

Chapter 4
Project Planning and Representation

Critical Questions

Why project planning is important?
What are the key components of project planning?
What is WBS (Work Breakdown Structure)? How it helps the project manager?
What are the different phases of project planning? What is the importance of each phase?
What is project representation? What is the significance of inconsistency and redundancy in project network?
What are the methods for checking consistency and redundancy in the project network?

Project Management in Practice

Bio Toilets in Indian Railways.

Bio toilet and green corridor is an indigenous and innovative development of technology to help the Indian railways fulfill its commitment in supporting and keeping the station premises and track clean. The primary rake with bio-latrines created with DRDO has been running in Bundelkhand Express since 18th January–2011. Bio-digester is given profluent which is released on target after bio de-degree. Direct vehicle from the latrine bowl to the tank supported by vacuum creation in the tank and pipeline and Waste is gathered at the end and handled; strong and fluid detachment is done in the actual tank.

J. J. Thakkar, *Project Management*, Management and Industrial Engineering,
https://doi.org/10.1007/978-981-15-3695-3_4

Key Project Management Challenges:

Time:

Under the Integrated Railway Modernization Plan, the Indian railways were to install 5,000 coaches by 2010. Until March 2006, only 261 coaches had been fitted, according to an audit report by the Comptroller and Auditor General of India. The target to install 'discharge free' toilets in all coaches was extended to 2011–13.

Cost:

Although it will save water, the bio-latrine includes some significant downfalls. The railroads would need to spend Rs 8 lakh as hardware cost and Rs 1.5–2 lakh as operations cost each year for each coach.

Performance:

Welding failure can cause significant performance challenges for bio-toilet. Maintenance of the septic tank is also an essential factor for their performance. Garbage dumped in the toilets of the train by passengers used to older toilet designs, blocks the toilet and creates a big problem, requiring additional money and maintenance.

Source ***Indian Railways' experimentation with eco-friendly toilets by R K Srinivasan, DownToEarth, Monday 31 March 2008*** (https://www.downtoearth.org.in/coverage/indian-railways-experimentation-with-ecofriendly-toilets-4386), ***accessed on 12th May 2021***.

4.1 Project Planning

Project planning is a procedural step in project management, where the required documentation is created in order to ensure successful project completion. Documentation includes all the necessary actions to define, prepare, integrate, and coordinate additional plans. The project plan clearly explains how the project is to be executed, monitored, controlled, and closed.

The process of project planning cannot be standardized as projects differ in terms of scope and scale. However, in general, project planning can be defined as "*the process of defining key milestones of projects starting from conceptualization to completion of the project including the factors such as time, cost and resources.*" A typical cycle of project planning, starting from identifying stakeholder requirements to controlling the project through an appropriate tracking mechanism, is illustrated in Fig. 4.1.

The critical components of project planning include:

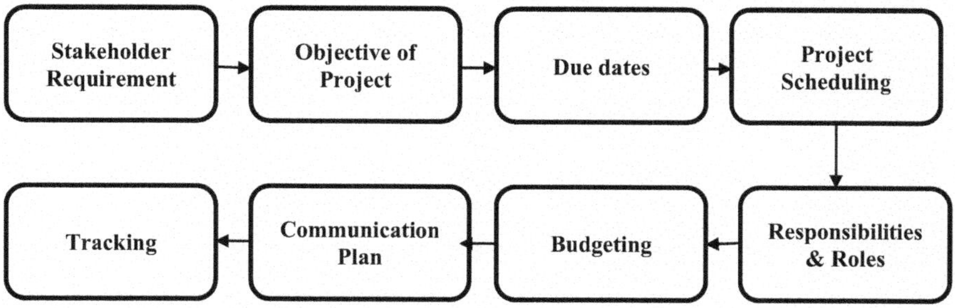

Fig. 4.1 A typical project planning cycle

(1) **Overview of the project**: This is the most primitive stage of project planning, where a bird's-eye view of the project is developed.
(2) **Project scope statement**: This stage defines the detailed project objectives and directs the project manager and team in the desired direction. It helps to minimize issues arising because of scope creep.
(3) **Work breakdown structure (WBS)**: It provides a hierarchical description to ensure the timely accomplishment of work toward achieving the targeted project deliverables. It helps develop a shared understanding of the scope of the project.
(4) **Project scheduling**: It defines the list of activities, duration, resources required, expected deliverables, and project milestones. It helps stakeholders develop clarity on project start and finish dates and plan a necessary course of action if there is any deviation from the schedule.
(5) **Resource and personnel management**: This stage involves the planning of all the resources, including raw material, labor, skilled personnel, technology, infrastructure, etc., to ensure timely allocation and to maximize efficiency.
(6) **Risk management**: Risk is inevitable irrespective of the scope and scale of the project. However, the severity of the risk differs depending upon the complexity of the project. Risk management intends to identify, evaluate, and mitigate the potential risks for achieving the intended outcomes in the stipulated time. An appropriate risk management mechanism helps project management oversight.

To appreciate the project planning process, we may consider an example of an R&D project that involves the construction of a modern warship with technological advancements. The relevance of various components of project planning for this can be summarized as in Table 4.1.

4.2 Work Breakdown Structure (WBS)

It is a tool to represent as deliverable-oriented, hierarchical decomposition of the work to be executed by the project team to accomplish the project objectives and create the required deliverables. The WBS breaks project tasks/goals into much smaller

Table 4.1 R&D project: construction of a modern warship

S. No.	Component	Remarks
1	Overview/summary/title	In the planning stage, we name the project and summarize the requirements. For example, the project name could be Warship 2020. The summary would include the technology intended for use in propulsion, weapons, other aspects or specifications like the speed to be achieved by the ship, the combat role of the ship, etc.
2	Project scope/scope statement	This stage involves listing all the physical and non-physical entities which form part of the research and construction of the new warship like the lab equipment, test equipment for research and experimentation, the engines, AC plant, weapons to be fitted, etc. Thus, it is a statement of the scope of everything required to complete the project
3	Work breakdown structure (WBS)	This stage involves a detailed and structured breakdown of all the activities, starting from initial research and testing to complete the ship's construction
4	Schedule/project scheduling	At this stage, the schedule for every activity (experimentation of new propulsion system, factory trials, fitment on the ship, etc.) would be decided, and network from project start to culmination would be made using a PERT chart or CPM method
5	Resource and personnel management	This stage involves planning resources and personnel required to develop new technology, construction, and fitting jobs. For example, resources like raw material, machinery, lab equipment; and personnel like scientists, vendors, skilled labor, managers, etc., are planned. This stage also plans for overall costing due to the required resources and personnel
6	Risk analysis/risk management	This stage involves contingency planning. For example, precautions and actions needed in case of a fire or a gas leak, or the actions necessary when a required resource is not available, which might otherwise risk the overall project/objective are planned and analyzed at this stage

targets distributing the overall project. The hierarchical representation of the WBS aids in understanding the project deliverables better and leads to accomplishing the objectives at every stage to provide the desired output. WBS interfaces with various project planning and scheduling activities such as responsibility assignment matrix, network scheduling, costing, risk analysis, organizational structure, coordination of objectives, and control (including contract administration).

Similar to a hierarchy tree, the oldest generations are positioned at the top, while the younger generations are positioned at the bottom. The critical attributes of WBS include:

(1) WBS has a minimum of three decomposition levels, and the levels may increase depending on the complexity of the project goal. The first level of the WBS is called the control account. It is a management control point where scope, budget, cost, and schedule are integrated and compared with the earned value to measure performance.

(2) The second level, just below the control account, is known as a planning package. It describes the context of the work to be completed without any detailed scheduled activities.
(3) The lowest level of the WBS is known as a work package. At this level, the cost and duration can be estimated/managed for the task. Work packages can be scheduled, cost assessed, monitored, and controlled.
(4) Along with the WBS, a WBS dictionary is also created to add further detail to the WBS components. The WBS dictionary is progressively elaborated as more details are acquired through the planning processes. It includes information such as the description of work, responsible organization, required resources, scheduled milestones, and other detailed information linked to the WBS.
(5) The project team develops a WBS. One of the essential features of WBS is that it explains the deliverables and not the sub-complexities attached with all the respective deliverables.
(6) WBS provides a well-thought-out visualization of the project deliverables by decomposing the project goal into smaller and manageable sub-divisions. It projects a clear picture concerning the main objective and the various associated sub-objectives, which also need to be detailed and divided into many sub-projects and sub-sub-projects so that each of the mentioned tasks can be detailed and monitored to the minute aspect of it. The decomposition of the scope of work into tasks is extended until it is acceptable to the stakeholders for planning, delivery, and control. Since the whole team is involved in creating the WBS, it provides a team buy-in or shared ownership of the project while keeping the communication between the project team and other stakeholders healthy.
(7) The most important output of creating a WBS is the scope baseline. A scope baseline is an approved version of a scope statement, the WBS, and its associated WBS dictionary. The scope baseline can only be modified through formal change control procedures. Whenever the scope of the project needs to be altered, the WBS and its dictionary serve as a basis for comparison.
(8) WBS links all the project's significant stakeholders with each minute detailing, tasking, and monitoring aspects of various sub-projects and sub-sub-projects so that any immediate change and course correction can be taken at any level and stage achieved for the betterment of tasking and overall project completion. The WBS assumes linkage with various elements of project planning and execution, including scope management, schedule management, financial management, resource allocation, and risk assessment
(9) The WBS system provides a hierarchical breakdown of the total scope of the project to the team. This linkage helps the team to accomplish the project objectives at various stages and thus provide the required deliverables. It also shows the interconnection and dependency of each of the big/significant tasks on the different smaller and multiple projects at the ground or shop floor level.

(10) The control account of the WBS is evaluated along with various milestones in the project. Finally, it is compared to earned value for understanding the performance of the cost incurred in the project.
(11) The WBS and schedule linkage here keeps every manager and supervisor aware of their tasks and deadlines. It follows up with each person associated with achieving the target so that the overall project can move/traverse and the cardinal targets/tasks of sub-projects can be achieved.
(12) WBS linkage here keeps the top executives and managers aware of personnel and resource status at the ground/tasking/supervising/executive level. Proper planning and forecasting can be done for effective execution and task accomplishment. Thus, HR and resource management are two major stakeholders for keeping any project floating after the mission is realized and finalized.
(13) WBS linkage keeps every stakeholder updated, as any contingency and emergency can pop up at any point of working time. The contingency plan gives people an option to move to plan B and progress with respective tasking and following the deadlines. The risk associated with each tasking and sub-project linked to an effective tasking needs to be visualized. Everyone needs to understand the complexities and dangers associated with such an emergency.

The steps involved in the creation of WBS include:

(1) Enlisting the task breakdown in successive levels
(2) Identifying data for each work package
(3) Reviewing work package information
(4) Calculating the costs associated with each of the work packages
(5) Scheduling the work packages
(6) Monitoring the actual resource consumption
(7) Monitoring the project schedule.

4.3 Work Breakdown Structure (WBS): Illustrative Examples

Example 3.1 WBS for new product development—Construction of a warship (refer Fig. 4.2).

Example 3.2 WBS for IT Project: Commissioning of Integrated Platform Management System (IPMS) for remote machinery monitoring and control (refer Fig. 4.3).

Example 3.3 WBS for competitive exam project (refer Fig. 4.4).

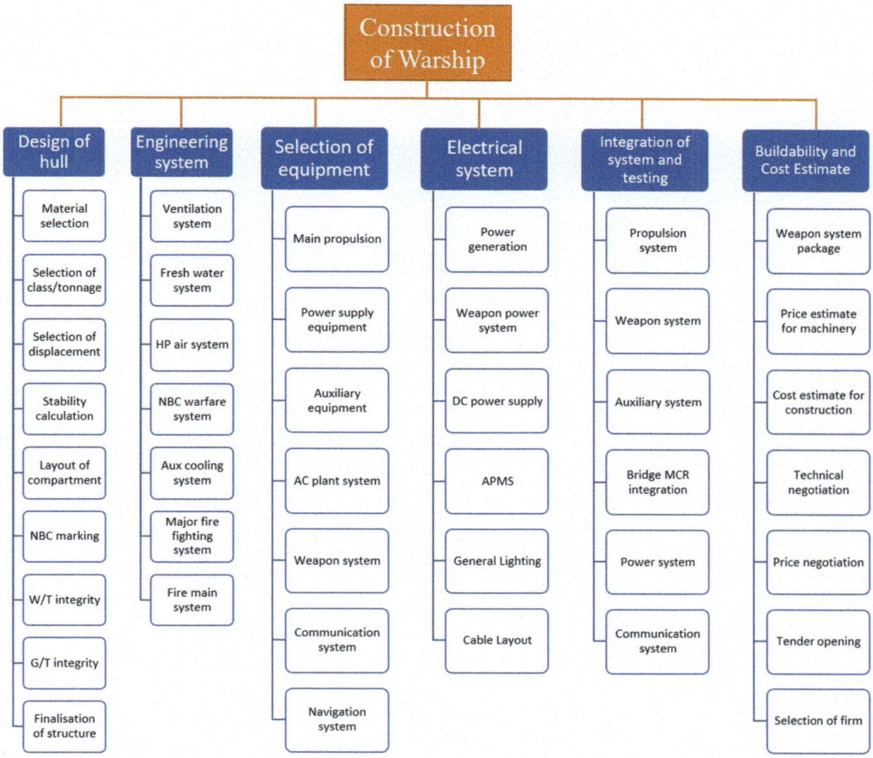

Fig. 4.2 WBS for new product development

4.4 Phases in Project Planning

The importance and need of project planning were discussed in the previous section. This section explains the various phases of project planning and their relevance in a highly complex and uncertain project environment.

Phase 1: Summary

In this phase of project planning, a summary of the entire project is made to have an overall picture of the project and its impact on the organization. Suppose the project is in a highly complex and uncertain environment. In that case, the inclusion of an uncertainty profile based on the complexities involved based on the geography, environment, economy, and other factors, will result in a better pre-planning process for the subsequent phases.

Phase 2: Objectives

This is the phase of project planning, where the goals/objectives of the project are laid out in a descriptive and detailed manner based on the project's scope statement. Most

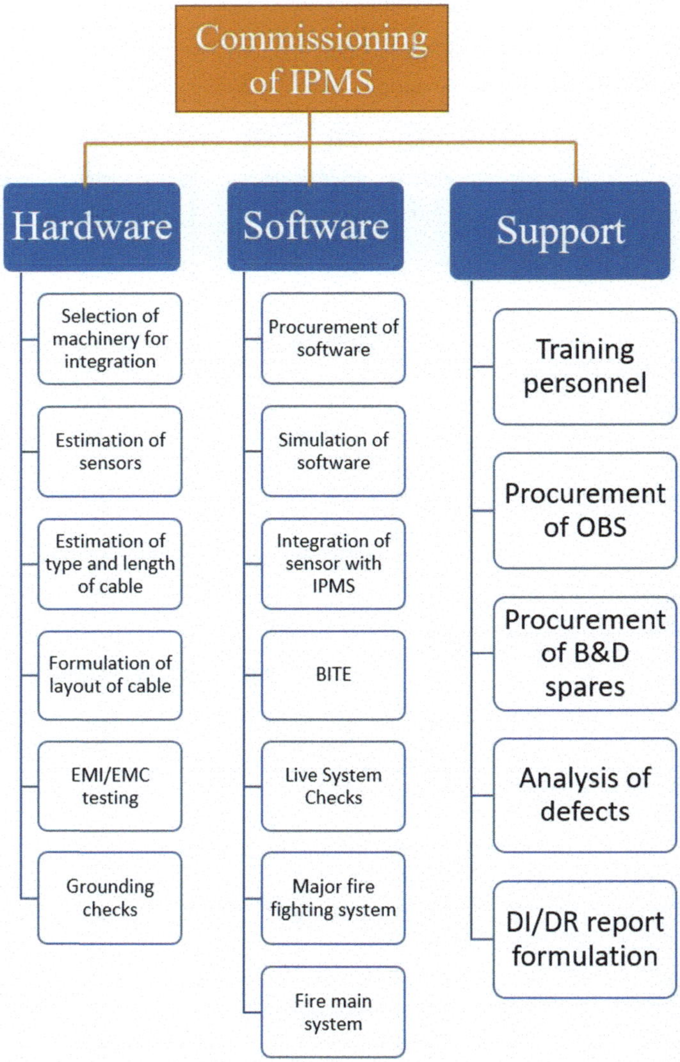

Fig. 4.3 WBS for IT project

of the failures in a highly complex and uncertain project environment are due to vague objectives or lack of clarity on the specifications or details of the goals/objectives that are to be achieved through the project. Therefore, a proper study and in-depth understanding of the objectives and consideration of all the sub-goals involved in achieving these objectives are carried out during this project planning phase. This is a crucial factor to perform better in a complex and uncertain project environment.

Phase 3: Contractual aspects

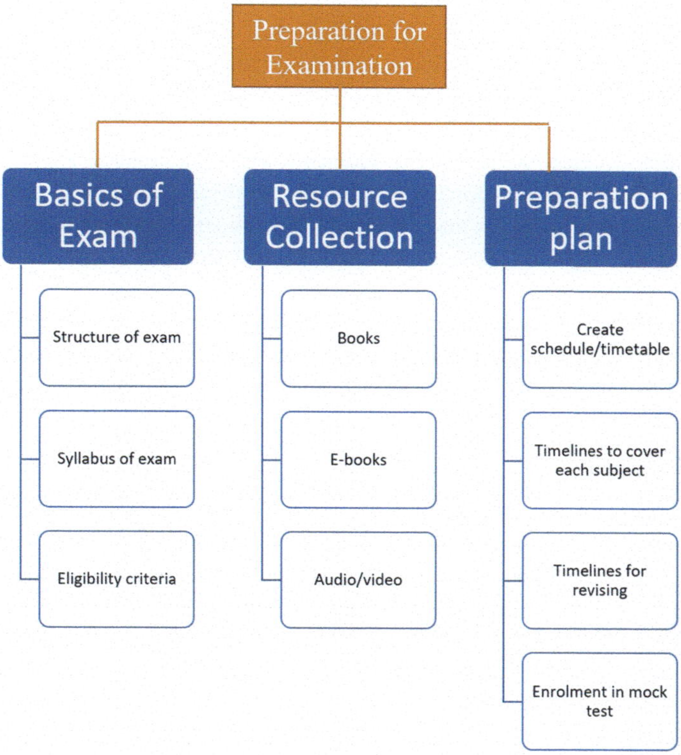

Fig. 4.4 WBS for competitive exam project

This is the most crucial phase in project planning. The success of a project is as good as the quality and specificity of a contract. The contract should be made only after a complete and proper understanding of the objective phase. There must be a lot of study on the customer profile, environment, geographical location, and resource availability before formulating a contract to deal with uncertainty and complexity in the project.

Phase 4: Schedule/Project Scheduling

This phase of project planning includes allocating timeframe to each activity of the project from the start to its termination and assigning each exercise's responsibility to the team's concerned team. This caters to the planning of the time-duration of each activity in the project. However, the approach to be considered at this phase for a highly complex and uncertain project environment is different from conventional methods. In general, the schedule is considered for a static environment. In uncertainty and complexity, the environment is dynamic, and the planned schedule might not be accurate to meet the project deadlines. Hence, many iterations are to be made before considering the final path of the activities and suitable flexibility is to be catered while scheduling for a complex and uncertain environment.

Phase 5: Resource and personnel management

This phase involves two activities—resource planning and personnel planning. Resource planning involves planning budget, overall cost monitoring/controlling, availability/unavailability of all deliverables, and resources required during the project keeping in view the time–cost trade-offs. Personnel planning involves planning skilled personnel selection, training, and identifying suitable persons for the right job. In a highly complex and uncertain environment, resource leveling should be made, considering the time–cost and performance trade-off. Also, there must be a flexibility of utilizing more resources in the occasion of high uncertainty in the project. In addition, personnel planning should cater to the selection of a project manager and team. It is a proven fact that a good project manager and a capable team can handle uncertainty and complexity in a skilled and efficient manner, thereby resulting in good project management.

Phase 6: Risk analysis

This phase of project planning includes planning how to deal with favorable and unfavorable eventualities of a project. Some of the eventualities include natural calamities, contractual labor strikes, unexpected technical breakthroughs, and so on. Uncertainty is bound to be present in this phase. No matter how perfect your plan, these crises might not be subdued completely. However, early planning may cater to some of the uncertainty and would cause lesser damage. During the initial phase of project planning, the uncertainty profile would help in better risk planning, thereby helping the project perform in a complex and uncertain environment.

Even after laying down a thorough project planning process, in case of a highly complex and uncertain project environment, it finally comes to the efficiency, skills, and capability of a project manager to quickly respond to inadvertent situations through his knowledge and social abilities. The problem is different and unique for every project manager as uncertainty and complexity are both random. Hence, the approach to these situations is also unique to every project manager.

4.5 Project Network Representation: Activity on Arc (A-O-A) and Activity on Node (A-O-N)

Project network representation is a convenient way to show activities and their precedence to the whole project. This helps the project manager in four ways: (a) responsibility allocation, (b) definition of subcontracting units, (c) role of different players, and (d) basic scheduling and establishment of work timetables. This helps in critical path determination and to apply selective management control for project execution. An accurate network representation also helps the project team's inappropriate allocation of resources. The resource planning for the project includes:

– Project crashing with time–cost trade-offs

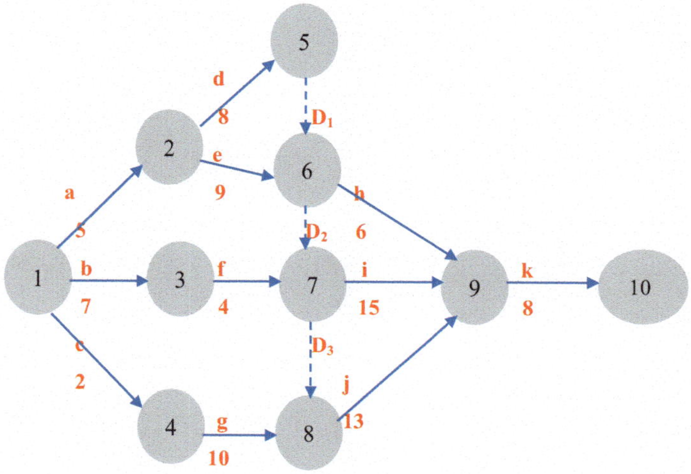

Fig. 4.5 A-O-A representation

- Resource aggregation
- Resource leveling
- Limited resource allocation.

The project network is thus a standard vehicle for planning, communicating, and implementing the project right from its inception.

There are two ways to represent a project in network structure.

- Activity on Arc (A-O-A): Arrow diagrams, event-oriented networks
- Activity on Node (A-O-N): Precedence networks, activity-oriented networks.

A-O-A representation (refer Fig. 4.5):

A-O-N representation (refer Fig. 4.6):

Develop A-O-A and A-O-N project representation for the dataset given in Table 4.2.

A-O-A representation of the project network for the dataset given in Table 4.2 is presented in Fig. 4.7.

A-O-N representation of the project network for the dataset given in Table 4.2 is presented in Fig. 4.8.

4.6 Consistency and Redundancy in Project Network

(1) The interdependence among project activities can be obtained from lists of jobs and their precedence relations. Consistency in a project network refers to the feasibility of generating a proper network from the given precedence

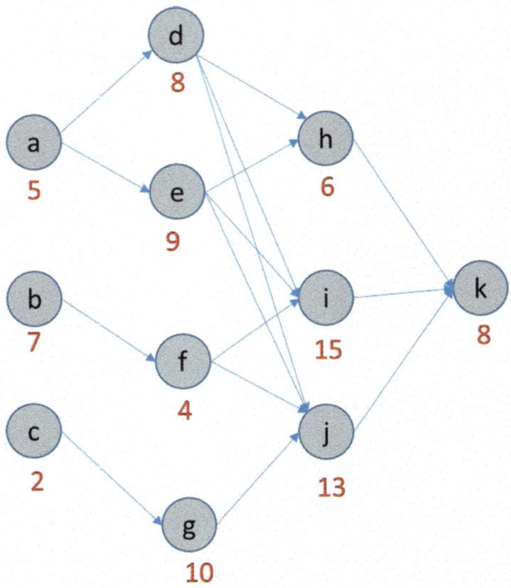

Fig. 4.6 A-O-N representation

Table 4.2 Project activity successor–predecessor relationship

Activity	Predecessor	Time (days)
A1	–	8
A2	–	6
A3	A1	4
A4	A1	7
A5	A2, A3	4
A6	A4, A5	5
A7	A3, A4, A5	2
A8	A4, A6	4

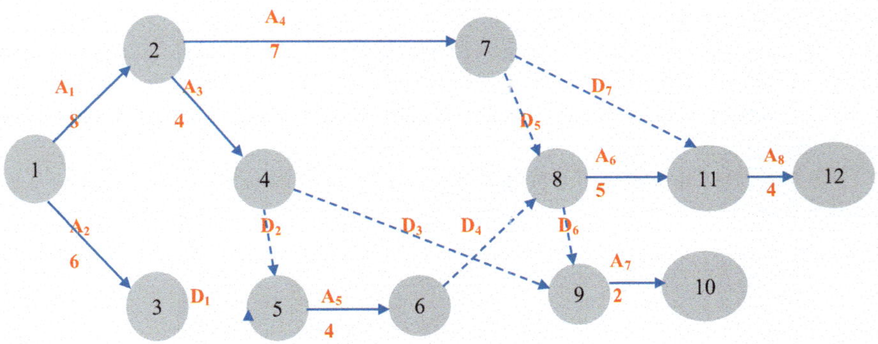

Fig. 4.7 A-O-A representation of project activities listed in Table 4.2

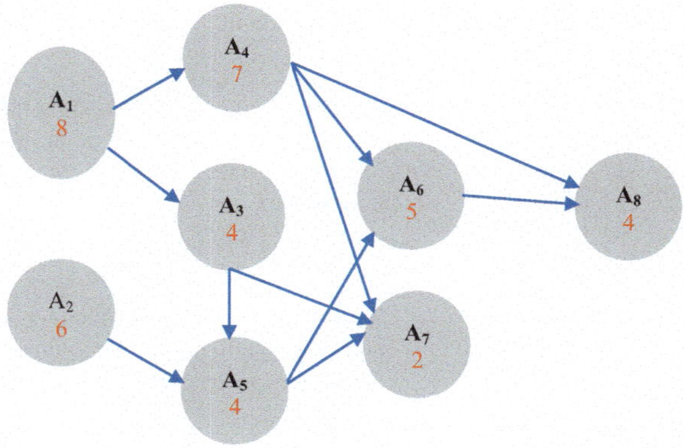

Fig. 4.8 A-O-N representation of project activities listed in Table 4.2

relations. By consistency, we mean that a project network should not contain any cycles or loops. In a consistent network, all the arrows define different paths without creating any loops. An inconsistent network with loops implies that a particular activity should be completed before it starts, which is logically inconsistent. Consistency checks broadly include checks that remove any kind of cycle within the project network. The various methods for checking project network consistency include:

(a) Topological ordering
(b) Fulkerson rule
(c) Marimont procedure
(d) Matrix squaring method

(2) Redundancy is the presence of any unnecessary information or additional information in the predecessor set. The redundancy check specifies the unnecessary arcs which can be omitted. Therefore, a systematic review needs to be carried out in a specified manner to remove redundancy from the project network.

(3) For timely and successful completion of a project, it becomes necessary for the project team to identify and accomplish the mission of critical tasks. Spending time, resources, and cost on unnecessary and illogical activities will adversely affect the project's completion. Therefore, the effective assessment of the project network, along with the consistency of network nodes and arcs and removal of redundancies from the project network, becomes the primary objective of the project planner and executives. Consistency needs to be maintained during all phases of project planning. From defining the project's scope to building the WBS, from planning for schedule and resources to planning for risk management, the project team needs to demonstrate consistency and avoid any kind of redundancy possible. Most importantly, the analysis of consistency and redundancy should be exercised at the WBS level, where a lot of sub-projects and sub-sub-projects exist. Clearing the clutter and prioritizing the tasks becomes mandatory for describing the process and selecting a hassle-free path of well-defined arcs and nodes in the network.

Table 4.3 Commissioning of a diesel alternator onboard a ship project

Activity	Description	Activity predecessor
A1	Selection of make, model, and rating of a diesel alternator depending on the power consumption and distribution requirements	–
A2	Completion of factory acceptance trails (FATS) and testbed trials	A1
A3	Initiating and installation of components and sub-components at the designated place at the selected machinery compartment	A2
A4	Completion of all alignment checks between engine and alternator. Followed by the commencement of fuel/oil flushing checks for initial tuning/setup checks	A3
A5	Commencement of harbor acceptance trials (HATs) and complete load trials	A4
A6	Completion of sea acceptance trials (SATs) and final acceptance of DAs	A5

Network representation (Activity on Arc (AOA))

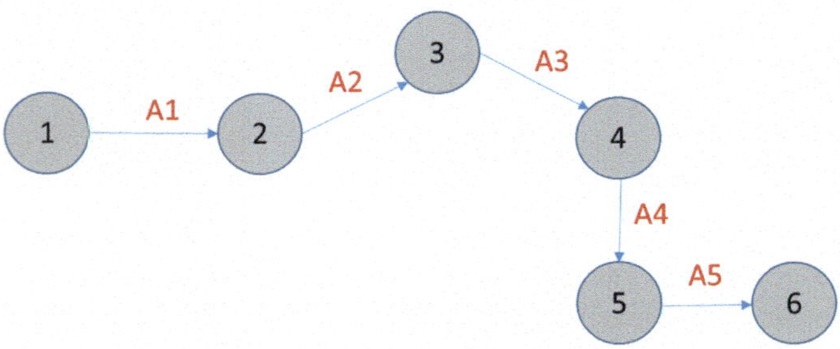

(4) The example under consideration is a project for the commissioning of a Diesel Alternator onboard a ship. The project details are given in Table 4.3.

4.7 Methods for Checking Consistency in Project Network

The methods used for consistency checking in the project network are explained below.

4.7.1 *Topological Ordering*

The order of activities is such that each activity appears only after its predecessors have appeared. Topological ordering is employed to order the activities topologically. The steps in topological ordering are as follows:

(a) Select a job that has no predecessors and place it on top of the list
(b) Delete the job that has just been placed from the predecessor list of remaining jobs
(c) Identify new sources, i.e., jobs without predecessors so generated that can be placed in the list and then go back to STEP 2
(d) Repeat the above steps until
(e) There are no sources on the list of remaining jobs which show the presence of inconsistency.
OR
(f) All the jobs have been placed on the list. It indicates that all jobs have been topologically ordered, and there are no inconsistencies in the network.

4.7.2 *Fulkerson Rule*

This is a procedure in which the number of nodes of a given network are arranged so that for every activity going from node i to j, the node number i is always less than the node number j. We continue to keep sequentially numbering the nodes until we either finish numbering all the nodes (indicating consistency) or we are left with unnumbered nodes (indicating inconsistency). The steps in the Fulkerson rule are as follows:

(a) Identify the source nodes in a given network and begin numbering them sequentially from number 1.
(b) For each numbered node, delete the outgoing arcs and then identify new sources
(c) Continue to number the newly discovered source nodes sequentially
(d) Keep repeating the above steps unit.

(i) All nodes are numbered, indicating a consistent network
OR
(ii) Absence of source in unnumbered nodes indicating an inconsistent network.

4.7.3 *Marimont Procedure*

Marimont procedure is applied to a project network. We first identify the source and sink nodes in this procedure and delete all incoming and outgoing arcs. The process is repeated until all nodes and arcs have been removed (indicating consistency) or have a residual sub-network (indicating inconsistency).

Steps in Marimont procedure:

(a) Identify the source and sink nodes
(b) Disconnect source with the network by deleting the source node and all the outgoing arcs
(c) Disconnect sink node by deleting the sink node and all the incoming arcs
(d) Repeat steps 1–3 on the remaining network until.
(e) All nodes and arcs have been disconnected, which indicates consistency
 OR
(f) We are left with a residual sub-network which indicates inconsistency.

4.7.4 Matrix Squaring Method

The matrix squaring method is used to identify inconsistencies in an adjacency matrix. An adjacency matrix is a representation of a project network in the form of a matrix. It is a square matrix ($n \times n$) of 0 s and 1 s, where n is the number of nodes in the network. In the matrix squaring method, we multiply the matrix M with itself at most n-1 times. If the matrix is consistent, it must be nilpotent of index "n" or less. For example, M is nilpotent of index k if $M^k = 0$ but $M^{(k-1)}$ is not equal to 0.

4.8 Illustrative Application of Project Network Consistency Methods

A Textile Mill inspected by OSHA (Occupational Safety and Health Administration) was found to violate several safety regulations. Thus, the OSHA inspectors ordered the mill to make some existing machinery safer (e.g., add safety guards), purchase some new machinery to replace older, dangerous machinery, and relocate some machinery to provide safer passageways. The machinery project details are given in Table 4.4.

Topological Ordering

Activity and activity predecessor of machinery alternation project (see Table 4.4) are separately presented in Table 4.5.

The steps in topological ordering are demonstrated as follows.

- We start by taking the source activities and adding them to a list. In the above example, activities a, b, and c are source activities as they have no predecessors. Therefore, we add these three activities to our list and delete them from the remaining jobs.

Table 4.4 Machinery alteration project

Activity	Description	Activity predecessor	Time
a	Order new machinery	–	5
b	Plan new physical layout	–	7
c	Determine safety changes in existing machinery	–	2
d	Receive equipment	a	8
e	Hire new employees	a	9
f	Make plant alternations	b	4
g	Make changes in existing machinery	c	10
h	Train new employees	d, e	6
i	Install new machinery	d, e, f	15
j	Relocate old machinery	d, e, f, g	13
k	Conduct employee safety orientation	h, i, j	8

Table 4.5 Predecessor and successor for machinery alteration project

Activity	Activity predecessor
a	–
b	–
c	–
d	a
e	a
f	b
g	c
h	d, e
i	d, e, f
j	d, e, f, g
k	h, i, j

- List = {a, b, c}
- Remaining jobs = {d, e, f, g, h, I, j, k}.

- Now we look at the remaining jobs list and identify source activities. We see activities d, e, f, and g are now the sources since their predecessors have already been added to the list and removed from the remaining jobs. So now the list and remaining jobs become:

  - List = {a, b, c, d, e, f, g}
  - Remaining jobs = {h, i, j, k}.

- Looking at the remaining jobs, we notice that h, i, and j are now sourced as all their predecessors d, e, f, and g have already been added to the list and are not part of the remaining jobs. Therefore, we can add h, i, and j to the list and remove them from the remaining positions.

- List = {a, b, c, d, e, f, g, h, i, j}
- Remaining jobs = {k}.

- We follow the same process for activity k: add it to the list and remove it from the remaining jobs since all its predecessors have already been added to the list and removed from the remaining jobs.
- Final list of activities = {a, b, c, d, e, f, g, h, I, j, k}.
- Since we were able to add all the activities in a topographical order, we can confirm that the project network is consistent.

Fulkerson Rule

The network diagram for the project is presented in Fig. 4.9.

We start by identifying the source nodes, i.e., nodes without predecessors. We can see nodes a, b, and c are source nodes without any predecessors from the network. We number them 1, 2, and 3 and then delete the outgoing arcs from them (as shown in Fig. 4.10).

Moving forward with our iteration, we notice that nodes d, e, f, and g are now source nodes because they do not have any predecessor nodes. Therefore, we can sequentially number them and also remove all outgoing arcs from these nodes. Thus, the network now looks as presented in Fig. 4.11.

Now we are left with nodes h, i, and j as source nodes. We will iterate the same process of sequentially numbering them and remove all outgoing nodes and continue to node k. The network now looks like this:

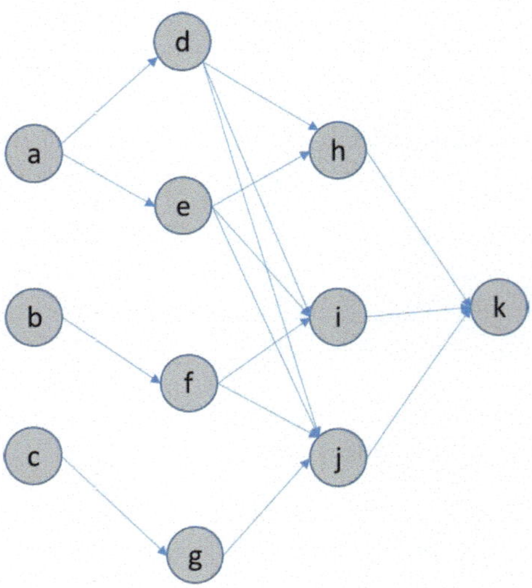

Fig. 4.9 Network diagram for machinery alteration project

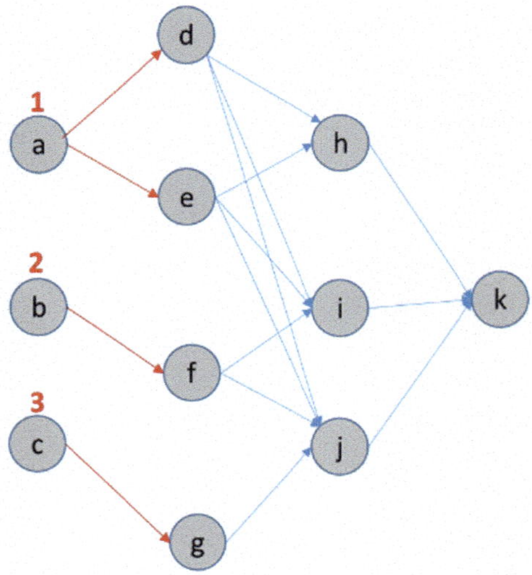

Fig. 4.10 Application of Fulkerson's rule for machinery alteration project (iteration 1)

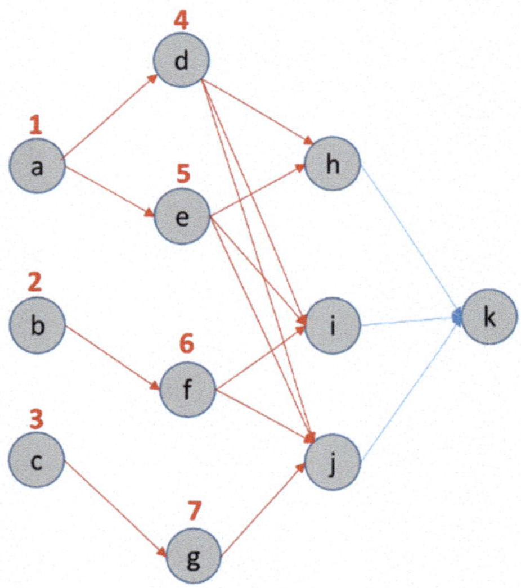

Fig. 4.11 Application of Fulkerson's rule for machinery alteration project (iteration 2)

Since we could number all nodes in the network (see Fig. 4.12) while keeping i < j for all arcs (i, j), we can confirm that the network is consistent.

Marimont's Procedure

Starting with the project network (see Fig. 4.13), we first identify the source and sink nodes.

Nodes a, b, and c are source nodes, and node k is the sink node. Now that we know the source and sink nodes, we can disconnect them by removing the source

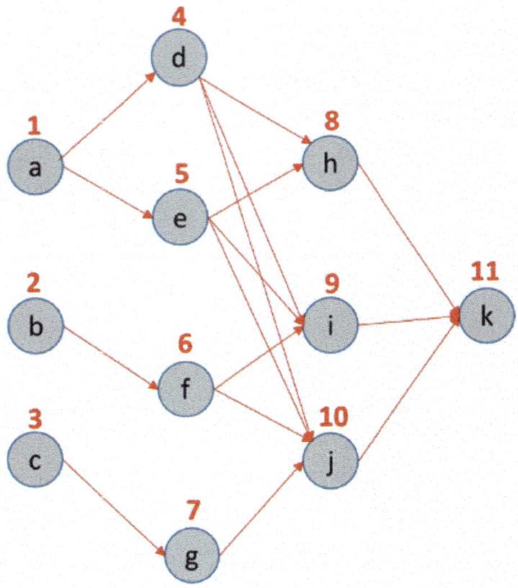

Fig. 4.12 Application of Fulkerson's rule for machinery alteration project (iteration 3)

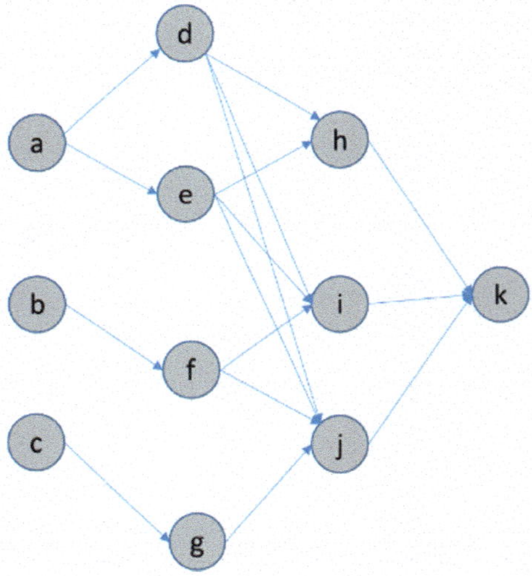

Fig. 4.13 Network diagram for machinery alteration project

node and all the arcs emanating from them and removing the sink node and all the arcs coming into it. The derived network is presented in Fig. 4.14.

We are now left with a sub-network with nodes d, e, f, g, h, i, and j. We can say that nodes d, e, f, and g are sources and nodes h, I, and j are sinks from this sub-network. Therefore, we can disconnect these nodes and the outgoing arcs from sources and incoming arcs to sinks. The derived network is presented in Fig. 4.15.

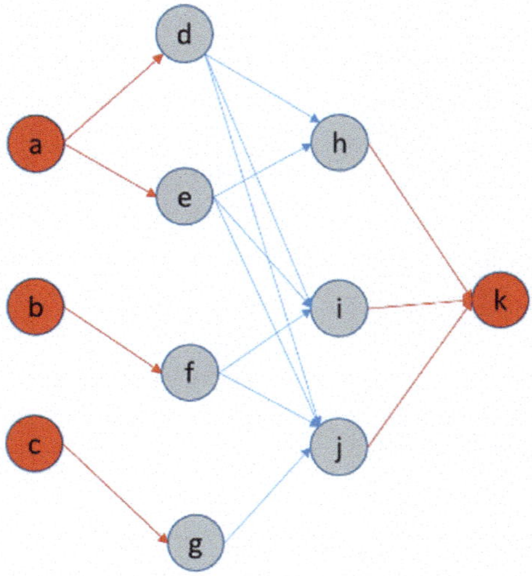

Fig. 4.14 Application of Fulkerson's rule for machinery alteration project (iteration 1)

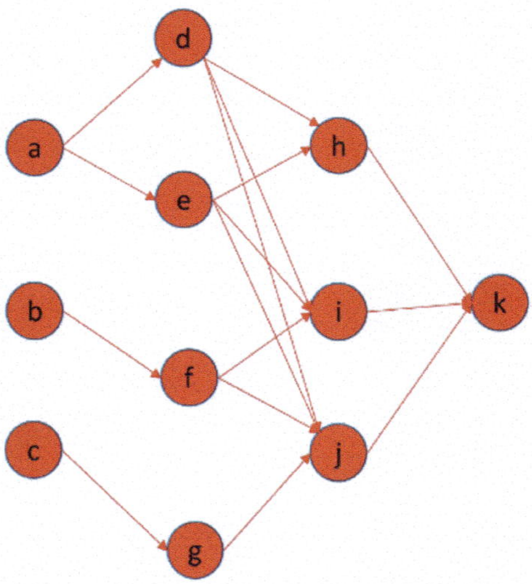

Fig. 4.15 Application of Fulkerson's rule for machinery alteration project (iteration 2)

Since we successfully disconnected all nodes and arcs without leaving behind a residual sub-network, we can confirm that the network is consistent.

Matrix Squaring Procedure

The adjacency matrix M for the given network is below.

	a	*b*	*c*	*d*	*e*	*f*	*g*	*h*	*i*	*j*	*k*	
A	0	0	0	1	1	0	0	0	0	0	0	2
B	0	0	0	0	0	1	0	0	0	0	0	1
C	0	0	0	0	0	0	1	0	0	0	0	1
D	0	0	0	0	0	0	0	1	1	1	0	3
E	0	0	0	0	0	0	0	1	1	1	0	3
F	0	0	0	0	0	0	0	0	1	1	0	2
G	0	0	0	0	0	0	0	0	0	1	0	1
H	0	0	0	0	0	0	0	0	0	0	1	1
I	0	0	0	0	0	0	0	0	0	0	1	1
J	0	0	0	0	0	0	0	0	0	0	1	1
K	0	0	0	0	0	0	0	0	0	0	0	0
	0	0	0	1	1	1	1	2	3	4	3	

M² becomes

0	0	0	0	0	0	0	2	2	2	0
0	0	0	0	0	0	0	0	1	1	0
0	0	0	0	0	0	0	0	0	1	0
0	0	0	0	0	0	0	0	0	0	3
0	0	0	0	0	0	0	0	0	0	3
0	0	0	0	0	0	0	0	0	0	2
0	0	0	0	0	0	0	0	0	1	0
0	0	0	0	0	0	0	0	0	0	0
0	0	0	0	0	0	0	0	0	0	0
0	0	0	0	0	0	0	0	0	0	0
0	0	0	0	0	0	0	0	0	0	0

M^3 is

A	*b*	*c*	*d*	*e*	*f*	*g*	*h*	*i*	*j*	*k*
0	0	0	0	0	0	0	0	0	0	9
0	0	0	0	0	0	0	0	0	0	4
0	0	0	0	0	0	0	0	0	0	2
0	0	0	0	0	0	0	0	0	0	0
0	0	0	0	0	0	0	0	0	0	0
0	0	0	0	0	0	0	0	0	0	0
0	0	0	0	0	0	0	0	0	2	0
0	0	0	0	0	0	0	0	0	2	0
0	0	0	0	0	0	0	0	0	2	0
0	0	0	0	0	0	0	0	0	2	0
0	0	0	0	0	0	0	0	0	2	0

M^4 is

A	*b*	*c*	*d*	*e*	*f*	*g*	*h*	*i*	*j*	*k*
0	0	0	0	0	0	0	0	0	0	0
0	0	0	0	0	0	0	0	0	0	0
0	0	0	0	0	0	0	0	0	0	0
0	0	0	0	0	0	0	0	0	0	0
0	0	0	0	0	0	0	0	0	0	0
0	0	0	0	0	0	0	0	0	0	0
0	0	0	0	0	0	0	0	0	0	3
0	0	0	0	0	0	0	0	0	0	3
0	0	0	0	0	0	0	0	0	0	3
0	0	0	0	0	0	0	0	0	0	3
0	0	0	0	0	0	0	0	0	0	3

M^5 is

$$
\begin{matrix} A & b & c & d & e & f & g & h & i & j & k \end{matrix}
$$

$$
\begin{vmatrix}
0 & 0 & 0 & 0 & 0 & 0 & 0 & 0 & 0 & 0 & 0 \\
0 & 0 & 0 & 0 & 0 & 0 & 0 & 0 & 0 & 0 & 0 \\
0 & 0 & 0 & 0 & 0 & 0 & 0 & 0 & 0 & 0 & 0 \\
0 & 0 & 0 & 0 & 0 & 0 & 0 & 0 & 0 & 0 & 0 \\
0 & 0 & 0 & 0 & 0 & 0 & 0 & 0 & 0 & 0 & 0 \\
0 & 0 & 0 & 0 & 0 & 0 & 0 & 0 & 0 & 0 & 0 \\
0 & 0 & 0 & 0 & 0 & 0 & 0 & 0 & 0 & 0 & 0 \\
0 & 0 & 0 & 0 & 0 & 0 & 0 & 0 & 0 & 0 & 4 \\
0 & 0 & 0 & 0 & 0 & 0 & 0 & 0 & 0 & 0 & 4 \\
0 & 0 & 0 & 0 & 0 & 0 & 0 & 0 & 0 & 0 & 4 \\
0 & 0 & 0 & 0 & 0 & 0 & 0 & 0 & 0 & 0 & 0
\end{vmatrix}
$$

M^6 is

$$
\begin{vmatrix}
0 & 0 & 0 & 0 & 0 & 0 & 0 & 0 & 0 & 0 & 0 \\
0 & 0 & 0 & 0 & 0 & 0 & 0 & 0 & 0 & 0 & 0 \\
0 & 0 & 0 & 0 & 0 & 0 & 0 & 0 & 0 & 0 & 0 \\
0 & 0 & 0 & 0 & 0 & 0 & 0 & 0 & 0 & 0 & 0 \\
0 & 0 & 0 & 0 & 0 & 0 & 0 & 0 & 0 & 0 & 0 \\
0 & 0 & 0 & 0 & 0 & 0 & 0 & 0 & 0 & 0 & 0 \\
0 & 0 & 0 & 0 & 0 & 0 & 0 & 0 & 0 & 0 & 0 \\
0 & 0 & 0 & 0 & 0 & 0 & 0 & 0 & 0 & 0 & 0 \\
0 & 0 & 0 & 0 & 0 & 0 & 0 & 0 & 0 & 0 & 0 \\
0 & 0 & 0 & 0 & 0 & 0 & 0 & 0 & 0 & 0 & 0
\end{vmatrix}
$$

For M^6, we get a 0 matrix. Since 6 is $<$ 11 (number of nodes). We can further say that the matrix is nilpotent of index six, which is less than the *n*. Therefore, we can confirm that the network is consistent.

4.9 Redundancy Checking in Project Network

In any given list of project predecessors, sometimes the predecessors are over-specified and may be pruned to yield a more compact network. As a result, there is unnecessary additional information which only adds clutter to the network. It is thus useful to check for any redundancies after the topological ordering of activities has been performed. A tubular method is widely used to represent an immediate and distant predecessor. The cell in which immediate and distant predecessors coincide with each other is called a "redundancy (extra information)" in the network. This needs to be eliminated for appropriate project planning.

4.9.1 Redundancy Checking for Project Network Presented in Table 4.3

	A1	A2	A3	A4	A5	A6	A7	A8
A1								
A2								
A3	x							
A4	x							
A5		x	x					
A6				x	x			
A7			x	x	x			
A8				x		x		

This network has two redundancies, as indicated by redline.

4.9.2 Redundancy Checking for Project Network Presented in Table 4.5

	a	b	c	d	e	f	g	h	i	j	k
a											
b											
c											
d	x										
e	x										
f		x									
g			x								
h				x	x						
i				x	x	x					
j				x	x	x	x				
k								x	x	x	

There is no redundancy in the network.

Project Management in Practice

Scope of Electrified Double Railway Track Project

The Government of Malaysia undertook a Electrified Double Railway Track Project. The project is being executed by IRCON International Ltd. (IRCON), a Public Sector Unit incorporated by the Government of India. IRCON is executing the project on a 'Design and Build' contract.

Key project elements:

- Infrastructure works comprising soil improvement works, earthwork and related retaining structures, drainage works, bridges, station buildings and yards, staff quarters, etc.
- Electrified double railway track in all respects with a total track length of 215 km.
- The system works of signaling, communication, and track electrification.
- Landscaping of area within the Railway reserve during and upon completion of construction.
- Environmental impact analysis mitigation/measures.

Scope of work of IRCON:

- Detailed planning, design, construction, testing, commissioning, and maintenance of completed works for two years.
- All procurements and manufacture, insurance, and handling of all local and imported material necessary for satisfactory completion of the works.

The designed track alignment runs parallel to an existing railway track, but the proposed track crisscrosses the current operational track at 23 locations. Laying of the new track and its commissioning needs to be done without disrupting services on the existing track, and this was a project execution challenge. The new double-track is as far as possible accommodated, in the existing railway land. However, to improve track geometry for the new design speeds and realignment requirements (due to crisscrossing of the existing and new tracks to be laid), acquisition of more land is needed. Land acquisition was under the scope of the client.

Takeaways:

- A 'Client driven' approach ensures effective project management.
- The Government being the client for significant engineering and construction projects can help evolve a project management culture. This helps to enhance the overall efficiency of various projects executed in the country.

Source An international railway project: a project management case study, by Vanita Ahuja and Hitesh Khanna, September 2013, https://doi.org/10.13140/2.1.1098.0487. (https://www.researchgate.net/publication/264697324), *accessed on 20th June 2021.*

Project Management in Practice
Development of a new propulsion system for spacecraft.

1. Satellite launches involve many sub-projects like construction of launch complex, development of propulsion system for positioning of the rocket used for launching the vehicle, development of propulsion system for the satellite launch vehicle, and development of avionics systems the final project objective, i.e., launch of a satellite vehicle. One such sub-project is the case study under consideration.
2. **Summary/Overview**: The project aims to develop a new diesel engine with higher power and rated load, which would be used in transporter vehicles to transport spacecraft to the launch site. This is an enhancement from previous launches as this time; the transporter vehicle requires higher horsepower and torque to transport new space crafts with heavier weights.
3. **Objective**: The previous engine used for the old satellite launches was a sixteen-cylinder diesel engine with a rated power of 1840 kW and 2467 horsepower. The project aims to develop a new eighteen-cylinder diesel engine with a rated power of 2150 kW and 2884 horsepower.
4. **Project Scope/Scope statement**: The scope of the project includes the following activities or resources.

    (a) Resources are required to study the previous version of the diesel engine and its design, specifications, performance, and ergonomics
    (b) Resources for R&D on the engine to be developed, including design considerations like power to ratio, volumetric efficiency, brake horsepower
    (c) Resources for hardware and software required to carry out step (a) and step (b)
    (d) Resources for finalization of design and model to be tested like special tools, testing apparatus
    (e) Cost for factory acceptance trials, dummy weight trials, and final trials of the equipment
    (f) Cost for procurement of spares and components for assembly of the engine and its auxiliaries
    (g) Cost of labor for fitment of engine and related auxiliary systems.

5. **Work breakdown structure**: The work breakdown structure for this project is illustrated in Exhibit 4.1.
6. The work breakdown structure above represents the structural breakdown of activities. The stakeholders for the above activities and their roles are as defined here. The assumption for this case study is that the organization follows a matrix organizational structure. The overall head of the organization is a Project Director (PD). A Deputy General Manager heads the project under consideration for this case study (Propulsion, Gearing, and shafting), considered the overall project manager for this sub-project. He is assisted by Manager (R&D), Manager (Planning), Manager (Material), Manager (Engineering), and

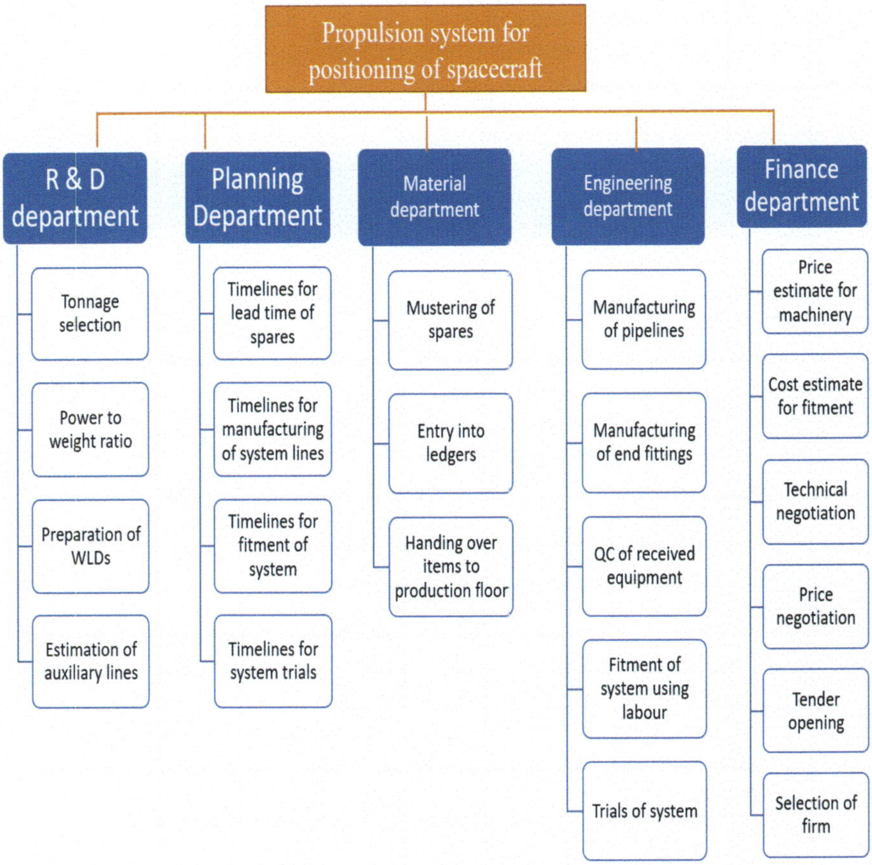

Exhibit 4.1 WBS for the development of propulsion system project

Manager (Finance), who in turn are assisted by deputy managers, and happen to be qualified engineers. The careful selection of exact individuals who are to be appointed to these posts would be made during the personnel planning phase. So it is evident that the prime stakeholder for this project is the project manager responsible for the PD. The other stakeholders are the managers and deputy managers as per authority and tasks assigned to them through delegation. The roles of the individual stakeholders are already defined in the work breakdown structure in Exhibit 4.1.

7. **Schedule/Project Scheduling**: The key issues which could be points of conflict for scheduling are considered for brainstorming before arriving at the final schedule as follows:

   (a) The ready availability of hardware, software, and test equipment required to carry out design calculations, simulations, and final design, including iterative tests

(b) The decision to use indigenous or imported equipment is based on the availability and procurement timelines.
(c) Availability of ready-to-use spares to carry out trials, fitment, installation, setting to work, and commissioning the equipment
(d) Availability of manpower based on geographical location and the number of holidays due to local festivals
(e) Availability of unique resources (specific metals or material required for construction) within planned timelines
(f) Slack/buffer time to be considered in case of failure of trials, unavailability of resources, manpower, and spares.

8. Based on the above factors and brainstorming, the list of activities with successor–predecessor relationships has been formulated and is tabulated in Exhibit 4.2.
9. The final A-O-A and A-O-N network diagrams of the project based on the activity schedule are presented in Exhibits 4.3 and 4.4.
10. By applying consistency checks like Fulkerson's rule or Marimont procedure, it is evident that the networks presented in Exhibits 4.3 and 4.4 are consistent.
11. The project schedule is followed by the resources/personnel planning where the cost analysis and appointment of apt individuals to the stakeholder posts as defined above would be made. For example, Mr. X would be appointed as DGM (PGS) due to his specialization in working with diesel engines during the portfolios held by him at deputy manager and manager levels. Similarly, Mr. A, Mr. B, Mr. C, and Mr. D would be appointed as Manager R&D, Manager Planning, Manager Engineering, and Manager Material due to their vertical specialization in respective fields.

Exhibit 4.2 Predecessor–successor relationships

Activity	Description	Activity predecessor	Time in weeks
a	Selection of tonnage (Design stage)	–	2
b	Market Survey	–	2
c	Design finalization of engine	a	3
d	Finalisation of OEM	b	4
e	Preparation of WLDs	c, d	2
f	Placing PO for Designed engine	e	1
g	QC and muster of spares	f	1
h	Testbed trials	g	2
i	Contracting for casual labor for fitment	g	1
j	Fitment of equipment at a site	h, i	3
k	Commissioning of engine and auxiliaries	j	2

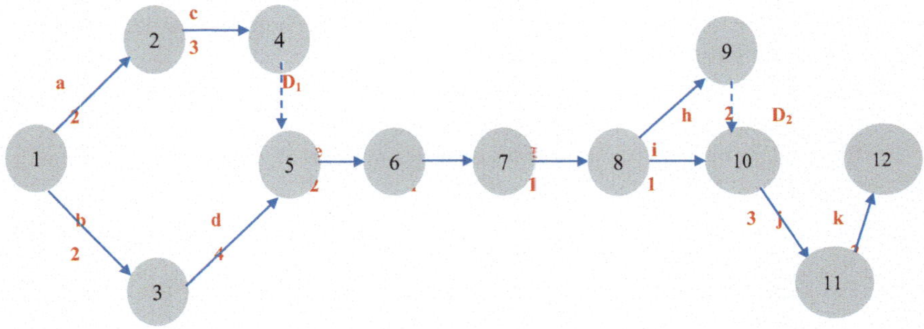

Exhibit 4.3 A-O-A network for the project

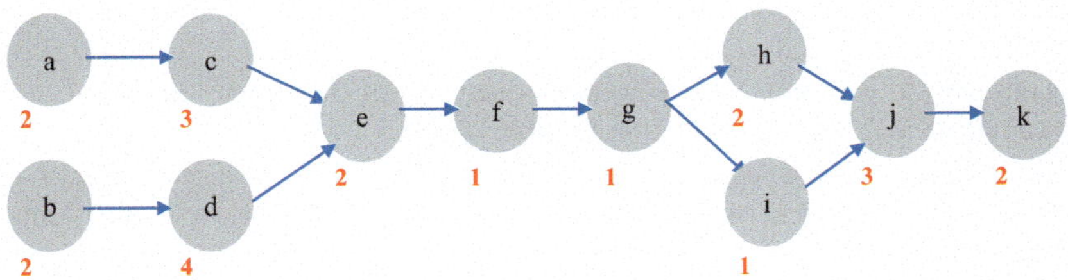

Exhibit 4.4 A-O-N network for the project

Exhibit 4.5 Risk management for spacecraft positioning vehicle

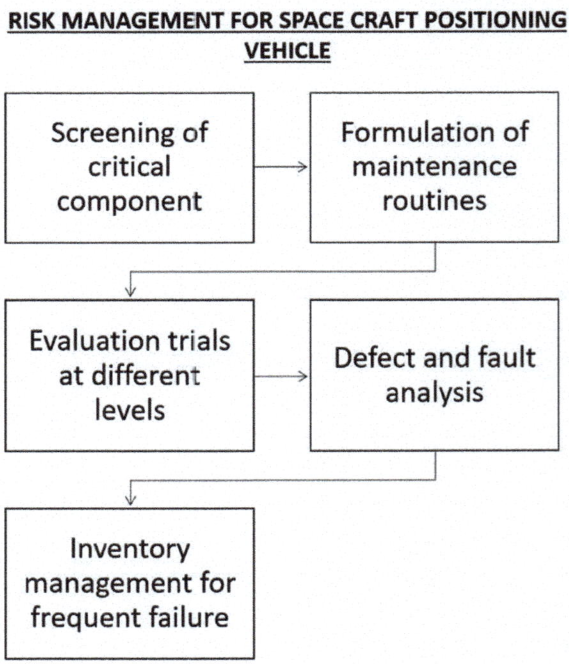

12. **Risk management**: The final phase in the project planning for this project is risk analysis or risk management. Exhibit 4.5 has been developed as a risk management methodology for the project under consideration.

The above case study covers all the phases involved in project planning for a hypothetical sub-project "The development of a new propulsion system for transporter vehicle used in spacecraft transport." Project planning is a crucial activity or stage of project management. However, it is pertinent to mention that the key to this project's success lies in selecting the Project Manager (DGM PGS, Mr. X), who can plan and push the project in a dynamic environment.

Questions for Class Discussion:

1. Why is an error-free project representation necessary for the case organizations?
2. What are the key insights a project manager and project team derive from a well-presented project network? How does this help in project planning and execution?
3. What is the role of the project network and defining roles and allocating resources for the various phases of the project? What are the implications of this on project completion time?

Summary

Project planning cannot be defined in a unique way as every project is unique in its own way and the plan for each project would be different.

The WBS breaks project tasks/goals into much smaller targets distributing the overall project. The hierarchal representation of the WBS aids in understanding the project deliverables better and leads to accomplishment of the objectives at every stage in order to provide the desired output.

The basic relation about a project can be obtained from lists of jobs and their precedence relations.

There are two ways to represent project network: (1) Activity on Arc (AOA) and (2) Activity on Node (AON)

Consistency in project network refers to feasibility of generating proper network from the given precedence relations. By consistency we mean that a project network should not contain any cycles or loops.

In any given list of project predecessors, sometimes the predecessors are over-specified and may be pruned to yield a more compact network. This is called redundancy in project network.

Questions for Discussion

1. What is project planning? What are the components of project planning? Illustrate the importance of each of the components of project planning for a typical R&D project in a Tabular format.

2. What is a work breakdown structure? What is its importance? What are the linkages of work breakdown structure with various elements of project planning and execution?
3. Develop work breakdown structures (WBS) for at least three different projects such as IT Project, New Product Development Project, Preparing for competitive exam project.
4. What are the different phases in project planning? Explain the importance of each phase for a highly complex and uncertain project environment.
5. What do you mean by consistency and redundancy of a project network? What is its impact on project planning and execution? Explain the practical relevance of consistency and redundancy with a suitable example.
6. Discuss the various methods for checking project network consistency—Topological ordering, Fulkerson rule, Marimont procedure, Matrix Squaring method.
7. A flyover construction project executed. You want to develop an error-free comprehensive network for this project. Demonstrate an application of all the methods explained in Question 6 for consistency checking in the project network for the example given below.

Activity	Predecessors	Expected duration
A	–	4
B	–	6
C	A	6
D	A	8
E	A	8
F	B, C	6
G	D, F	15
H	D, F	4
I	E, H	8

8. Develop A-O-A and A-O-N project representation for the data set given in Question 7.
9. Consider a typical R&D project involving the following activities. First, develop A-O-A and A-O-N project representation for the data set given below.

Job	Predecessor	Duration (Days)
A	–	5
B	A	6
C	A	4
D	B	5
E	C	3

(continued)

(continued)

Job	Predecessor	Duration (Days)
F	D, E	5
G	C	5

10. Check the redundancy of project networks given in Question 7 and Question 9 using the Tabular method.

Web-based Exercise

Study the details of a Metro Project on the Internet. Develop a detailed work breakdown structure and define the various project modules. Enlist at least 15 critical activities with precedence relationships to successfully execute the Metro Project and develop A-0-A and A-O-N representations.

Group Project

Do it in a Group of Four. Undertake a detailed brainstorming session on the project "Organizing Picnic." Enlist all the necessary activities with precedence–successor relationships. Apply procedures for consistency and redundancy check and develop an error-free comprehensive project network. Document at least ten critical decisions you can take using this network details to plan and execute the Picnic successfully.

Chapter 5
Project Scheduling

Critical Questions

What is project scheduling? State its importance.
What are the project scheduling approaches?
How project scheduling can be conducted using CPM and PERT?
What are the errors in PERT?
What is GERT?

Project Management in Practice

Naini-Itarsi Rail Project.

Electrical locomotives are powerful and operate at high speeds. This feature implies that fewer locomotives are needed in big passenger trains or heavy haul freight trains, resulting in lower maintenance costs. Electrical locomotives also employ fewer components than diesel locomotives, further reducing maintenance costs. In addition, electric locomotives lose less power in high gradient terrain and warm weather, leading to lower energy consumption. Faster acceleration and deceleration makes electric locomotives particularly suitable for handling urban/suburban transport. Electrical locomotives are more energy-efficient because of easy control and the ability to regenerate electricity during braking. Electrical locomotives are also more environment-friendly as they generate much fewer fumes and are relatively less noisy than diesel locomotives.

J. J. Thakkar, *Project Management*, Management and Industrial Engineering,
https://doi.org/10.1007/978-981-15-3695-3_5

Key Project Management Challenges

Time:

In 2012, electrification projects of the Indian Railways (IR) worth 27 billion were in progress. IR had electrified 17,809 km of its railway routes by the end of the Xth Five-Year Plan. The target for the XIth Five-Year Plan (2007–2012) was 3,500 km. However, a study found that pollution created by electricity generation in thermal power plants was more harmful to the environment than the pollution by diesel locomotives. These factors were brought before the court, and the settlement of disputes was done. This caused the project to halt various times.

Cost:

Among other target documents laid down, an electrification target of 14,000 km in this decade for 126 billion was planned by IR, and the Itarsi-Naini project was one of them. Included in the signaling costs incurred in 2015–2016 was 50 million for upgrading to light emitting diodes (LED) lighting to improve reliability and visibility, which would have to be incurred even if electrification was not undertaken. Although the budget was allocated on time, the projects were frequently delayed by two years, with the longest delay being four years. Electrical engineering cost 4600.00, signaling cost 662.80, telecom cost 452.60, civil engineering cost 340.00, and total capital expenditure 6005.4. All these allocations were delayed by IR, which caused the project to be delayed further.

Performance:

Although most electrical systems have been using overhead wire lines for electrical energy supply, countries differ drastically in terms of the choice of voltage (ranging from 600 to 25,000 V) and type of current output (DC—direct current or AC—alternating current). Since 1957, the railway electrification (RE) system has been standardized to 25,000 V (25 kV) for all future projects. The electrified engines run very fast, at almost double the speeds of the diesel engines.

Source The Naini-Itarsi Railway Electrification Project by Ravi Anshuman, Narang Tapsi, Anand Sharma, Nov 1, 2012, HBS Case: #IMB395-PDF-ENG (https://www.hbsp.harvard.edu/product/IMB395-PDF-ENG).

5.1 Importance of Project Scheduling

A schedule is the conversion of a project action plan into an operating timetable.

Project scheduling is a mechanism to communicate what tasks need to get done and which organizational resources will be allocated to complete those tasks in what timeframe.

A project schedule is a document collecting all the work needed to deliver the project on time.

It is an integral part of the project planning process. It is not possible to run a project without having a proper plan. Scheduling plans the timelines, delivery, and availability of project resources (personnel, inventory, or capital). The projects executed without a detailed schedule encounter uncertainties and chaos.

The complex nature and multifaceted range of activities make project management highly complex. It is impossible to derive an ideal schedule for a project that does not experience any fluctuations under organizational and environmental uncertainties. Project scheduling is not an exact process; however, an absence of a schedule can make the project deviate from its milestones, fail to extend the expected deliverables and break down. The project schedule is used in the approximation of the resources needed and the duration required. This is an essential part of the project plan as it indicates how the project will progress. Projects must be scheduled carefully to arrange and finish them in a timely, quality, and financially responsible manner. Therefore, effective project scheduling plays a vital role in ensuring the success of the project. For keeping projects on track, one needs to fix realistic time frames, allot resources appropriately, and manage quality to reduce the errors of the product. This ensures better cost control and enhances customer satisfaction. The essential elements of the project schedule are illustrated in Fig. 5.1. This includes project scheduling designs (which primarily deal with the policy formulation and creating procedures), activities identification, resource allocation, calculating activity duration, dependencies creation, and analysis of time and cost constraints.

It helps to oversee how the undertaking is advancing and whether any issues are to be managed or if the customer should be told about a time overrun in delivery (Fig. 5.2).

Figure 5.2 is a basic Gantt chart used for monitoring and control of project activities. It shows the tasks in a security and access control program. The works are outlined in two sections. Each task represents a yellow triangle to indicate the beginning date of the work and a green triangle facing down to indicate the date of completion. The sub-contractors responsible for the project are shown in Fig. 5.2 (in the column named R-E-S-P).

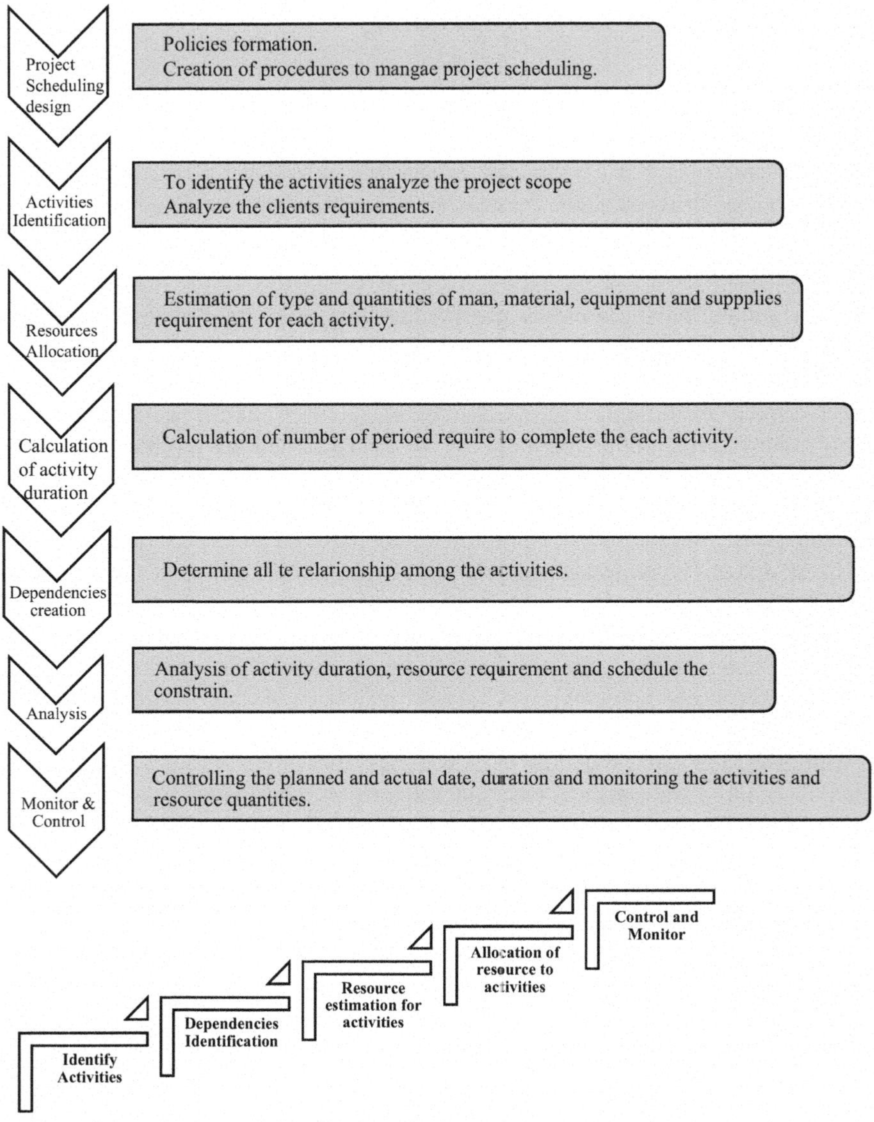

Fig. 5.1 Project scheduling

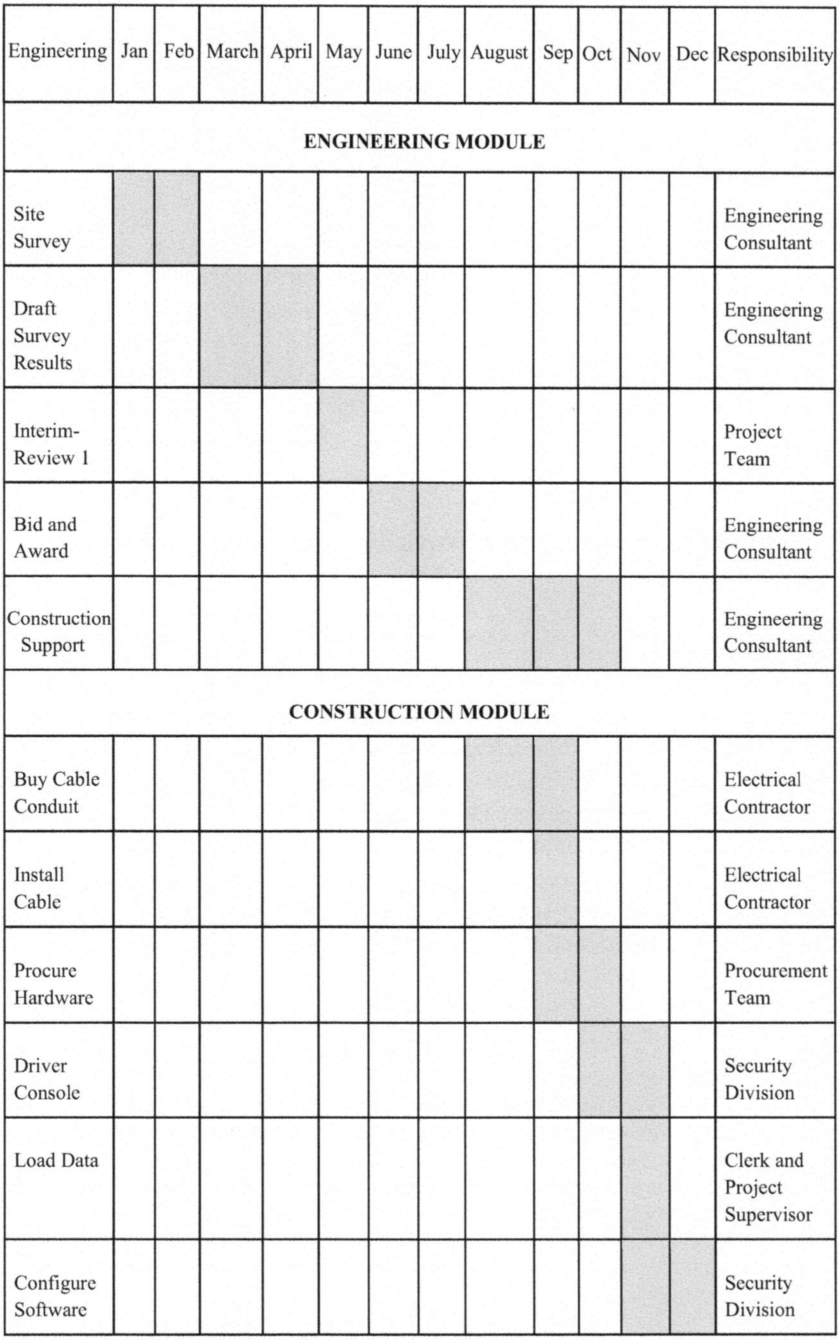

Fig. 5.2 Project security and control system

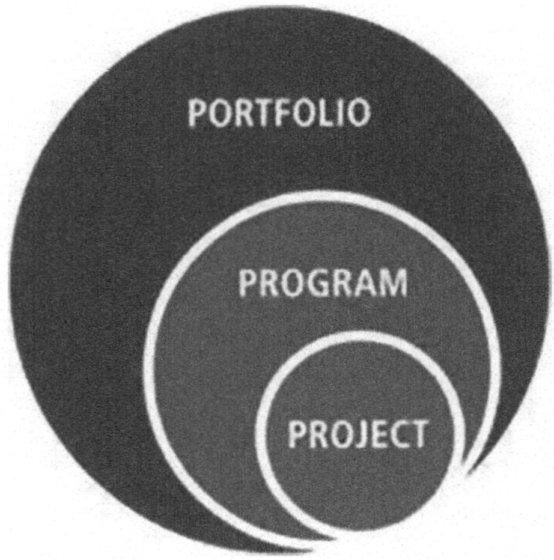

Fig. 5.3 Key outcomes of the project schedule for a project management team

5.2 Project Scheduling and Work Breakdown Structure (WBS)

Project Scheduling.

The schedule is the instrument that shows the work to be carried out, the resources required for performing the work, and the duration of the work. In addition, it indicates the work that needs to be performed on time. Without having a complete plan, the project manager will be unable to estimate the total effort required, regarding cost and assets, essential to convey the project.

The project schedule should be checked, edited, and updated by the committee associated with the project, keeping everybody educated on the general project status. The key outcomes of project schedules are presented in Fig. 5.3. These three categories are interconnected, and if one of these collapses, the others will be affected. If the life cycle of a project extends, it will alter both the program and portfolio schedule.

Project	"A temporary endeavor undertaken to create a unique product, service or result." Efficiently and effectively handling a project indicates taking advantage of tools and techniques to utilize resources ideally. Projects are smaller units that form the basis of the programs and portfolios. All projects have unique start and finish dates. A project manager can also decide to terminate the project on achieving the target or when the target cannot be achieved or when the need to complete the target diminishes

(continued)

(continued)

Program	"A group of related projects managed in a coordinated way to obtain benefits and control not available from managing them individually. Programs may include elements of related work outside the scope of discrete projects in the program." According to its definition, a program is a heap of projects that run at the same time. Overseeing different projects simultaneously is not a simple job, and therefore, scheduling techniques are constantly used to control the advancement of the program. Inside a program, it is beneficial to utilize scheduling methods to differentiate projects running admirably from those facing delays and challenges. Using scheduling procedures, it is conceivable to move resources among different projects to streamline their improvement
Portfolio	"A collection of projects or programs and other work that are grouped to facilitate effective management of that work to meet strategic business objectives. The projects or programs of the portfolio may not necessarily be interdependent or directly related." The portfolios usually do not have a due date. However, the associated programs and projects concerned must be completed on time

Due dates and time limits are the most imperative limitations that influence the notoriety of an organization. Conveying on-time projects, items, or administrations implies being dependable and experienced. Postponements or disappointments can happen, but it is conceivable to anticipate and oversee them through definite planning methods. Promising clients the conveyance of the task at a particular date and not having the capacity to submit it is synonymous with inconsistency and the absence of booking devices. Outcomes will be punishments, loss of clients, and also salaries.

Work Breakdown Structure (WBS).

The components of a schedule start with a Work Breakdown Structure (WBS). The WBS is a hierarchical representation of all the work in the project. The objective is fulfilling the expectations, and hence, the work should be performed.

The right way to implement a work breakdown structure is by organizing the steps in the project scope by starting from the mission statement of the project at the top of the hierarchy. You can then move on to the main deliverables and outline how each of them will be accomplished.

5.3 Components of Project Scheduling

In project management, scheduling refers to listing of activities, deliverables, and milestones. These issues are often evaluated by resource allocation, cost planning, working periods, and other issues incorporated into the project calendar when the conditions are booked. In addition, components in the timetable may be strongly recognized with a list of work requisites (WBS) terminal details, statement of work, or contract data requirements.

> Project managers must develop a full understanding of all the key constraints affected by deliverable timeliness versus delay.

If you want to create a culture of timeliness, be careful about declaring due dates for deliverables.

Before designing a project timetable, the schedule maker must have a work lapse (WBS), a projected assessment for each project, and a resource list with access to each resource.

To create a project plan, the following should be completed:

- The purpose of the project
- A series of actions
- Five project phases (survey, definition and planning, opening, performance, closing)
- Work dependencies map
- Critical path analysis
- Project milestones.

The schedule records the activities to be finished amid the project and distinguishes what resources such as workforce and materials are essential to complete the activities. This typically comes as a statement of work (SOW) or scope statement, a detailed description of what will be proficient amid the project's life. A schedule can be as basic as a hand-written outline or a complex chart made utilizing a software application like Microsoft Project. Every individual assignment distinguished in the schedule is given a length comprising of a begin and end time. Resources would then be relegated to the activities, and milestones can be given more focus inside the project plan.

Milestones: Milestone tools use tool management. These points may indicate announcers, among others, requiring the start and end date of the project, external study or input, and budget checks. In many cases, milestones are not complete. Instead, they focus on the key progress points that need to be achieved successfully.

Milestones are the most valuable features in a table. They are significant events with zero periods that help measure the progress of a project, which will help senior management. The schedule can provide senior management with an idea about how the project will progress. They can confirm that they are reaching milestones within the acceptable time.

Using Milestones in Scheduling

Milestones can add significant value to planning. When programmers evaluate and review coordinates with the technique (PERT) or the critical path method (CPM), it allows the project managers directly or indirectly to determine more accurately whether the project is on schedule. By controlling milestones-related dates, you can determine the path for more critical gaps in the entire project. Each table interval is calculated on slack/float. This section of the schedule indicates timetable issues in the intervals and allows for a better view of efficient performance. They usually only improve in the course of the path and ignore non-critical acts. It is common to move from critical activities to critical actions to ensure the meeting of milestones. This seems to be a plan in the project when some projects are ignored.

Activity: It is an action that requires a deadline to work toward a specific period or work-related goals. This is a minor factor that acts as a means of different diverging components of a project. It can be broken into assignments, which must have a limited start and end date or a deadline. One or more of the jobs are functioning under the operation. The work ends when all tasks of a particular task are completed. Finally, works can be combined to create.

Completion of tasks usually requires the integration of others. Several individuals must meet a general goal of the joint involvement of integrated human contact time, energy, effort, effectiveness, and coordination. Coordination efforts are necessary to achieve solidarity as a larger workforce is involved in doing the work. The integration allows you to complete large tasks.

Deliverable: Deliverables are components of output within the scope of a project. They are the result of objectively focused work performed within the project process. They can be internal or external. A deliverable that is internal to your organization is not visible to anyone external to the organization. A deliverable that is external to the organization is to be done for a client, customer, stakeholder, or revenue. Either way, it usually means that the deliverable is due on a specific date.

5.4 Critical Path

The critical path is the longest sequence of the project's plan, which must be completed on time to end the project on the due date. If there is any delay on the critical path, then the entire project will be delayed by that much time.

The most important way to help manage any project is to use selective control through the critical path. If the critical path is identified, you can see that efforts cannot be compromised.

A critical path is a series of collaborative actions or tasks. This is the longest route (i.e., the longest path) of the project to end. At the same time, the long path is the shortest way to complete a plan. If there is a delay in the process of the path, the entire plan gets delayed.

Importance of Using the Selective Control in The Project Management

ABC analysis (or choice inventory control) is an analytical technique widely used for applying selective control using Pareto's (80–20) principle. ABC analysis is divided into three sections, namely "A items" (very tight control), "B items" (less tightly controlled), and "C items" (least restrictive controls).

ABC analysis provides a mechanism for identifying substantially impacted items at the total cost of goods, providing a mechanism for identifying and focusing on a few vital items for better monitoring and control.

ABC analysis suggests that a company's inventory is not of equal value. Thus, inventory value is a group of three groups (A, B, and C).

"A" products are very important for a company. Due to the high value of these "A" products, frequent rating analysis is required. In addition, a company must select a bar sequence (e.g., "just-in-time") to avoid an excess capacity.

A critical path-based project scheduling and monitoring work on the same concept as that of selective control. A project manager cannot focus on all the project activities, and hence, identifying a critical path helps the manager focus critically on select activities out of many.

5.5 Project Scheduling with Critical Path Method (CPM) and Program Evaluation and Review Technique (PERT)

The activity times can be deterministic or probabilistic.

Deterministic time: (as in CPM): When previous experience yields reasonably accurate estimates of activity duration, e.g., construction activity, market surveys, a single time estimate is used for each activity. This is taken from experts who have prior knowledge and experience of the activity.

Probabilistic time (as in PERT): When there is uncertainty in times, for instance, in R&D activities, new activities are carried out for the first time. Three time estimates (optimistic, most likely, and pessimistic) are commonly used for each activity based on the group's consensus.

CPM and PERT are the most widely used approaches for project scheduling. The following are the most critical differences between PERT and CPM:

(1) PERT is a technique that controls planning and time, unlike CPM, which is a means of controlling costs and time
(2) A deterministic model is used in CPM. Instead, the PERT uses probability models
(3) PERT consists of three time estimates, i.e., optimistic time (t_o), pessimistic time (t_p), and most likely time (t_m). On the other hand, there is only one estimate in the CPM
(4) PERT deals with unpredictable actions, but CPM deals with predictable activities
(5) PERT is used to review the nature of work which is non-repetitive. On the contrary, the CPM involves repetitive work.

5.6 Critical Path Determination: Path Enumeration Method

The CPM provides a structured approach to project planning, which allows identifying the most essential steps for effective execution of the project.

Breaking the Work

The project needs to be broken down into a set of modules and tasks or activities. The activities should not be broken into too small or too large segments. If the activities are too large, it will create conflict in responsibility, duplication of resources, and budget allocation.

Organize Tasks with Prerequisites

It is essential to deduce an accurate successor–predecessor relationship among the activities. This ensures the smooth conduct of the project through appropriate monitoring and control mechanisms.

Estimated Duration of the Tasks Involved

A project team must estimate the activity timings as accurately as possible based on the experience and consensus of the team members. The factors such as reliability of suppliers and contractors, the complexity of the project, experience of the project team in handling such projects play an important role in estimating activity or task duration.

Find the Longest Path

The longest path decides the completion of the project. A critical path of the project and activities on this path need to be carefully monitored to complete the project. It also sets the basis for deciding the reward or penalty clause in project contracts.

Example Consider a network of a typical project "construction of industrial shed" as given in Fig. 5.4. First, we wish to identify the critical path (the longest path in the network using the path enumeration method).

The project network of "construction of industrial shed" has three paths connecting "start of the project," i.e., activity "A" and "end of the project," i.e., activity "K." The activities "A" and "K" indicate the start and end of the project and have zero activity time.

The path lengths (time in months) for all three paths can be calculated by summing up the activity timings on a particular path as given below.

Path 1: A-B-C-E-J-K = 17 months.

Path 2: A-B-C-D-J-K = 19 months.

Path 3: A-B-C-F-G-H-I-J-K = 38 months.

The longest path in the network is "Path 3." Hence, it is critical. A project manager and team need to be extremely critical about activities on this path as any delay in the activity of this path will increase the project duration.

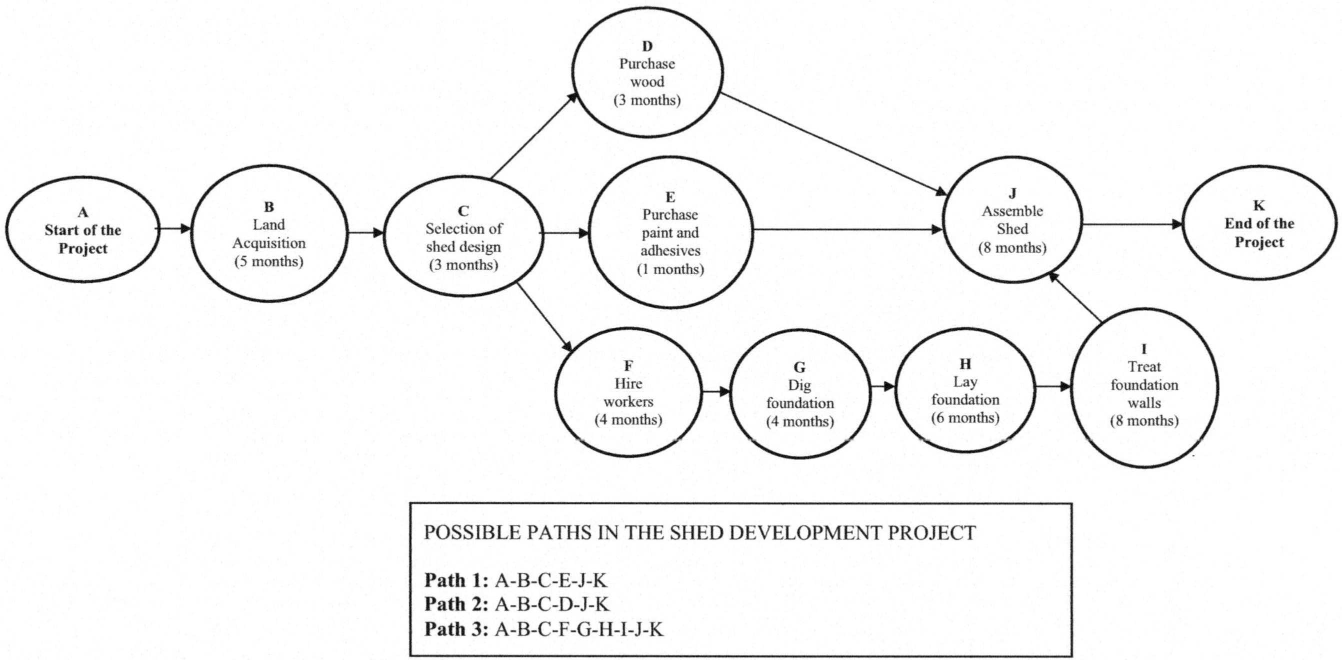

Fig. 5.4 Network of "construction of industrial shed" project

Table 5.1 Project activities and predecessors

Activity	Predecessor	Time (days)
A1	–	8
A2	–	4
A3	A1	3
A4	A1	9
A5	A2, A3	5
A6	A4, A5	6
A7	A3, A4, A5	3
A8	A4, A6	4

5.7 Project Scheduling with CPM: Illustrative Example

Mr. Abhishek is leading a construction project for an infrastructure company. He has identified the set of activities within the WBS as follows. He has also clearly understood the successor–predecessor relationship among the activities, as indicated in Table 5.1.

Tasks:

Task 1	Draw the A-O-A and A-O-N networks with a single source and sink
Task 2	Use critical path method (CPM) and conduct forward and backward pass computation. Determine the critical path
Task 3	Compute various floats—total float, free float, safety float, independent float. Provide a detailed interpretation of each float

Solution:

The project representation on activity on arc (A-O-A) and activity on node (A-O-N) is illustrated in Figs. 5.5 and 5.6, respectively.

A-O-A network:

A-O-N network:

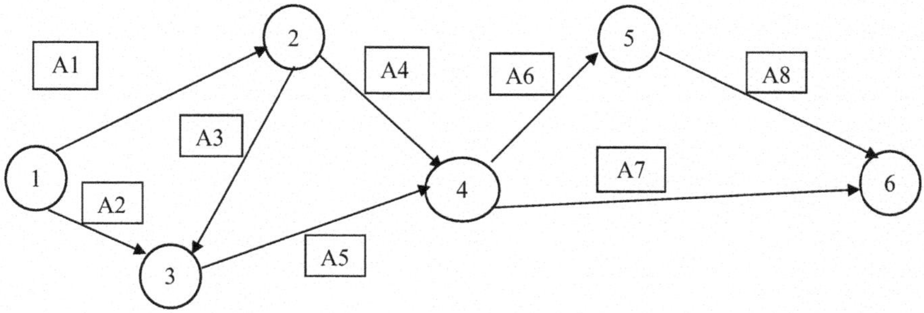

Fig. 5.5 A-O-A representation of project in Table 5.1

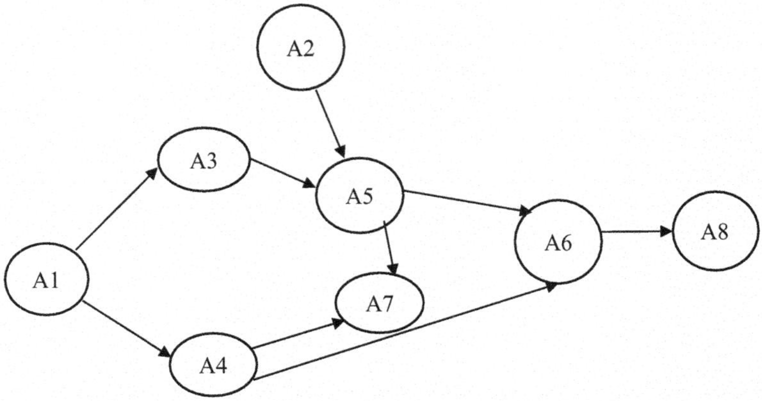

Fig. 5.6 A-O-N representation of project in Table 5.1

Forward and Backward Pass Computations:

Forward pass and backward pass computations help to perform critical path analysis. The procedure for performing forward and backward pass computations is given in Table 5.2.

Table 5.2 Forward and backward pass computations

<table>
<tr><td colspan="2">Forward Pass Computation
Forward pass determines earliest start (ES) and earliest finish (EF) time by moving forward through the project network diagram</td></tr>
<tr><td>Earliest Start Time Rule:
• If an activity has only a single immediate predecessor, its ES equals the EF of the predecessor
• If an activity has multiple immediate predecessors, its ES is the maximum of all the EF values of its predecessors
• ES = Max {EF of all immediate predecessors}</td><td>Earliest Finish Time Rule:
• If an activity has only a single immediate predecessor, its ES equals the EF of the predecessor
• If an activity has multiple immediate predecessors, its ES is the maximum of all the EF values of its predecessors
• ES = Max {EF of all immediate predecessors}</td></tr>
<tr><td colspan="2">Backward Pass Computation
Backward pass performs the latest finish (LF) and latest start (LS) calculations by moving backward through the project network</td></tr>
<tr><td>Latest Finish Time Rule
• If an activity is an immediate predecessor for just a single activity, its LF equals the LS of the activity that immediately follows it
• If an activity is an immediate predecessor to more than one activity, its LF is the minimum of all LS values of all activities that immediately follow it
• LF = Min {LS of all following immediate activities}</td><td>Latest Start Time Rule
• The latest start time (LS) of an activity is the difference between its latest finish time (LF) and its activity time
• LS = LF − Activity time</td></tr>
</table>

Table 5.3 ES, EF, LS, and LF computations

	ES	EF	LS	LF
A1	0	8	0	8
A2	0	4	8	12
A3	8	11	9	12
A4	8	17	8	17
A5	11	16	12	17
A6	17	23	17	23
A7	17	20	24	27
A8	23	27	23	27

The forward pass (ES and EF) and backward pass (LS and LF) computations for the project activities indicated in Table 5.1 are summarized in Table 5.3.

From Table 5.3, the activities which have ES-LS or EF − LF = 0 are the activities on critical path. Hence, the critical path is A1-A4-A6-A8 (duration = 27 days).

FLOAT COMPUTATIONS:

An activity, in general, has both predecessors and successors. Thus, each of the four kinds of float depends on how it accommodates the activity.

Total float = LS − ES = LF-EF of activity.

Safety float = Total float − Slack on preceding node.

Free float = Total float − Slack on succeeding node.

Independent float = Max (0, Total float − Slack on preceding and succeeding nodes).

Slack on preceding node = Max (LF of predecessors) − ES.

Slack on succeeding node = LF − Min (ES of successors).

Interpretation of Floats:

The floats can be interpreted for more significant practical insights using the matrix representation in Fig. 5.7.

For activity A2, total float = LS − ES = 8 − 0 = 8

$$\begin{aligned}\text{Safety float} &= \text{Total float} - \left[\text{Max (LF of predecessor)} - \text{ES}\right]\\ &= 8 - (0 - 0)\end{aligned}$$

Fig. 5.7 Float interpretation

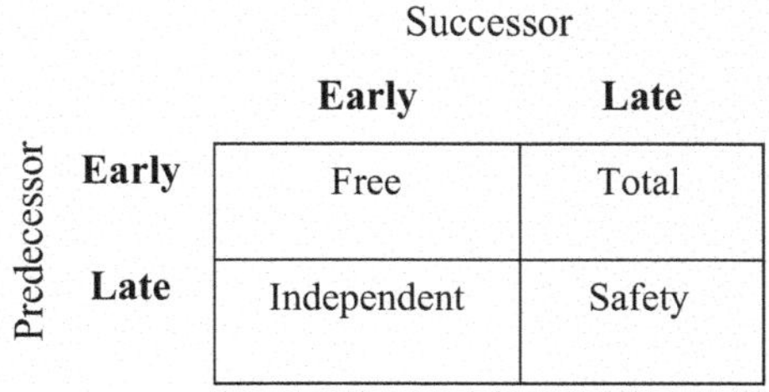

Table 5.4 Float computations

	Total	Safety	Free	Independent
A1	0	0	0	0
A2	8	8	7	7
A3	1	1	0	0
A4	0	0	0	0
A5	1	0	1	0
A6	0	0	0	0
A7	7	7	7	7
A8	0	0	0	0

$$= 8$$

$$\begin{aligned}\text{Free float} &= \text{Total float} - [\text{LF} - \text{Min (ES of successor of A2 is A5)}]\\ &= 8 - (12 - 11)\\ &= 7\end{aligned}$$

$$\begin{aligned}\text{Independent float} &= |\text{total} - (\text{free} + \text{safety})|\\ &= |8 - (8 + 7)|\\ &= 8\end{aligned}$$

The floats are calculated in Table 5.4 using the above expressions.

5.8 Project Scheduling with PERT: Illustrative Example

A petrochemical mill was inspected by OSHA and found to violate several safety regulations. In order to comply with OSHA standards, the mill was forced to make changes to existing machinery to make it safer (e.g., by installing safety guards); replace older, hazardous machinery with new machinery, and relocate machinery to give workers safer passages and access to the exits. OSHA permitted the mill only 35 weeks to make the changes; if they did not make the changes, the mill would be fined Rs. 50 lakh. The mill determined the activities in a PERT network that would have to be completed and then estimated the indicated activity times, as shown in Table 5.5.

Determine:

(a) Expected activity times
(b) ES, EF, LS, LF, and critical path
(c) Total float, free float, safety float, independent float
(d) Expected project duration and variance
(e) What is the probability that the project will be completed in 15 weeks?
(f) Compute probabilities that the project is completed within 75% and 125% of the critical duration
(g) What is the probability that the mill will be fined Rs. 50 lakh?

Solution:

(a) Expected Activity times

Formula:

$$\text{Expected time(te)} = \frac{to + 4tm + tp}{6}$$

$$\text{For activity A, te} = \frac{2 + 4 * 3 + 4}{6} = 3$$

The expected activity durations for other activities are calculated similarly, as shown in Table 5.6.

Table 5.5 Project activities and predecessors for the petrochemical mill

Activity	Description	Activity predecessor	Time estimates (in weeks)		
a	Order new machinery	–	2	3	4
b	Plan new physical layout	–	2	6	8
c	Determine safety changes in existing machinery	–	1	3	5
d	Receive equipment	a	4	12	25
e	Hire new employees	a	5	7	12
f	Make plant alternations	b	10	17	25
g	Make changes in existing machinery	c	5	10	14
h	Train new employees	d, e	2	3	7
i	Install new machinery	d, e, f	1	6	6
j	Relocate old machinery	d, e, f, g	2	8	12
k	Conduct employee safety orientation	h, i, j	2	4	8

Table 5.6 Expected activity durations

Activity	Expected time
A	3
B	5.67
C	3
D	12.83
E	7.5
F	17.16
G	9.83
H	3.5
I	5.167
J	7.67
K	4.33

(b) ES, EF, LS, LF, and Critical path

A-O-N representation of the project details given in Table 5.5 is presented in Fig. 5.8. The forward and backward pass computations of ES, EF, LS, LF are summarized in Table 5.7.

The activities for which ES-LS or EF-LF becomes zero are the activities with zero slack (as highlighted in Table 5.7) and hence considered activities on the critical path. It is evident from Table 5.7 that the critical path for the project is B-F-J-K.

(iii) Float Computations

$$\text{Total float} = \text{LS} - \text{ES} = \text{LF} - \text{EF}$$

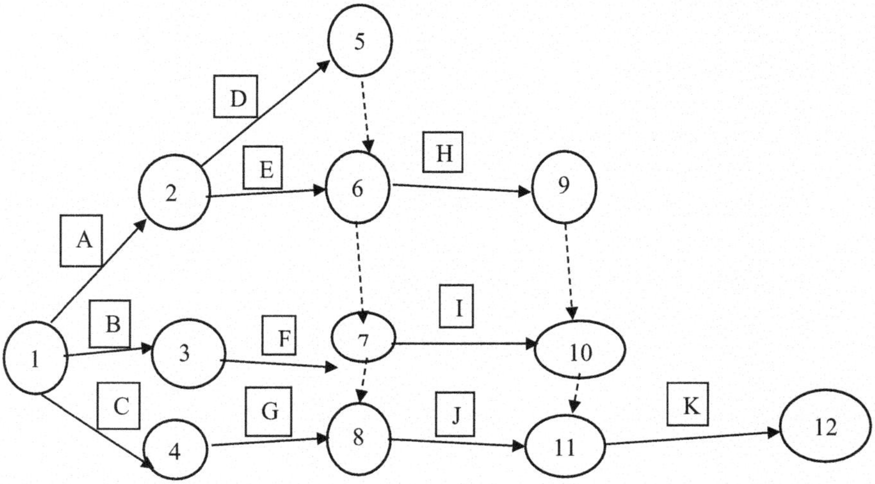

Fig. 5.8 A-O-N representation of the project in Table 5.5

Table 5.7 ES, EF, LS, LF

Activity	ES	EF	LS	LF
A	0	3	2.67	5.67
B	0	5.67	0	5.67
C	0	3	10	13
D	3	15.83	10	22.83
E	3	10.5	15.33	22.83
F	5.67	22.83	5.67	22.83
G	3	12.83	13	22.83
H	15.83	38.125	27	30.5
I	22.83	28	25.33	30.5
J	22.83	30	22.83	30.5
K	30.5	34.83	30.5	34.83

Note Highlighted activities are critical path activities with zero float

$$\text{Safety float} = \text{Total} - [\max(\text{LF of the predecessor}) - \text{ES}]$$

$$\text{Free float} = \text{Total} - [\text{LF} - \min(\text{ES of successor})]$$

$$\text{Independent Float} = \text{Total Float} - \text{Tail Event Slack}$$

The floats are calculated in Table 5.8.

Table 5.8 Total, safety, free and independent floats

Activity	Total float	Safety float	Free float	Independent float
A	2.67	2.67	0	0
B	0	0	0	0
C	10	10	0	0
D	7	4.33	0	2.67
E	12.33	9.66	5.33	2.67
F	0	0	0	0
G	10	0	10	0
H	11.17	4.17	11.17	4.17
I	2.5	2.5	2.5	2.5
J	0	0	0	0
K	0	0	0	0

(iv) Expected project duration and variance

$$\begin{aligned}\text{Expected project duration} &= \text{Sum of duration of the activities on the critical path}\\ &= \text{Sum of}(\text{B} + \text{F} + \text{dummy} + \text{J} + \text{K})\\ &= 5.67 + 17.16 - 7.67 + 4.33\\ &= 34.83 \text{ weeks}\end{aligned}$$

$$\sigma = \sqrt{12 + 2.52 + 02 + 1.672 + 12}$$

Standard deviation = 3.322.
Variance = (standard deviation)^2 = 3.3222 = 11.

(e) What is the probability that the project will be completed in 15 weeks?

$$\begin{aligned}z &= \frac{X - \mu}{\sigma}\\ z &= \frac{15 - 34.83}{3.322}\\ &= -5.969\end{aligned}$$

Therefore, the probability is approximately zero (from the normal distribution table).

(f) Compute probabilities that the project is completed within 75% and 125% of the critical duration.

i. $x = 0.75 \times 34.83 = 26.12$

Therefore,

$$\begin{aligned}z &= \frac{x - \mu}{\sigma}\\ &= \frac{26.12 - 34.83}{3.3222}\\ &= -2.622\end{aligned}$$

Now, probability corresponding to z = -2.622 is 0.44%.

ii. x = 1.25 × 34.83 = 43.53

Therefore,

$$\begin{aligned}z &= \frac{x - \mu}{\sigma}\\ &= \frac{1.25\mu - \mu}{3.3222}\\ &= 2.622\end{aligned}$$

Now, the probability corresponding to z = 2.622 is 99.56%.

(g) What is the probability that the mill will be fined Rs. 50 lakh?

$$z = \frac{35 - 34.83}{3.322}$$
$$= 0.0511$$

Therefore, probability corresponding to $z = 0.0511$ is 52%
So, the probability of 50 lakh fine will be (100 − 52)% = 48%

Project Management in Practice

What are the challenges in CPM and PERT for real-life projects?

- Not all precedence relationships can be anticipated before a project begins; this is especially true for research and development types of projects which involve new technology and unknown problems
- Predecessor relationships are assumed to be firm—that is, a given job must be completed in its entirety before any successor jobs can begin. Frequently, predecessor relationships are "soft."
- Some precedence relationships are contingent upon the outcome of previous activities. For example, if a test phase for a new missile guidance system is successfully completed, the next activity may be to manufacture the system. But if the test fails, the system may go back to the drawing boards. Neither PERT nor CPM allows for conditional activities and precedence ordering of this type
- Time estimates are subjective and depend both on the vagaries of the humans who produce these estimates and on the environment within which they operate (Pessimistic–Conservative–Realistic)
- System of rewards within which contractors operate–cost-plus contract/incentive-type contract lead to bias.

5.9 Graphical Evaluation and Review Technique (GERT)

Graphical evaluation and review technique (GERT) is a network analysis technique used in project management that allows probabilistic treatment to network logic and estimate activity durations. The technique was first presented in 1966 by Dr. Alan B. Pritsker of Purdue University and WW Happ. The key features of GERT include (i) probabilistic branching (stochastic models), (ii) network looping (feedback loops), (iii) multiple sink nodes (multiple outcomes), and (iv) multiple node realizations (repeat events), which are unavailable in PERT/CPM. These GERT features allow the user to model and analyze projects and systems of a very general form. Since many real-world system problems involve probabilistic occurrences, false starts, activity repetition, and multiple outcomes, GERT is an ideal tool for modeling and analysis. Typically, GERT helps to overcome the following limitations of PERT and CPM.

1. There are no alternative paths (i.e., all activities have to be performed)
2. There are no loops (i.e., it is not possible to repeat activity; next similar activity has to be added)
3. There are no decisions in the diagram (i.e., this means there are no "AND," "OR," "XOR" options that help to choose the proper path)
4. There is no scaling (i.e., it is impossible to replace some detailed group of activities with one summary task).

GERT uses activity-on-arrow notation only. That means that each activity is described on an arrow. The nodes are used to connect activities and determine the type and conditions of relations between them.

Each task has two parameters: duration and probability of appearance.

There are three logical operators in GERT which concern activities incoming to the node:

- "XOR": alternative (only one path possible)
- "OR": alternative (one or more paths can be performed)
- "AND": all paths have to be performed.

The most common is AND, which means that every incoming activity has to happen before the outcoming starts.

There are two types of relations that concern activities outcoming from the node. They are:

- Deterministic: Every outcoming activity has a probability equal to 1, which means that each activity will be performed
- Probabilistic: Each outcoming activity has some probability of appearance.

The symbols used in GERT are presented in Table 5.9. In GERT representation, each node consists of one logical operator and one relation.

Steps in GERT

The steps employed in applying GERT are:

Step 1: Convert a qualitative description of a system or problem to a model in network form.

Step 2: Collect the necessary data to describe the branches of the network.

Step 3: Obtain an equivalent one-branch function between two nodes of the network;

Step 4: Convert the equivalent function into the following two performance measures of the network:

(a) The probability that a specific node is realized
(b) The moment generating function (M.G.F.) of the time associated with an equivalent network.

Table 5.9 GERT symbols

Name	Symbol	Characteristics
EXCLUSIVE-OR	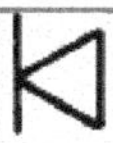	The realization of any branch leading into the node causes the node to be realized; however, one and only one of the branches leading into this node can be realized at a given time
INCLUSIVE-OR		The realization of any branch leading into the node causes the node to be realized. Thus, the time of realization is the smallest of the completion times of the activities leading into the INCLUSIVE-OR node
AND		The node will be realized only if all the branches leading into the node are realized. The time of realization is the greatest of the completion times of activities leading to the AND node
DETERMINISTIC		All branches emanating from the node are taken if the node is realized, i.e., all branches emanating from this node have a p-parameter equal to i
PROBABILISTIC		Exactly one branch emanating from the node is taken if the node is realized

Step 5: Make inferences concerning the system under study from the information obtained in Step 4.

GERT Representation

Figure 5.9 represents a brief schematic highlighting PERT/CPM and GERT differences and demonstrating the various GERT characteristics and attributes. The primary difference between PERT/CPM and GERT networks is that GERT has two types of nodes, deterministic and probabilistic. Node 3 in Fig. 5.8 (the identification number is on the right-hand side of the cone-shaped node) is a probabilistic node. Instead of one deterministic branch (arrow) as in PERT/CPM, there are multiple possible outcomes (success, failure, recycle), each with a probability of occurrence.

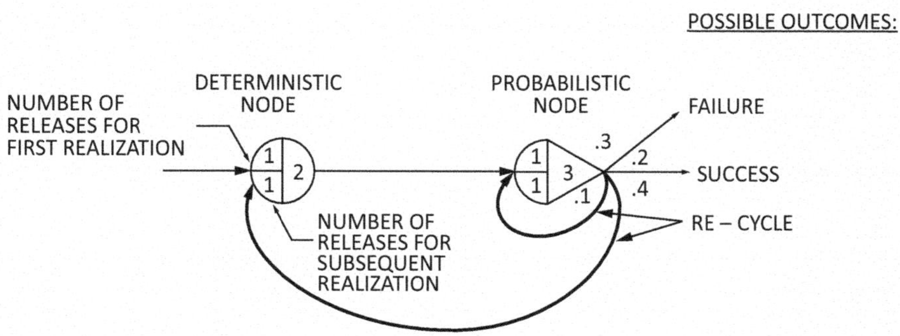

Fig. 5.9 GERT characteristics (Adapted from: Taylor, B. W. (1978). Project management using GERT analysis. *Project Management Quarterly, 9*(3), 15–20.)

Thus, a choice situation exists at a probabilistic node where one of several alternatives may be selected based on the associated probabilities. However, the sum of the probabilities for all activities emanating from a probabilistic node must be 1.00 (i.e., there is a 1.0 probability that one of the activities will be realized).

If the activity emanating from node three proceeds by looping back to node 2 occurs, this will cause activities 2–3 to be repeated. On the other hand, if the activity labeled "failure" was realized, the network might flow to a "sink" node that ends the network. Alternatively, if the activity labeled "success" is realized, the network might continue for several more activities before the network ended in another (different) "sink" node. The fourth activity at node 3 is activity 3–3, representing a self-loop back to the same node. These alternative activities reflect the feedback, multiple outcomes, and repeating activity characteristics of GERT.

Node 2 is deterministic as used in PERT/CPM. Because node 2 is deterministic, the probability of realization for activity 2–3 is 1.0. In both node 2 and node 3, the number in the upper left-hand quadrant represents the number of releases necessary for the first realization of the node (in both cases shown, only one activity release is required). The number in the lower left quadrant of each node is the number of activity releases needed for all subsequent realizations of the node.

GERT is relatively easy to use since it requires only that the project of interest be.

(1) Diagrammed in network form
(2) Converted to program input data describing the network
(3) Simulated using the prewritten GERTS-IIIZ simulation package.

By simulating the network, statistical data can be collected at different nodes for network duration and cost. The GERT simulation package can have nine different probability distributions for activity times: constant, normal, uniform, Erlang, lognormal, Poisson, beta, gamma, and the beta fitted to three parameters. Additionally, the GERT model can assign fixed and variable costs to network activities. This means that, in a fixed cost system, the cost is accumulated each time an activity occurs; in a variable cost system, the cost is calculated based on the amount of time it takes to complete the activity.

GERT has effectively applied to many systems problems, including product planning, research and development planning, market research, production planning, quality control, and manpower planning.

5.10 Errors in PERT

The most important error in PERT is that PERT calculates the maximum of expected length, but in reality, PERT should take the expected maximum lengths. For example, if there are three activities in a project, then the probability of completion should be $P(T) = p1 \times p2 \times p3$. This gives the actual probability of project completion time, but we accept only the probability of critical path, and other subcritical paths are neglected. This ignorance of subcritical paths induces error which makes PERT optimistic.

First, the errors may occur at the activity level. What this means is that we have made some assumptions about the activity distributions. For example, we may assume that activities follow beta distribution, but the actual distribution of the activity may be significantly different. Second, errors may occur because of aggregation. These are perhaps much more serious. This typically happens because of the primary focus on the critical path and neglecting other sub-critical paths in the network.

In summary, the following issues may be appropriately addressed to minimize errors in PERT-based project scheduling.

- In PERT, it is assumed that the activities are independent, but in a real-life project, some activities are dependent on other aspects
- PERT assumes that activities that are not on a critical path are ignored, but those activities were not on a critical path in real-life projects but were near critical
- It follows a beta distribution throughout, but it may not be true in real-life projects and might follow some other distribution
- Errors of aggregation mean that the focus is only on the critical path which connects the source to sink, but there may be other paths in real-life projects, and if they are ignored, we may not get the correct project completion date.

Project Management in Practice

Schedule delay in the construction project of Ash handling Plant in a Thermal Power Plant.

Infrastructure construction projects provide a major thrust to the Indian economy. However, many projects run behind schedule due to various reasons. Therefore, the risk factors which are critically responsible for the cause of such delays need to be identified and adequately assessed. This case develops the understanding of the reasons behind the schedule delay and the risk factors that lead to such delay in an infrastructure project—Ash handling Plant construction at a Thermal Power Plant.

Key Success Factors in Infrastructure Projects:

- More likely to succeed when the owner:
- Defines guarantee well
- Defines scope and quality very carefully
- Defines milestones meticulously
- Defines LD/penalty clauses
- Makes payment terms specific.
- The contractor also has ways to improve project success:
- Adopts similar terms and conditions as owner regarding quality, guarantee, etc., for sub-contracts/vendors
- Does not keep terms open-ended
- Coordinates vigilantly to reduce chances of errors at the site.

Different Systems of Ash Handling Plant

- Bottom ash system
- Ash water system
- Fly ash wet system
- Ash slurry system and ash disposal pipeline
- Dry ash system and ash silo.

Specification of Ash Handling Plant

- World's single longest conveying distance of 1.6 km for fly ash handling
- Technology—Dense-phase pneumatic conveying
- Installed capacity—6 × 40 tons per hour
- Units—1 × 500 MW.

The key issues in the project are investigated using SAP-LAP analysis as presented in Exhibit 5.1.

Exhibit 5.1 Scope and challenges in Ash handling project: SAP-LAP analysis

SAP ANALYSIS			
Stages			Issues
Situation	How did we reach the situation of risk assessment?		Schedule delay is an essential aspect of construction management Risk analysis is becoming a need for the identification of critical risk factors
	What is happening now?		Schedule delay is a main concern for the project management team Poor coordination among the project contractors
Actor	What are the world views?		Risk is an integral part of any construction project
	What roles and capabilities are exhibited?		Power plant constructions have a well-structured project team Project construction to be computerized
	In what domains is freedom of choice available?		Freedom is available with the choice of construction methodology
Process	What is being done?	What are the variables?	The project team, engineers, consultants, materials, equipment design
		What are the parameters?	Schedule, risk factors, engineering designs
		What can be changed?	Attitude toward risks The supply chain of the materials
	Why is it being done?		To assess the risk factors associated with the project
	How is it being done?	What else?	Through hiring external consultant and experts
		Why else?	There were no internal experts available
		How else?	Through quantitative risk assessment techniques

LAP ANALYSIS

Learning	**What are the key issues related to the situation?**	**The key issues related to the case study is the construction of projects**
	What are the key issues related to actor (s)?	• Technical risk, financial and economic risk, organizational risk, environmental risks, natural hazards • Knowledge about the critical risk factors that contribute toward the delay
	What are the key issues related to process(s)?	• The technical risk was the major issue also the environmental regulations • Development of better coordination between the owner and the sub-contractors
Action	What should/be done to improve the situation?	• Risk assessment and mitigation policy based on the identified critical risk factors • Management involvement in the risk assessment team
	What can be done to improve the actor?	• Better management and project management team • The company's objective should be proactive toward the risk
	What ought to be done to improve/implement the process(s)?	Better coordination between the project domains, risk ranking, and risk breakdown framework
Performance	What will be its impact on the situation?	• Timely finish of the project • Better implementation of the power plant • Return of investment will not be affected by the delay
	How will the actor(s) be affected?	• The actor will be in a better situation to analyze the entire project schedule • The construction team will be more professional in finishing the project
	How will the performance of the process(es) be affected?	• The process will be performing in a better way • Improvement in schedule and planning of the project • Construction time will be optimized

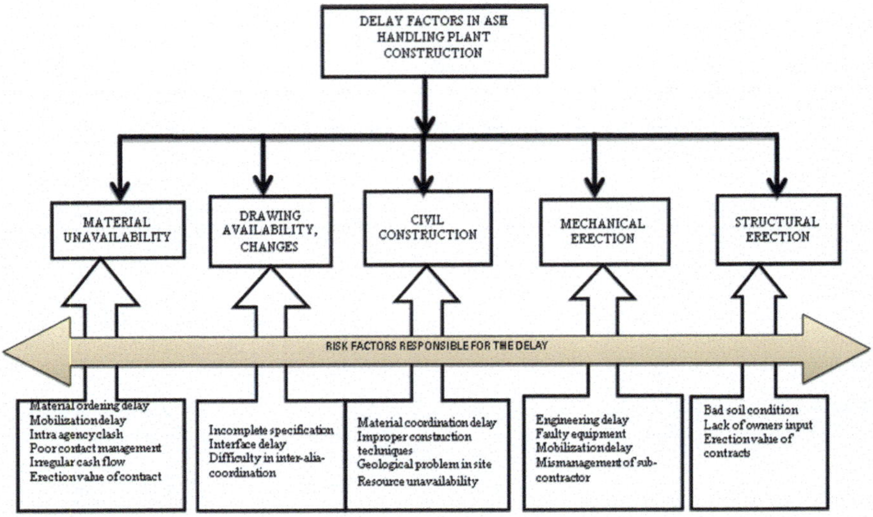

Exhibit 5.2 Risk factors in Ash handling plant construction project

Delay Factors in ash Handling Plant

A detailed analysis of risk factors causing a delay in the project is presented in Exhibit 5.2.

Key Recommendations to Reduce Schedule Delay in Ash Handling Infrastructure Project:

- The construction site should be handed over to the contractors at the beginning of the project
- All the peripheral amenities such as the supply of 440 kV electrical power supply, supply of water, and the site approach roads should be constructed by the client
- There should not be frequent changes in the design and specifications
- Bought-out items (BOI) should be finalized by the contractors as soon as they receive the general arrangement drawings, detailed drawings, part drawings, etc.
- Vendors for the BOI must be selected, and orders must be placed in time
- There should not be any delay in the manufacturing of significant parts
- Equipment mobilization should be started as soon as the site is handed over to the contractors
- Sub-contractor management is an area of concern for both clients NTPC and contractors.
- Health and safety at the site must be maintained properly to boost the morale of the construction engineers and workers
- Resource unavailability is a significant issue in every mega construction project; since there is no such specific solution to this problem, so the delay caused due to this factor needs to be considered during the project planning phase itself

- Delay due to mismanagement of sub-contractors can be reduced by setting up an individual project construction cell rather than mechanical erection, civil erection, contractor office, etc. Site-level decision-making must be encouraged
- All equipment must be checked periodically to stop malfunctioning and break down
- Political issues and land acquisition and environmental clearance are a major problem in West Bengal, reflected in the land acquisition of a village adjoining the ash pond. In reality, almost one year of work was stopped due to the land acquisition of the Shankarpur village. Discharge of ash slurry was not allowed in the Kanoi River due to environmental norms.

Question for Discussion

1. Why is project scheduling essential for the smooth execution of the project? What are the key benefits of a good project schedule for a typical infrastructure project?
2. What are the critical challenges in an infrastructure project? How can these be addressed at the stage of project planning and scheduling?
3. What is the role and importance of contractors in infrastructure projects? What are the critical criteria for which contractors should be evaluated?
4. What are the uncertainties associated with the Ash Handling infrastructure project? How will you address them using the critical information derived from SAP-LAP analysis?

Summary

A project can be viewed as a set of activities that are performed in a logically determined sequence. Thus, in dealing with a given project, first the activities are identified and precedence relationships are established. Scheduling gives the timelines, delivery, and availability of project resources (personnel, inventory, or capital).

Projects must be scheduled carefully to arrange and finish them in a timely, quality, and financially responsible manner.

Effective project scheduling has an important role in assuring the success of the project. For keeping projects on track, one needs to fix realistic time frames, allot resources properly, and manage quality to reduce the errors of the product.

The most important way to help manage any project is the critical path. If the critical path is identified, you can see clearly that efforts cannot be compromised. If there are any changes in the important path, the final date of the project will be affected. PERT and CPM are two network analysis techniques used in planning and controlling projects. They both use similar concepts. While CPM is deterministic, the PERT is probabilistic in nature. Using expected times, the critical path is obtained and critical activities are determined. The length of critical path indicates the expected duration of project. Further, the summation of the critical activities' variance yields variance of the project completion time. Its square root gives the standard deviation.

Milestones can add significant value to planning. When programmers evaluate and review coordinates with the technique (PERT) or the critical path method (CPM), it allows the project managers directly or indirectly to determine more accurately whether the project is on schedule.

Probabilities of completion of the project are calculated by considering the project completion time to be normally distributed with parameters mean and standard deviation as obtained above.

Most important error in PERT is that PERT calculates maximum of expected length, but in reality, PERT should take expected of maximum lengths.

GERT is used to model the stochastic project networks.

Questions for Discussion

1. What is the importance of project scheduling? Why is it essential for the effective execution of the project? Explain with a suitable example.
2. How do you relate project scheduling with Work Breakdown Structure? What are the key outcomes of the project schedule for a project management team? Illustrate your response for a typical construction project.
3. What are the various components of project scheduling? Explain each one in detail for an example of the construction project.
4. What is the critical path? What is its importance for applying selective control in project management? Explain your response to a case of the R&D project.
5. What is the difference between CPM and PERT? How do you see the utility of such techniques for determining a critical path for a typical real-life project of building a multi-specialty hospital? Discuss the practical relevance and consequences in detail.
6. What are the different approaches for determining critical path? Explain the essence of each one in brief with a suitable example.
7. Dr. Kothari is the lead consultant for a refinery project. He has identified the set of activities within the WBS as follows. He has also clearly understood the successor–predecessor relationship among the activities.

Activity	Predecessor	Time (days)
A1	–	10
A2	–	6
A3	A1	3

(continued)

(continued)

Activity	Predecessor	Time (days)
A4	A1	11
A5	A2, A3	5
A6	A4, A5	8
A7	A3, A4, A5	5
A8	A4, A6	7

Your Tasks:

Task 1	Draw the A-O-A and A-O-N networks with a single source and sink
Task 2	Use CPM (Critical path method) and conduct forward and backward pass computation. Determine the critical path
Task 3	Compute various floats—Total float, free float, safety float, independent float. Provide a detailed interpretation of each float

8. Consider a typical flyover construction project in Hyderabad. The list of activities with predecessors and time estimates are given in the table below. This is Public–Private Partnership (PPP) project, and it is expected to be completed in 32 months. There will be a revenue loss of Rs. 5 lakh every month if the project takes longer than 32 months. Mr. Raxroth, a project manager, is expected to carry out detailed project scheduling using PERT. He is interested in determining various parameters as a part of this.

Determine:

(a) Expected activity times
(b) ES, EF, LS, LF, and critical path
(c) Total float, free float, safety float, independent float
(d) Expected project duration and variance
(e) What is the probability that the project will be completed in 28 months?
(f) Compute probabilities that the project is completed within 75% and 125% of the critical duration.

Activity	Predecessor	Time estimates (in weeks)			Activity	Predecessor	Time estimates (in weeks)		
A	–	2	3	4	**I**	A, B, C, D	10	17	25
B	–	2	6	8	**J**	O, E, N	5	10	14
C	–	1	3	5	**K**	B, C, D	2	3	7
D	–	4	12	25	**L**	K	1	6	6
E	B, C, D	5	7	12	**M**	B, C, D	2	8	12
F	A, B, C, D	6	8	10	**N**	B, C, D	2	4	8
G	A, B, C, D	5	7	8	**O**	A, B, C, D	4	6	9
H	F, G, I	3	5	10					

(g) What is graphical evaluation and review technique (GERT) based on the project schedule? Illustrate the application of GERT for a hypothetical dataset/example.
(h) What do you mean by errors in PERT? Why is PERT optimistic?
(i) What is GERT? What are its advantages over CPM and PERT?
(j) What is the Theory of Constraints (TOC) and Critical Chain Project Management (CCPM)? Explain the key issues and components of each of these two methods?

Web-Based Exercise

Visit the various Internet links providing the details on the Aadhaar card implementation project. Develop a list of activities necessary for the successful implementation of the Aadhaar card system in India. Develop a project schedule and identify the critical path using forward and backward path computations for the assumed activity timings. Critically analyze the mistakes made by government and implementation agencies in executing the Aadhaar card project. Based on your schedule and understanding, develop an extended set of recommendations that may help implementation agencies overcome some of the shortcomings.

Group Project

Undertake a brainstorming session on "Organizing an International Conference on Project Management." Identify the list of activities and key milestones to be achieved. Consider one and half years as the total planning period. Develop a detailed schedule with necessary assumptions of activity timings, and identify the critical path. What are the key guidelines you would extend to the organizing committee to meet the deadline and ensure the hassle-free conduct of the event?

Chapter 6
Project Scheduling with Exact Distribution Method (EDM), Chance-Constrained Programming (CCP), and Simulation

Critical Questions

What are the limitations of CPM and PERT-based project scheduling?
What is exact distribution method for project scheduling? How it arrives at the better estimates?
How chanced-constrained programming (CCP) can help to overcome optimistic nature of PERT?
What are the steps in project simulation?
What are the key concerns and rationales in project simulation?
How do we execute project simulation?

Project Management in Practice

Bilgibeel Bridge

Known as the second-longest rail-cum-street Scaffold, the Bogibeel Bridge spreads over the Brahmaputra Waterway in the northeastern district of India. The 4.94 km connection is worked across the Dhemaji and Dibrugarh regions of Assam, India. It expects to diminish the train head out time between Delhi to Dibrugarh by around three hours. Likewise, it will directly network from Dibrugarh to Itanagar, lessening the movement distance by 705 km. The Bogibeel Scaffold is also perceived as the fifth-longest extension in India. The scaffold is developed as a two-path rail track on the lower deck, and the upper deck comprises three paths. The extension has been planned so that it can, without much of a stretch, vehicle armed force troops and tanks, and even warrior stream arrivals.

J. J. Thakkar, *Project Management*, Management and Industrial Engineering,
https://doi.org/10.1007/978-981-15-3695-3_6

Key Project Management Challenges

Time:

The venture got an endorsement from the government Bureau Panel on Financial Issues (CCEA) in September 1997, and development started in April 2002. For building the scaffold, the Indian Rail lines needed to develop the scope of the Brahmaputra Waterway from 10 to 5 km; this elaborate development of "manage bunds." The venture required around 505 hectares of land, which was gained across 19 towns. Significant earthworks and fortifying of the north and south dykes were finished by June 2011. The bridge was inaugurated in 2018, with construction costs of the project estimated at Rs 5,920 cr.

Cost:

HCC, in a joint endeavor with DSD Brouckenbau GmbH, Germany, and VNR Frameworks Ltd, got a 987 crore request from the Upper east Outskirts Railroad to build the superstructure of Bogibeel rail-cum-street Extension in November 2011. HCC's offer in a specific order is 51%, DSD Brouckenbau GmbH is 20%, and VNR Foundations' is 29%. Bogibeel rail-cum-street Scaffold.

Performance:

Since the Bogibeel Bridge is located in an earthquake zone, it is one of the first bridges to have a welded steel–concrete beams/trusses construction which holds the ability to withstand an earthquake of magnitude up to 7 on the Richter scale. The stiffest challenge was the construction of guide bunds on the turbulent Brahmaputra River within the short time of four to five months a year because of the rainy season and flood.

Sources

(a) *Bogibeel Bridge: Asia's Second Longest Rail-Road Bridge, Encardio Rite* (https://www.encardio.com/blog/case-study-bogibeel-bridge-asias-second-longest-rail-road-bridge/) *accessed on 1st June 2021.*
(b) *Bogibeel bridge: Challenges of bridging the Brahmaputra, The Hindu, June 25, 2014* (https://www.thehindu.com/news/national/other-states/bogibeel-bridge-challenges-of-bridging-the-brahmaputra/article6147290.ece) *accessed on 23rd May 2021.*
(c) *Bogibeel bridge: A dream come true for Assam and Arunachal Pradesh, India Today, December 24, 2018* (https://www.indiatoday.in/india/story/bogibeel-bridge-assam-arunachal-pradesh-1416655-2018-12-24).

6.1 Chance-Constrained Programming (CCP) and Simulation in Project Scheduling

Need for chance-constrained programming (CCP): The widely used methods for obtaining project completion time are program evaluation review technique (PERT) and critical path method (CPM). They both assume that the processing time as a whole cannot exceed the time taken by methods in the critical path. By contrast, in reality, the non-critical paths also have some probability of exceeding the time. Probabilistic constraints play a vital role in determining the real project completion time when the activity timings are uncertain.

The problems in PERT are as follows.

1. Ignoring non-critical paths
2. Computation of maximum expected path length, rather than expected maximum path length.

Chance-constrained programming considers the probabilities associated with all the paths and finds a solution inside the highest confidence space. The features of CCP are mentioned below.

1. The critical path in a project may be determined as a linear programming (LP) problem.
2. Primal–dual consideration of this linear programming formulation gives some valuable insight.
3. CCP is an extension of this LP framework to accommodate probabilistic activity duration.
4. CCP modifies dual in terms of event times and formulates the problem.

Once the probability distribution of all the paths is determined, CCP can help to determine the most likely solution. CCP approach minimizes the optimism associated with PERT analysis, and hence, it is preferred over PERT or CPM.

Need for Simulation in Project Scheduling: An appropriate option to explore when the project has a probabilistic nature of activities is a simulation. Unlike PERT, the simulation assumes no simplification. The key features of a simulation are summarized below.

1. Simulation permits different paths to become critical on different realizations rather than labeling one official critical path.
2. Simulation is more accurate compared to PERT because of no simplifying assumption.
3. The simulation generates information on the mean, variance, and distribution of the project completion times.
4. Simulation can give information on the likelihood of any path becoming critical, the probability distribution of nodes, slack or activity floats, etc.

A simulation is a collection of software and hardware systems, which are used to mimic the behavior of some entity or phenomenon. Simulation may be used to

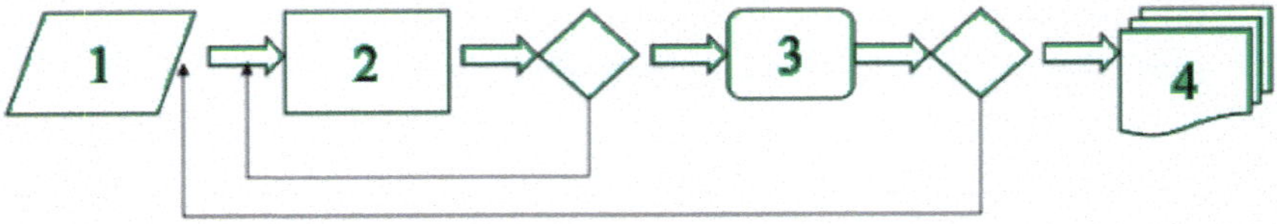

1. Input 2. Monte Carlo Sampling
3. Basic scheduling for each network realization 4. Final reports

Fig. 6.1 Simulation flow chart

analyze various theoretical models, which may be too difficult to grasp from a purely conceptual level.

The basic simulation approach collects all the data about the probability distribution of each activity in the project. It uses the Monte Carlo sampling technique to generate the distribution of each activity. When all activity times are known, basic scheduling consisting of a forward and backward pass is conducted on the network. This completes one realization of the project network of one simulation run. Finally, the required number of simulations is run. The flowchart of a simulation is illustrated in Fig. 6.1.

One advantage of simulation is that it can provide users with practical feedback when designing real-world systems. Because of no generalized assumption and a more realistic result, simulation is always preferred over PERT or CPM.

6.2 Error in PERT and PERT Optimism

Error in PERT: Program evaluation review technique (PERT) is an optimistic method to determine the project completion time. The primary reasons for optimism are indicated below.

1. Expected time or average time: Average time taken for the completion of the job or activity is denoted by t_e

$$t_e = \left(t_o + 4t_1 + t_p\right)/6$$

It is one of the most important assumptions in PERT that activities follow the beta distribution. This may not hold in a real-life situation.

1. PERT calculates the maximum expected time: PERT determines the critical path first and then calculates the project completion time according to the probabilities associated with the critical path only. Any other non-critical path with a certain probability of taking time more than the critical path is ignored. Hence, PERT always calculates the maximum of expected time, whereas our main focus

should have been determining the expected maximum of project completion time.

2. Ignorance of subcritical paths: If P1, P2, P3,…are probabilities associated with different paths in the network connecting source and sink, then $P(T) = \text{p1} \times \text{p2} \times \text{p3}$…this gives the actual probability of project completion time, but we are accepting only the probability of critical path, and other subcritical paths are neglected. This makes PERT optimistic. Ignorance of subcritical paths includes an error.

PERT is Optimistic: The precedence relationship of project activities can be completely represented by a non-cyclical network graph, in which each activity connects directly to its immediate successor. Activity times may be estimated either as single-point estimates or as three-point PERT estimates and are independent of each other. The reasons behind the optimistic nature of PERT are mentioned below.

1. Not all precedence relationships can be anticipated before a project begins; this is especially true for research and development types of projects which involve new technology and unknown problems.
2. Predecessor relationships are assumed to be firm. That is, a given job must be completed in its entirety before any successor jobs can begin. Frequently, predecessor relationships are "soft."
3. Some precedence relationships are contingent upon the outcome of the previous activities. For example, if a test phase for a new missile guidance system is completed, the next activity may be to manufacture the system. But if the test fails, the system may go back to the drawing boards. Neither PERT nor CPM allows for conditional activities and precedence ordering of this type.
4. Time estimates are subjective and depend both on the vagaries of the humans who produce these estimates and on the environment within which they operate (pessimistic, conservative, and realistic).
5. The system of rewards within which contractors operate is usually a cost-plus contract or incentive-type contract, leading to bias.

Most of the time, these assumptions never hold, and the calculation of project completion time is prone to errors. Hence, PERT is optimistic.

6.3 Exact Distribution Method for Project Scheduling Under Uncertainty: Illustrative Example

This section demonstrates the application of the exact distribution method for overcoming errors in PERT (optimism in PERT) with an illustrative example (for project network given in Table 6.1). In this method, we attempt to determine an exact distribution of time at each node while moving from source to sink to obtain the resulting exact distribution at the sink node. The approach of exact distribution method for determining project completion time is cumbersome and time-consuming.

Table 6.1 Project network

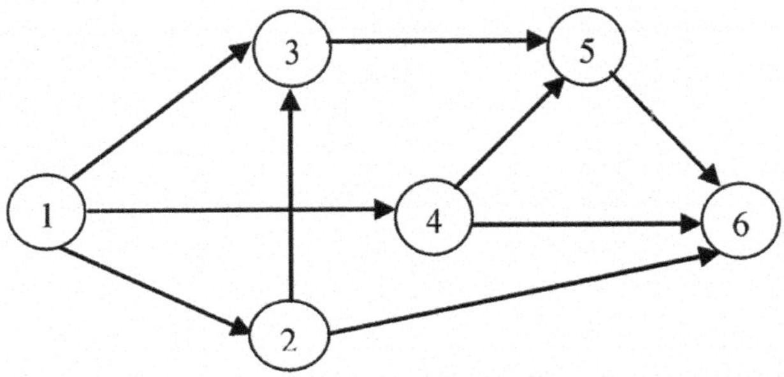

Activity	Optimistic-most likely–pessimistic time
1–2	2, 4, 8
1–3	1, 4, 6
2–3	3, 5, 9
1–4	1, 5, 8
4–5	2, 4, 7
4–6	1, 3, 7
3–5	2, 5, 8
2–6	3, 7, 9

Table 6.2 Identifying distribution at node 3 reaching via node 2

		1/3	1/3	1/3
		2	4	8
1/3	3	5	7	11
1/3	5	7	9	13
1/3	9	11	13	17

1/9	2/9	1/9	2/9	2/9	1/9
5	7	9	11	13	17

The distribution at node 3, while reaching node 3 via node 2 is tabulated in Table 6.2.

The resulting distribution at node 3 is identified in Table 6.3.

The distribution at node 5 via node 3 is determined in Table 6.4.

The distribution at node 5 via node 4 is determined in Table 6.5.

The resulting distribution of the network at node 5 is identified in Table 6.6.

The distribution at node 6 is determined in Table 6.7.

The Distribution at Node 6 Via Node 2 is Given in Table 6.8.

Table 6.3 Resulting distribution at node 3: maximum (via node 1, via node 1 and node 2)

		1/9	2/9	1/9	2/9	2/9	1/9
		5	7	9	11	13	17
1/3	1	5	7	9	11	13	17
1/3	4	5	7	9	11	13	17
1/3	6	6	7	9	11	13	17

2/27	1/27	6/27	3/27	6/27	6/27	3/27
5	6	7	9	11	13	17

Table 6.4 Distribution at node 5 while reaching node 5 via node 3

		2/27	1/27	6/27	3/27	6/27	6/27	3/27
		5	6	7	9	11	13	17
1/3	2	7	8	9	11	13	15	19
1/3	5	10	11	12	14	16	18	22
1/3	8	13	14	15	17	19	21	25

2/81	1/81	6/81	2/81	4/81	6/81	8/81	4/81	12/81	6/81	3/81	6/81	9/81	6/81	3/81	3/81
7	8	9	10	11	12	13	14	15	16	17	18	19	21	22	25

Table 6.5 Distribution at while reaching node 5 via node 4

		1/3	1/3	1/3
		2	4	7
1/3	1	3	5	8
1/3	5	7	9	12
1/3	8	10	12	15

1/9	1/9	1/9	1/9	1/9	1/9	2/9	1/9
3	5	7	8	9	10	12	15

Assuming the distribution of activity 5–6: (1, 2, 3) with equal probabilities (1/3, 1/3, 1/3), the distribution at node 6 via node 5 is given in Table 6.9.

The resulting final distribution at node 6 of the network is identified in Table 6.10 in two steps—via node 5, via node 4 and subsequently, via node 2 and maximum (via node 5, via node 4).

The project completion time based on the above resulting final distribution (with exact distribution method) is (summation of each time multiplied by its corresponding probability) = 17.96.

The activity completion times using PERT with the beta distribution are calculated in Table 6.11. The critical path is identified based on float calculations as shown in Table 6.12.

Table 6.12 indicates the activities with "zero" total float. This has helped to identify the critical path as: 1–2–3–5–6.

Project completion time using PERT = 4.33 + 5.33 + 5.00 + 2.00 = 16.66.

It can be seen that PERT ensures the completion of the project in less time (i.e., 16.66 months), while the exact distribution method gives 17.96 months as project completion time. This justifies that PERT is optimistic. This error in PERT may have serious consequences in delivering the project timely and designing contracts with incentive and penalty clauses.

Table 6.6 Resulting distribution at node 5: maximum (via node 3, via node 4)

		2/81	1/81	6/81	2/81	4/81	6/81	8/81	4/81	12/81	6/81	3/81	6/81	9/81	6/81	3/81	3/81
		7	8	9	10	11	12	13	14	15	16	17	18	19	21	22	25
1/9	3	7	8	9	10	11	12	13	14	15	16	17	18	19	21	22	25
1/9	5	7	8	9	10	11	12	13	14	15	16	17	18	19	21	22	25
1/9	7	7	8	9	10	11	12	13	14	15	16	17	18	19	21	22	25
1/9	8	8	8	9	10	11	12	13	14	15	16	17	18	19	21	22	25
1/9	9	9	9	9	10	11	12	13	14	15	16	17	18	19	21	22	25
1/9	10	10	10	10	10	11	12	13	14	15	16	17	18	19	21	22	25
2/9	12	12	12	12	12	12	12	13	14	15	16	17	18	19	21	22	25
1/9	15	15	15	15	15	15	15	15	15	15	16	17	18	19	21	22	25

6/729	6/729	33/729	21/729	24/729	78/729	64/729	32/729	141/729	54/729	27/729	54/729	81/729	54/729	27/729	27/729
7	8	9	10	11	12	13	14	15	16	17	18	19	21	22	25

Table 6.7 Distribution at node 6 while reaching node 6 via node 4

		1/3	1/3	1/3
		1	3	7
1/3	1	2	4	8
1/3	5	6	8	6
1/3	8	9	11	15

1/9	1/9	2/9	2/9	1/9	1/9	1/9
2	4	6	8	9	11	15

Table 6.8 Distribution at node 6 while reaching node 6 via node 2

		1/3	1/3	1/3
		3	7	9
1/3	2	5	9	11
1/3	4	7	11	13
1/3	8	11	15	17

1/9	1/9	1/9	3/9	1/9	1/9	1/9
5	7	9	11	13	15	17

6.4 Steps Involved in Chance-Constrained-Based Project Scheduling

The procedure involved in chance-constrained-based project scheduling is explained as below.

The critical path in a project may be determined as a linear programming problem. Primal–dual considerations of this LP formulation give some valuable insights. Chance-constrained programming is an extension of this LP framework to accommodate probabilistic activity durations. The steps involved in chance-constrained scheduling are mentioned below.

1. Dual Variable and Node or Event Times:

Dual variables are linearly related to the node or event times
$t_i = A + Bw_i$, where A is the start time and B a scaling parameter.
(With A = 0, B = −1, $w_i = -t_i$, and the dual variables are interpreted as the negative of the event times)

1. Modified Dual in terms of Event Times:

Minimize $t_n - t_1$
Subject to,
$t_j - t_i \geq t_{ij}$, for all (i, j) in arc set A (Fig. 6.2).

Table 6.9 Distribution at node 6 while reaching node 6 via node 5

		6/729	6/729	33/729	21/729	24/729	78/729	64/729	32/729	141/729	54/729
		7	8	9	10	11	12	13	14	15	16
1/3	1	8	9	10	11	12	13	14	15	16	17
1/3	2	9	10	11	12	13	14	15	16	17	18
1/3	3	10	11	12	13	14	15	16	17	18	19

		27/729	54/729	81/729	54/729	27/729	27/729
		17	18	19	21	22	25
1/3	1	18	19	20	22	23	26
1/3	2	19	20	21	23	24	27
1/3	3	20	21	22	24	25	28

6/2187	12/2187	45/2187	60/2187	78/2187	123/2187	166/2187	174/2187	237/2187	227/2187	222/2187
8	9	10	11	12	13	14	15	16	17	18

135/2187	162/2187	135/2187	135/2187	81/2187	81/2187	27/2187	27/2187	27/2187	27/2187
19	20	21	22	23	24	25	26	27	28

Table 6.10 Resulting final distribution at node 6: maximum (via node 5, via node 4, via node 2)

Maximum (via node 5, via node 4)

		6/2187	12/2187	45/2187	60/2187	78/2187	123/2187	166/2187	174/2187	237/2187	227/2187	222/2187	135/2187
		8	9	10	11	12	13	14	15	16	17	18	19
1/9	2	8	9	10	11	12	13	14	15	16	17	18	19
1/9	4	8	9	10	11	12	13	14	15	16	17	18	19
2/9	6	8	9	10	11	12	13	14	15	16	17	18	19
2/9	8	8	9	10	11	12	13	14	15	16	17	18	19
1/9	9	9	9	10	11	12	13	14	15	16	17	18	19
1/9	11	11	11	11	11	12	13	14	15	16	17	18	19
1/9	15	15	15	15	15	15	15	15	15	16	17	18	19

Maximum (via node 5, via node 4)

		162/2187	135/2187	135/2187	81/2187	81/2187	27/2187	27/2187	27/2187	27/2187
		20	21	22	23	24	25	26	27	28
1/9	2	20	21	22	23	24	25	26	27	28
1/9	4	20	21	22	23	24	25	26	27	28
2/9	6	20	21	22	23	24	25	26	27	28
2/9	8	20	21	22	23	24	25	26	27	28
1/9	9	20	21	22	23	24	25	26	27	28
1/9	11	20	21	22	23	24	25	26	27	28

(continued)

Table 6.10 (continued)

Maximum (via node 5, via node 4)

		162/2187	135/2187	135/2187	81/2187	81/2187	27/2187	27/2187	27/2187	27/2187
1/9	15	20	21	22	23	24	25	26	27	28

/19683 – >	36	90	315	543	624	984	1328	2056	2133	2043	1998	1215
	8	9	10	11	12	13	14	15	16	17	18	19

/19683 – >	1458	1215	1215	729	729	243	243	243	243
	20	21	22	23	24	25	26	27	28

Maximum (via node 2, maximum (via node 5, via node 4))

	/19683 –>	36	90	315	543	624	984	1328	2056	2133	2043	1998	1215
		8	9	10	11	12	13	14	15	16	17	18	19
1/9	5	8	9	10	11	12	13	14	15	16	17	18	19
1/9	7	8	9	10	11	12	13	14	15	16	17	18	19
1/9	9	9	9	10	11	12	13	14	15	16	17	18	19
3/9	11	11	11	11	11	12	13	14	15	16	17	18	19
1/9	13	13	13	13	13	13	13	14	15	16	17	18	19
1/9	15	15	15	15	15	15	15	15	15	16	17	18	19
1/9	17	17	17	17	17	17	17	17	17	17	17	18	19

(continued)

Table 6.10 (continued)

Maximum (via node 2, maximum (via node 5, via node 4))

	1458	1215	1215	729	729	243	243	243	243
	20	21	22	23	24	25	26	27	28
1/9	20	21	22	23	24	25	26	27	28
1/9	20	21	22	23	24	25	26	27	28
1/9	20	21	22	23	24	25	26	27	28
3/9	20	21	22	23	24	25	26	27	28
1/9	20	21	22	23	24	25	26	27	28
1/9	20	21	22	23	24	25	26	27	28
1/9	20	21	22	23	24	25	26	27	28

/177147 –>	72	306	945	4581	3744	8496	9296	20,368	17,064	26,496	17,982	10,935
	8	9	10	11	12	13	14	15	16	17	18	19

/177147 –>	13,122	10,935	10,935	6561	6561	2187	2187	2187	2187
	20	21	22	23	24	25	26	27	28

Table 6.11 Activity completion time

Activity	Optimistic-most likely–pessimistic time	Expected time	Variance
1–2	2, 4, 8	4.33	0.44
1–3	1, 4, 6	3.83	0.69
2–3	3, 5, 9	5.33	1.00
1–4	1, 5, 8	4.83	1.36
4–5	2, 4, 7	4.17	0.69
4–6	1, 3, 7	3.33	1.00
3–5	2, 5, 8	5.00	1.00
2–6	3, 7, 9	6.67	1.00
5–6	1, 2, 3	2.00	0.11

Table 6.12 Float calculations

Activity	Expected time	ES	EF	LS	LF	Total float
1–2	4.33	0	4.33	0	4.33	0.00
1–3	3.83	0	3.83	5.83	9.66	5.83
2–3	5.33	4.33	9.66	4.33	9.66	0.00
1–4	4.83	0	4.83	5.66	10.49	0.00
4–5	4.17	4.83	9.00	10.49	14.66	5.66
4–6	3.33	4.83	8.16	13.33	16.66	8.50
3–5	5.00	9.66	14.66	9.66	14.66	0.00
2–6	6.67	4.33	11.00	9.99	16.66	5.66
5–6	2.00	14.66	16.66	14.66	16.66	0.00

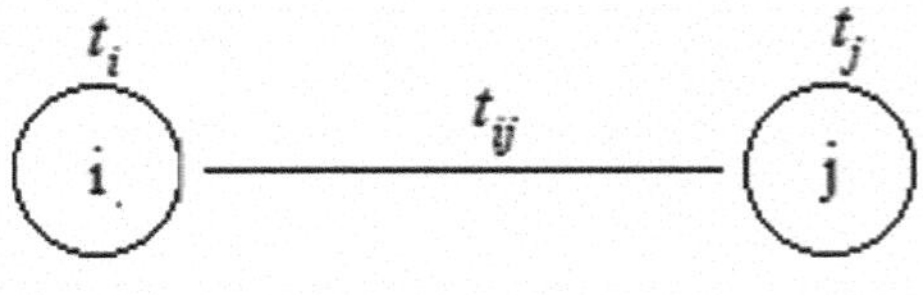

Fig. 6.2 Event times and activity duration

When activity duration t_{ij} is a random variable, the constraints may or may not be satisfied if a fixed value is substituted for t_{ij} on the right-hand side.

Fig. 6.3 Activity Distribution cdf (f) and cdf (F)

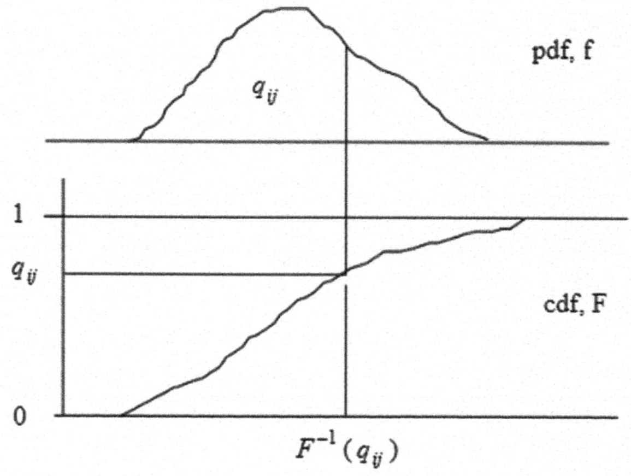

1. Chance-Constrained Programming Formulation:

Determine the node times to
Minimize $t_n - t_1$
Subject to,
Probability $[(t_j - t_i) \geq t_{ij}] \geq q_{ij}$, For all (i, j) in the arc set A.
[q_{ij} is the probability that the node times t_i and t_j would accommodate the activity ij].

1. Activity Constraints:

If the activity (ij) follows a probability distribution with a given pdf f and cdf F, then the constraint
Probability $[(t_j - t_i) \geq t_{ij}] \geq q_{ij}$ is equivalent to
$(t_j - t_i) \geq \mathrm{F}^{-1}\,(q_{ij})$
This is owing to the monotonic nature of the cumulative density function.

1. Activity Distribution (Fig. 6.3):

6.5 Steps Involved in Project Simulation

The generic steps involved in Monte Carlo simulation are mentioned below sequentially.

Step 1: Construction of a Simulation Universe

Construct a simulated "universe" of some randomizing mechanism (of cards or dice etc.), whose composition is similar to the universe, whose behavior we wish to investigate. The term "universe" refers to the system relevant for a single sample of an event.

Step 1 includes two operations:

A. Decide which symbols will stand for the elements of the universe you will simulate.
B. Determine whether the sampling will be with or without replacement.

Critical thinking is required to determine the appropriate real universe whose properties interest us.

- **Step 2: Specify the Rules in the Universe**

Specify the procedure that produces a pseudo-sample that simulates the real-life sample in which we are interested. That is, specify the procedural rules by which the sample is drawn from the simulated universe. These rules must correspond to the behavior of the real universe in which you are interested. To put it another way, the simulation procedure must produce simple experimental events with the same probabilities that the simple events have in the real world.

- **Step 3: Describe Composite Events**

If several simple events must be combined into a composite event, and if the composite event was not described in the procedure in step 2, describe it now.

- **Step 4: Calculate the Probability**

Calculate the probability of interest from the tabulation of outcomes of the resampling trials.

Procedure for Project Simulation with Monte Carlo Simulation

Step 1: **Determine the Probability Distribution of Activity Times**

– Distribution of each of the activity times is determined using experience, past data, or expert opinions.

Step 2: **Determine the Number of Simulation Runs**
This depends on the following parameters:

- Statistic of interest
  – Variance
  – Mean
  – Criticality
- Precision
- Confidence

Estimation of Project Variance

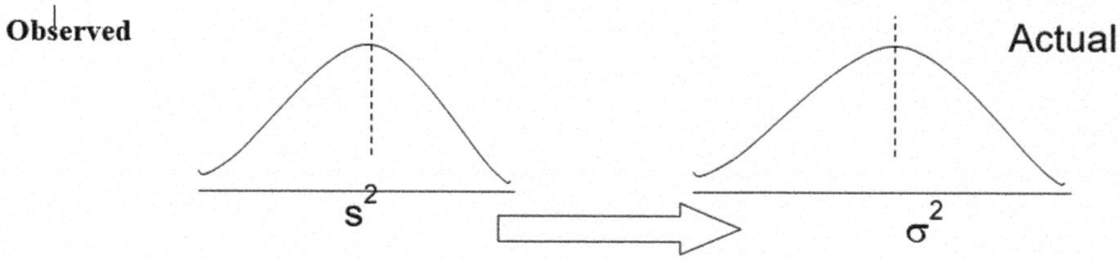

$$\Pr(0.98\,\sigma^2 \le s^2 \le 1.02\,\sigma^2) \ge 0.99$$

Precision ± 2%

Confidence 99%

$\dfrac{Ks^2}{\sigma^2}$ has a Chi-squared Distribution with (K-1) degrees of freedom

For large K $\approx$ N (k-1, 2(k-1))

or

$$\frac{\frac{Ks^2}{\sigma^2}-(k-1)}{\sqrt{2(k-1)}} \text{ has a N (0,1)}$$

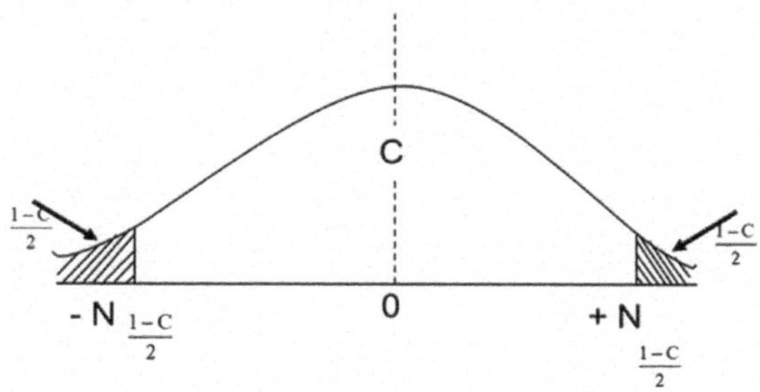

$$\Phi\left(\frac{1-Pk}{\sqrt{2(k-1)}}\right) \le \frac{1-C}{2}$$

$$\frac{1-Pk}{\sqrt{2(k-1)}} \le -N_{\frac{1-C}{2}}$$

From Normal Tables

Estimation of Project Mean Duration

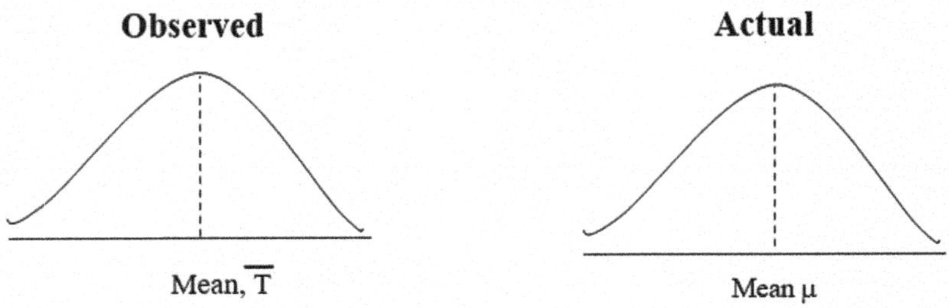

- The project distribution is assumed normally distributed
- The observed sample mean (of K realizations) follows a t-distribution with (K - 1) degrees of freedom.

For large K, this t-distribution converges to the normal. It is desired to estimate μ to within 0.01σ of its true value.

$$\Pr\left[(\mu - 0.01\sigma) \le \overline{T} \le (\mu + 0.01\sigma)\right] \ge 0.99$$

Precision Confidence

$$\Pr\left[\frac{-0.01\sigma}{\sigma/\sqrt{k}} \le \frac{\overline{T}-\mu}{\sigma/\sqrt{k}} \le \frac{0.01\sigma}{\sigma/\sqrt{k}}\right] \ge 0.99$$

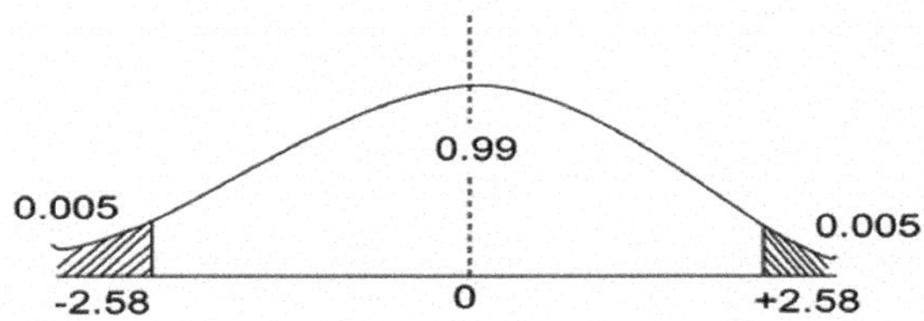

$$\Phi(-0.01\sqrt{k}) \le 0.005$$
$$0.01\sqrt{k} \ge 2.58$$
$$k \ge 66{,}564 \text{ samples}$$

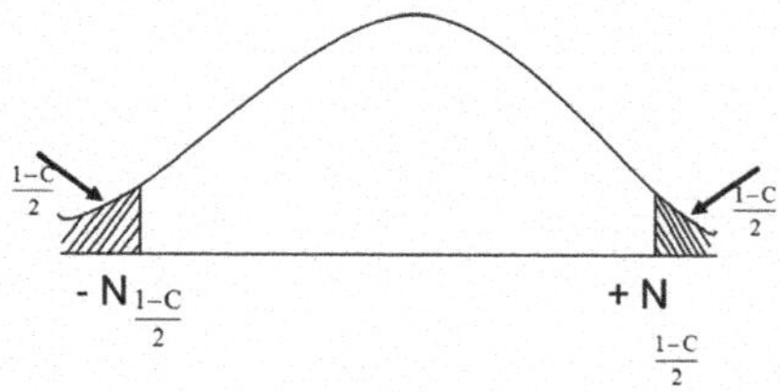

$$\Phi\left(-P\sqrt{k}\right) \le \frac{1-C}{2}$$

$$P\sqrt{k} \ge N_{\frac{1-c}{2}}$$

$$k \ge \left(\frac{N_{\frac{1-c}{2}}}{P}\right)^2$$

Estimation of Criticality Index

- P_a = Probability that activity is critical
- P = Precision
- Probability of getting exactly r success in K trials (simulation run)
  Binomial distribution
- r_a = No. of times the activity is critical
- K = Number of realization (number of simulation run)
- Every time evaluate activity critical or not.
- For large K, binomial distribution may be approximated to a normal distribution. When binomial is approximated to normal, the mean and variance are:

  Mean = KP_a
  Variance = $KP_a\,(1 - P_a)$

- When we want in fraction, divide mean by K and variance by K^2

$$Mean = P_a$$

$$Var = \frac{P_a(1-P_a)}{K}$$

$$P_a \approx N\left(P_a, \frac{P_a(1-P_a)}{K}\right)$$

$$\mu = r_a$$

r_a is the observed value, and it should be close to a true value.

$$P_r(r_a - P \le P_a \le r_a + P) \ge 0.99$$

$$1 - 2\phi\left(1 - \frac{0.01}{\sqrt{\frac{P_a(1-P_a)}{K}}}\right) \ge 0.99$$

Proof:

- $P_r(r_a - P \leq P_a \leq r_a + P) \geq 0.99$

  here $\mu = r_a$ ……..observed value

$$\Pr\left(\frac{r_a - 0.01 - r_a}{\sqrt{\frac{P_a(1-P_a)}{K}}} \leq \frac{P_a - r_a}{\sqrt{\frac{P_a(1-P_a)}{K}}} \leq \frac{r_a + 0.01 - r_a}{\sqrt{\frac{P_a(1-P_a)}{K}}}\right)$$

$$= 1 - 2\phi\left(1 - \frac{0.01}{\sqrt{\frac{P_a(1-P_a)}{K}}}\right) \geq 0.99$$

Under Worst-Case Analysis

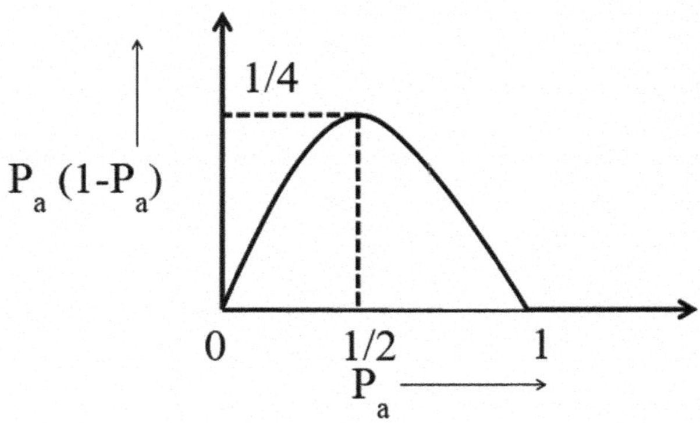

If we are conservative and substitute the worst possible value of P_a that is 50% = 1/2.

The worst value of the above expression is 1/4 at Pa = ½

$$\varphi\left(\frac{-0.01\sqrt{K}}{\sqrt{P_a(1-P_a)}}\right) \leq 0.005$$

$$\varphi\left(-0.02\sqrt{K}\right) \leq 0.005$$

$$\frac{\sqrt{K}}{50} \geq 2.58$$

$$K \geq 16641$$

This worst case can be generalized in terms of P and K.

$$\emptyset\left(-2P\sqrt{\mathrm{K}}\right) \leq (1 - C/2)$$

K based on a critical index (number of simulation run).

***Step 3* : Select a random number string** using the random number table or any other random number generator.

***Step 4* : Assign two-digit random numbers** in a sequence of the random number string to the activities.

***Step 5* : Determine the activity times using the random numbers** and distribution of activity times as follows. Let r be the random number of an activity

$$F(x) = r/100$$

***Step 6* : Apply the usual CPM approach** to determine the critical path and project completion time for the current network realization.

***Step 7* : Repeat Step 3 to Step 6** till the number of simulation runs calculated in Step 2 is achieved.

6.6 Illustrative Example: Chance-Constrained Programming (CCP) for Project Scheduling

DFM is a pharmaceutical company interested in setting up a manufacturing unit in Kolkata. The project manager has conceived the relationships (as shown in Table 6.13) to execute the project. He assumed that each activity has a triangular distribution with a, m, b values of 6,8,10 use chance-constrained programming to determine all the node times such that for all jobs, the probability of completion is.

(A) 90%
(B) 95%
(C) 99%.

Compare these results with the corresponding values under standard PERT assumptions showing how optimistic PERT is under these circumstances.

Table 6.13 Project network relationships

Job	Predecessors
A	–
B	–
C	A
D	A
E	A
F	B,C
G	D,F
H	D,F
I	E,H

Triangular Distribution:

pdf,

$$\begin{aligned} f(x) &= \frac{2(x-a)}{(b-a)(m-a)} \quad \text{for } a \le x \le m \\ &= \frac{2(b-x)}{(b-a)(b-m)} \quad \text{for } m \le x \le b \end{aligned}$$

cdf,

$$\begin{aligned} F(x) &= \frac{(x-a)^2}{(b-a)(m-a)} \quad for\ a \le x \le m \\ &= 1 - \frac{(b-x)^2}{(b-a)(b-m)} \quad for\ m \le x \le b \end{aligned}$$

Therefore,

$$\begin{aligned} F^{-1}(x) &= a + \sqrt{F(x)(b-a)(m-a)} \quad for\ a \le x \le m \\ &= b - \sqrt{(1-F(x))(b-a)(b-m)} \quad for\ m \le x \le b \end{aligned}$$

Network (A-O-A) (as given in Fig. 6.4).

Expected time of each activity = (6 + 8 + 10)/3 = 8

Variance = $((10–6)^2 + (10–8)^2 + (8–6)^2)/36 = 0.67$.

For 90% probability of completion, the calculations are shown in Table 6.14.

$x = F^{-1}(x) = 6 + \sqrt{0.9(10-6)(8-6)} = 8.68$ for $6 \le x \le 8$: Conditions is not satisfied.

$x = F^{-1}(x) = 10 - \sqrt{(1-0.9)(10-6)(10-8)} = 9.11$ for $8 \le x \le 10$: Conditions is satisfied.

z = 1.282.

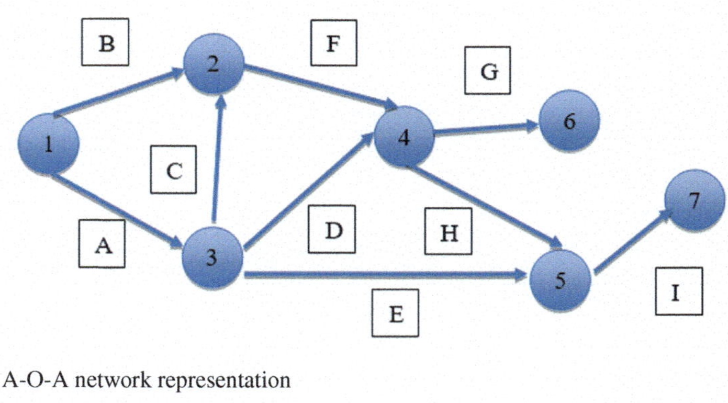

Fig. 6.4 A-O-A network representation

Table 6.14 CCP versus PERT calculations for 90% probability

Node	Expected Time	Variance	CCP	PERT
1	0	0	0	0
2	16	1.34	18.22	17.48
3	8	0.67	9.11	9.05
4	24	2.01	27.33	25.81
5	24	2.01	27.33	25.81
6	32	2.68	36.44	34.10
7	40	3.35	45.55	42.34

Table 6.15 CCP versus PERT calculations for 95% probability

Node	Expected Time	Variance	CCP	PERT
1	0	0	0	0
2	16	1.34	18.72	17.90
3	8	0.67	9.36	9.35
4	24	2.01	28.08	26.33
5	24	2.01	28.08	26.33
6	32	2.68	37.44	34.69
7	40	3.35	46.80	43.01

For 95% probability of completion, the calculations are shown in Table 6.15.

$x = F^{-1}(x) = 6 + \sqrt{0.95(10-6)(8-6)} = 8.75$ for $6 \le x \le 8$: Conditions is not satisfied.

$x = F^{-1}(x) = 10 - \sqrt{(1-0.95)(10-6)(10-8)} = 9.36$ for $8 \le x \le 10$: Conditions is satisfied.

z = 1.645.

For 99% probability of completion, the calculations are shown in Table 6.16.

$x = F^{-1}(x) = 6 + \sqrt{0.99(10-6)(8-6)} = 8.81$ for $6 \le x \le 8$: Conditions is not satisfied.

$x = F^{-1}(x) = 10 - \sqrt{(1-0.99)(10-6)(10-8)} = 9.72$ for $8 \le x \le 10$: Conditions is satisfied.

z = 2.326.

Table 6.16 CCP versus PERT calculations for 99% probability

Node	Expected time	Variance	CCP	PERT
1	0	0	0	0
2	16	1.34	19.44	18.69
3	8	0.67	9.72	9.90
4	24	2.01	29.16	27.30
5	24	2.01	29.16	27.30
6	32	2.68	38.88	35.81
7	40	3.35	48.60	44.26

6.7 Illustrative Example: Monte Carlo Simulation for Project Scheduling

A residential construction project has six major activities. The activity timings are uncertain, and the cost associated with optimistic estimates of the PERT is significant. A project analyst has decided to simulate the project. He made certain assumptions about the most appropriate distribution for each of the activities. The project network structure and activity distribution details are given below. You are expected to simulate the project.

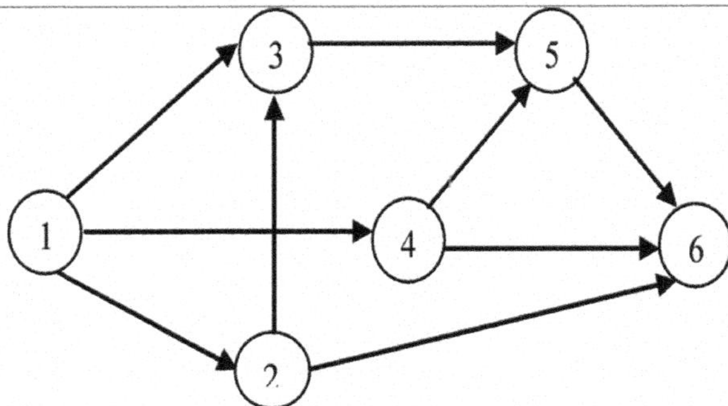

Activities	Distribution
(1–2)	4, 6, 10, discrete with equal probability
(1–3)	The uniform distribution between 2,7
(1–4)	5, 10, 12, triangular distribution
(2–3)	Uniform distribution between 10,20
(2–6)	Exponential distribution with mean 8
(3–5)	4, 6, 10, discrete with probabilities 0.3, 0.3, 0.4
(4–5)	5, 9, 14, triangular distribution
(4–6)	Uniform distribution between 6,18
(5–6)	4, 8, 10, triangular distribution

Your Tasks:

Task 1	Use standard PERT assumptions to compute the expected project duration, expected floats for all activities, and node slacks. Also, compute the criticality indices for all activities
Task 2	Use Monte Carlo sampling from a string of 2 digit random numbers (578, 918, 451, 857, 594) to generate activity durations in **TWO NETWORK REALIZATION** and compute the project duration, activity floats, and node slacks
Task 3	Compare with results obtained in Task 1? Is PERT optimistic?—Comment
Task 4	Compute the number of simulation runs required to estimate the project means with a confidence of 97% and accuracy of 2%

Table 6.17 Float calculations

Activity	Expected time	ES	EF	LS	LF	Total float
(1–2)	6.67	0	6.67	0	6.67	0.00
(1–3)	4.50	0	4.50	17.17	21.67	17.17
(1–4)	9.00	0	9.00	9.33	18.33	9.33
(2–3)	15.00	6.67	21.67	6.67	21.67	0.00
(2–6)	8.00	6.67	14.67	28.00	36.00	21.33
(3–5)	7.00	21.67	28.67	21.67	28.67	0.00
(4–5)	9.33	9.00	18.33	19.34	28.67	10.34
(4–6)	12.00	9.00	21.00	24.00	36.00	15.00
(5–6)	7.33	28.67	36.00	28.67	36.00	0.00

Critical path: 1–2–3–5–6
Therefore, project completion time = 36.00

Task 1: Float calculations (as given in Table 6.17).

Task 2:

Network Realization 1 for random numbers: (57 89 18 45 18 57 59 45 78) is given in Table 6.18.

Therefore, critical path: 1–2–3–5-6.

Project completion time = 34.88.

Network Realization 2 for random numbers: (91 84 51 85 75 94 57 89 18) is given in Table 6.19.

Therefore, critical path: 1–2–3–5-6.

Project completion time = 38.50.

Activities 1–2, 2–3, 3–5, and 5–6 are critical for both the simulation runs. The criticality index for these activities is 2 and 0 for all other activities.

Task 3:

Expected project completion time based on two network realizations = (34.88 + 38.50)/2 = 36.69 > 36 (Pert).

The difference in expected project completion time using simulation and PERT approach is not significant for only two runs. Still, if more simulation runs are performed, then the optimistic nature of PERT will be evident.

Task 4:

Let k be the number of simulation runs required for estimating the project variance (i.e., statistics of interest is variance).

$$\emptyset\left(\frac{1-0.02k}{\sqrt{2(k-1)}}\right) \geq (1-0.97)/2$$

Table 6.18 Network realization 1

Activity	Random number (r)	Distribution	Actual time	ES	EF	LS	LF	Total float
(1–2)	57	4, 6, 10, discrete with equal probability	6	0	6	0	6	0.00
(1–3)	89	Uniform distribution between 2, 7	6.45	0	6.45	14.05	20.50	14.05
(1–4)	18	5, 10, 12, triangular distribution	7.50	0	7.50	9.30	16.80	9.30
(2–3)	45	Uniform distribution between 10, 20	14.50	6	20.50	6	20.50	0.00
(2–6)	18	Exponential distribution with mean 8	1.58	6	7.58	33.30	34.88	27.30
(3–5)	57	4, 6, 10, discrete with probabilities 0.3, 0.3, 0.4	6	20.50	26.50	20.50	26.50	0.00
(4–5)	59	5, 9, 14, triangular distribution	9.70	7.50	17.2	16.80	26.50	9.30
(4–6)	45	Uniform distribution between 6,18	11.40	7.50	18.90	23.48	34.88	15.98
(5–6)	78	4, 8, 10, triangular distribution	8.38	26.50	34.88	26.50	34.88	0.00

$$\left(\frac{1-0.02k}{\sqrt{2(k-1)}}\right) \leq -2.170$$

$$0.0004k^2 + 4.30k - 3.34 \leq 0$$

$$k \geq 10751$$

Table 6.19 Network realization 2

Activity	Random number (r)	Distribution	Actual time	ES	EF	LS	LF	Total float
(1–2)	91	4, 6, 10, discrete with equal probability	10	0	10	0	10	0.00
(1–3)	84	Uniform distribution between 2, 7	6.20	0	6.20	22.30	28.50	22.30
(1–4)	51	5, 10, 12, triangular distribution	9.22	0	9.22	18.68	27.90	18.68
(2–3)	85	Uniform distribution between 10, 20	18.5	10	28.50	10	28.50	0.00
(2–6)	75	Exponential distribution with mean 8	11.09	10	21.09	33.49	44.58	23.49
(3–5)	94	4, 6, 10, discrete with probabilities 0.3, 0.3, 0.4	10	28.50	38.50	28.50	38.50	0.00
(4–5)	57	5, 9, 14, triangular distribution	9.60	9.22	18.82	28.90	38.50	19.68
(4–6)	89	Uniform distribution between 6, 18	16.68	9.22	25.90	27.90	44.58	18.68
(5–6)	18	4, 8, 10, triangular distribution	6.08	38.50	44.58	38.50	44.58	0.00

Project Management in Practice

Project simulation for a project—"Constructing Flyover"

A company with long experience in executing many projects for the Andhra Pradesh state government has acquired expertise in building roads, bridges, flyovers, and other government infrastructure projects. The company has been given the project to construct a 5.5-km flyover that would bypass a busy corridor to provide a smooth passage to the traffic for longer distances in Bangalore. The location for the flyover is from X landmark to Y landmark; it would be the two-point connector and will not branch to other locations. The company has assigned a senior project engineer and three project engineers who would all be reporting to the project manager. The project engineers have been responsible for surveying the locations and clearing out technical details for the project after discussion with the team. Based on a technical outline of the plan, the project manager sits with the team and specifies all major, minor phases. The technical experts and some experienced supervisors are asked to give opinions about their assessment of the time needed for different phases of the project. The past project data is used to take into consideration any undesirable situations and other possibilities. The project manager combines all the bits of information and determines the activity time distributions. Again, the project manager takes expert advice for any unaccounted risk factors and then finally fixes the distributions for the activity times (Exhibit 6.1).

Exhibit 6.1 Flyover construction project details

S. No.	Activities	Predecessors	Activity times (in months)
1	Alignment of the flyover is laid out, and the pier and abutment locations are located	–	Uniform distribution 2, 4
2	Excavated to the appropriate depth and driving pile	1	Uniform distribution 1, 2
3	Subgrade material is compacted and leveling, or rat slab is built	2	Uniform distribution 3, 4
4	Concrete placed on the bottom portion of the foundation, within formwork, or using the ground as a form	3	Uniform distribution 2, 4
5	Pier column or abutment concrete placed	4	Uniform distribution 2, 3
6	Building pier caps	5	Uniform distribution 4, 6
7	Building abutments	5	Uniform distribution 4, 5
8	Placement of steel or precast girders, or box girders, etc.	6, 7	Uniform distribution 2, 5
9	Building deck with either precast slabs or by placing a concrete deck	8	Uniform distribution 3, 6
10	Building railings, roadway, guardrail, and paint pavement markings	9	Uniform distribution 1, 4

Network (A-O-N):

The A-O-N network representation of the flyover construction project is given in Exhibit 6.2.

Simulation Runs:

Network Realization 1 for random numbers: (57 89 18 45 18 57 59 45 78 91) is given in Exhibit 6.3.

Usage for the Company:

It is recommended that an adequate number of simulation runs is necessary to arrive at a reliable estimate of activity times and overall project completion time. A reliable

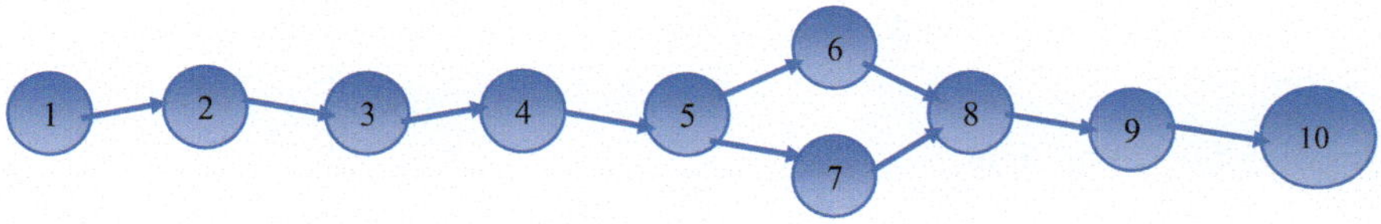

Exhibit 6.2 Network representation (A-O-N) for flyover construction project

Exhibit 6.3 Network realization for flyover construction project

Activity	Random numbers	Activity times (in months)	Actual time	ES	EF	LS	LF	Total float
1	57	Uniform distribution 2, 4	3.14	0	3.14	0	3.14	0.00
2	89	Uniform distribution 1, 2	1.89	3.14	5.03	3.14	5.03	0.00
3	18	Uniform distribution 3, 4	3.18	5.03	8.21	5.03	8.21	0.00
4	45	Uniform distribution 2, 4	2.9	8.21	11.11	8.21	11.11	0.00
5	18	Uniform distribution 2, 3	2.18	11.11	13.29	11.11	13.29	0.00
6	57	Uniform distribution 4, 6	5.14	13.29	18.43	13.29	18.43	0.00
7	59	Uniform distribution 4, 5	4.59	13.29	17.88	13.84	18.43	0.55
8	45	Uniform distribution 2, 5	3.35	18.43	21.78	18.43	21.78	0.00
9	78	Uniform distribution 3, 6	5.34	21.78	27.12	21.78	27.12	0.00
10	91	Uniform distribution 1, 4	3.73	27.12	30.85	27.12	30.85	0.00

Critical path: 1–2–3–4–5–6–8–9–10

Project completion time = 30.85 months (assuming 30 days/month)

estimate of activity times and project completion time can help minimize underestimating or overestimating required resources, which helps to stick to the schedule and budget. The organization can take the help of the above approach, replace the activity time distributions with their distributions based on experience from other projects, and thus execute the project without much deviation from the resource and time allocation plan.

Pitfalls:

1. The above approach may not be useful if an adequate number of simulation runs are not performed.
2. Simulation may give unrealistic results if the activity time distributions are not appropriately determined.
3. The approach may not take into consideration some disrupting events with less probability.

Question for Discussion:

1. What should be considered a basis by the case organization to determine the nature and distribution of activities?
2. Why do you feel that the simulation approach can help the case organization better plan the project compared to traditional methods like CPM and PERT?
3. What are the essential precautions you would like to recommend to the case organization executing a project using a simulation approach?

Summary

CCPM and simulation are seen as better alternatives to PERT to minimize optimism involved in PERT analysis.

PERT ignores the subcritical path and this induces error in PERT analysis.

PERT is optimistic in nature.

Exact distribution method explores all the paths and provides accurate estimate of project completion time but it is cumbersome in nature.

Monte Carlo simulation for project scheduling is the most widely used approach for estimating project completion time under uncertainty.

Monte Carlo simulation allows all the paths equally to be a candidate for critical path and hence minimizes bias in PERT analysis.

There are three statistics for determining simulations runs with a given degree of precision and accuracy. This can be calculated for statistics of interest (i) variance, (ii) mean, and (iii) criticality index.

Questions for Discussion

1. What is the need for chance-constrained programming (CCP) and simulation in project scheduling?
2. What is an error in PERT? Why is PERT optimistic?
3. Use the exact distribution method for the network given below and determine the project completion time.

[*Hint*: Find exact distribution at each node and then find resulting distribution. Continue up to the last node (node 6) and get the resulting final distribution. Use this final distribution to determine project completion time. Assume equal probability (1/3) for each time (optimistic-most Likely–pessimistic) of an activity]

Also, use the PERT approach with forwarding and backward pass computation and find project completion time. Compare the project completion times and comment—"How much is the PERT optimistic?"

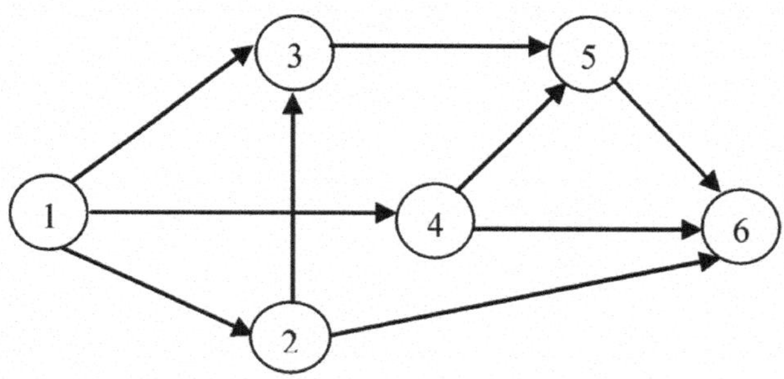

Activity	Optimistic-most likely–pessimistic time
1–2	3, 5, 9
1–3	2, 5, 8
2–3	3, 6, 9
1–4	3, 6, 8
4–5	4, 4, 7
4–6	2, 3, 8
3–5	2, 6, 8
2–6	3, 7, 9

4. Enlist the steps (complete procedure) involved in chance-constrained-based project scheduling.
5. Enlist the steps (complete procedure) involved in project simulation with the Monte Carlo simulation approach
6. Mr. Singh, a project manager for an infrastructure company executing the construction of a multi-specialty hospital, has conceived the relationships (as given in table below) for executing the project. He assumed that each activity has a triangular distribution with a, m, b values of 8, 10, 12 use chance-constrained programming to determine all the node times such that for all jobs, the probability of completion is

   (A) 90%
   (B) 95%
   (C) 99%

Compare these results with the corresponding values under standard PERT assumptions showing how optimistic PERT is under these circumstances.

Job	Predecessors
A	–
B	–
C	A
D	A
E	A
F	B, C
G	D, F
H	D, F
I	E, H

7. A company has been given the contract to construct a restaurant at Delhi Airport. This has six major activities. The activity timings are uncertain, and the cost associated with optimistic estimates of the PERT is significant. A project analyst has decided to simulate the project. He made certain assumptions about the most appropriate distribution for each of the activities. The project network structure and activity distribution details are given below. You are expected to simulate the project.

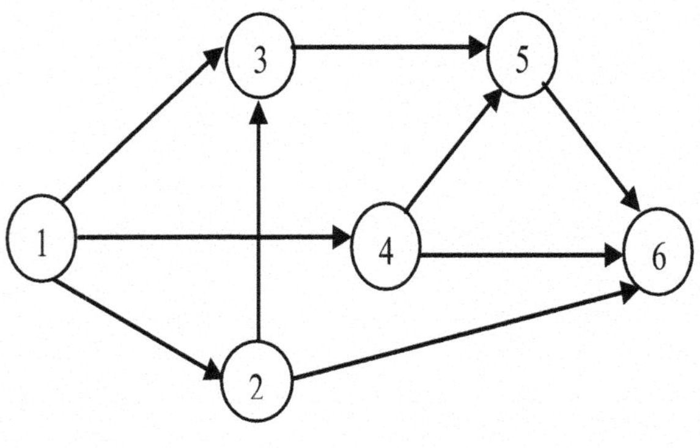

Activities	Distribution
(1–2)	Uniform distribution between 4,8
(1–3)	6,8,10, discrete with equal probability
(1–4)	5, 10, 12, triangular distribution
(2–3)	Uniform distribution between 12,28
(2–6)	Uniform distribution between 10, 20
(3–5)	4, 6, 10, discrete with probabilities 0.3, 0.3, 0.4
(4–5)	5, 9, 14, triangular distribution

(continued)

(continued)

	(4–6)	4, 8, 10, triangular distribution
	(5–6)	Exponential distribution with mean 10

Your Tasks:

Task 1	Use standard PERT assumptions to compute the expected project duration, expected floats for all activities, and node slacks. Also, calculate the criticality indices for all activities
Task 2	Use Monte Carlo sampling from a string of 2 digit random numbers (678, 818, 551, 858, 516) to generate activity durations in TWO NETWORK REALIZATION and compute the project duration, activity floats, and node slacks
Task 3	Compare with results obtained in Task 1? Is PERT optimistic?—Comment
Task 4	Compute the number of simulation runs required to estimate the project means with a confidence of 97% and accuracy of 2%

Group Project

Study the nature and characteristics of the project executed by the Government of India project construction of "Statue of Unity." What are the key milestones a project management team would have envisaged? Undertake a detailed brainstorming session in a group of four students and enlist all the necessary activities and milestones for this project. Assume appropriate distributions for the activities and carry out a detailed project simulation. What is the realistic time in which the government should have completed this project? What do you think as the critical shortcomings of the project which might have escalated project completion time and cost?

Chapter 7
Project Scheduling with Theory of Constraints (TOC) and Critical Chain Project Management (CCPM)

Critical Questions
What are the key principles of TOC and CCPM?
How TOC helps to address the critical constraints of the system?
Why CCPM-based project management is more effective?
What are the key implementation challenges in TOC- and CCOPM-based project management?

7.1 Theory of Constraints (TOC) and Critical Chain Project Management (CCPM)

The theory of constraints is a means of identifying a key objective factor (i.e., blocking) that leads to accomplishing the task and imposes control and restrictions on the system. In production, control is called a crisis. It adopts a scientific approach to progress. Each complex system, including manufacturing processes, considers several connected operations, one of which operates as a controller in the entire system (i.e., the control function is "the weakest link in the chain"). TOC works on the three essential measures which control the performance of an organization. It includes throughput, operational expense, and inventory. "Throughput" is the rate at which the system generates output. "Operational Expense" is the expenditure of the system to turn inventory into throughput. "Inventory" is the stock built by the system by procurement.

"TOC (control theory)" is a method of increasing performance by managing "constraints" (bottlenecks). A constraint is anything that prevents the system from achieving more of its goal. Constraints can be internal (e.g., market demands from the system that it can deliver) or external (e.g., the system can produce more than

J. J. Thakkar, *Project Management*, Management and Industrial Engineering,
https://doi.org/10.1007/978-981-15-3695-3_7

the market will bear). A constraint always prevents the system from getting more throughput (typically, revenue through sales). The constraint is not just a breakdown in the equipment, people, or policy. This is a theoretical basis of supply chain management, managing the cash flow-based profit, and commercial processes' decision on the business process.

Critical chain project management (CCPM) program and planning strategy emphasize an effective utilization of the necessary resources (people, equipment, indoor space) to manage the project specifications. The method was introduced by Eliyahu M. Goldratt in his book "Critical Chain" in the year 1997.[1] CCPM is a TOC-based project planning technique. It differs from the traditional methods derived from the critical path and PERT algorithms, emphasizing work sequence and planning. The traditional methods (PERT and CPM) focus on the critical path, but in CCPM, the focus is on the critical chain. In the critical path, activities can have an early start or late start, early finish or late finish, whereas in the critical chain, activities have only a start and an end. If there is no connection for resources, then the critical chain is the same as the critical path. Typically, when there are resource contentions in a project, the critical chain needs to be found considering the impact of statistical fluctuation and dependent events. The CCPM is extensively used for having 10–50% faster and/or cheaper programs than traditional methods (i.e., CPM, PERT, Gantt, etc.). It has been observed that the projects generally end up at 222% of the scheduled time, 189% of the original budget cost, and 30% of the times; they are canceled before the completion. CCPM attempts to improve the performance of these traditional statistics.

Typically, a project gets delayed because of seven reasons: (i) Murphy's law (the worst happens when it was least desired!), (ii) bad multitasking (carrying out several activities at the same time); (iii) student's syndrome (tendency to keep the activity completion pending till the last moment), (iv) task dependence (delays are passed on from one task to another, seldom the benefit of early completion passed on), (v) Parkinson's law (work fills up the available time. Since there are no rewards for early completion, extra effort is not made to complete an activity early); (vi) self-protection (an early finish calls for less allotted time later); and (vii) dropped baton (early finish of an activity may not ensure the early start of next activity if people are not ready).

The smooth and cost-effective execution of the project is achieved using the theory of constraints (TOC). The key components of TOC include:

- Five focus steps (a time to identify and delete obstacles)
- Thought processes (problems analysis and solving tools)
- Process accounting (performance measurement management and guidance management).

[1] Goldratt, Eliyahu M. Critical Chain. Great Barrington, MA: North River Press, 1997.

7.1.1 What Are the Problems with Traditional Methods (CPM and PERT) of Project Scheduling?

1. **Inflated estimates of the project managers**

There is always a significant difference in the projected completion time of the activity, and the actual time it takes. To overcome this difficulty, project managers add a huge contingency in projecting an activity completion time, and this leads to a project which will take much more time than required. CCPM uses the concept of project buffers to overcome this difficulty.

2. **Procrastinating work till it becomes urgent**

There are two types of people with their reporting tendency in the project. One type of people will ensure timely completion of the project but the other type (maybe about 10%) will delay things till the last moment and report late completion. This indicates the procrastination behavior of human beings. This means 99% of the activities should finish on time or earlier than expected, and 1% of the activities are delayed until they become urgent. This is typically called a "students syndrome" (trying to complete the entire syllabus of the exam at the last moment). This human nature of delaying the things till it gets urgent leads to the well-known phenomena "work expands to fill available time." The traditional methods of project planning suffer because of this problem.

3. **Manipulating resource or time savings**

The traditional methods focus on the sequences and scheduling of the activities to ensure the project's timely completion. A typical human behavior emerges when it comes to reporting the savings in resource or time because the activity is completed early and with less consumption of material, human, and other resources. This happens because there is no reward associated with the early completion of the task. On the contrary, you may get fewer resources and time next time because you have completed the work early.

4. **Strict requirement for merging the activities to satisfy predecessor-successor relationships**

The project has a set of interdependent activities which are taking place simultaneously. Typically, it has been observed that many things get piled up and start surfacing out toward the end of the project. This is attributed to only one problem in traditional project scheduling methods that is the "need of merging the activities at a particular node and satisfying predecessor–successor relationships." For example, suppose four activities merge at a particular node and two of them are completed early. If one is completing on time and the last is delayed, then the succeeding activity can not start until all the activities get completed. This means the project manager can not avail of the benefits emerging from early or on-time completion of three activities out of four.

5. **Delays caused because of multitasking**

Usually, "multitasking" is considered beneficial for improving the availability of resources and the system's productivity. Typically, in the context of the projects, it is not true. The projects get delayed because of multitasking. This can be explained simply. For example, we are working on three activities in a typical project. Each of the activities is expected to take one month to complete. When you opt for multi-tasking, you are dividing your day into three segments and as a result, each of the activities will take three months to complete. This will result in budgeting for three months for an activity that should take only one month to complete. This leads to the fact that the total time needed for each task is longer than it ought to be too.

7.1.2 Key Issues in the Application of TOC and CCPM

The theory of constraints (TOC) can be used to analyze business and identify possible weaknesses that might cause the company to perform below optimum levels. While the theory can be a useful tool for improving business performance, one has to be aware of its limitations before implementing its recommended solutions.

Theory	The theory of constraints maintains that a few major constraints keep complex processes such as business functions from reaching optimized performance. Managers have to identify the constraints that affect their business and work to mitigate their impact or remove them. There is always one constraint that has the biggest effect. When a business deals with it, another constraint becomes the most important. Managers can reduce the effects of constraints as they identify them and use the theory of constraints to guide them in continuously improving company performance
Identification	A key process in the theory of constraints is the identification of constraints that reduce performance. A major challenge involves how to identify those constraints. The theory might either work on a constraint that is caused by another constraining factor, or it may focus on a constraint that is irrelevant to the present production bottlenecks This limitation of the theory could encourage you to spend time and waste resources on problems that are not critical to the success of the company

(continued)

(continued)

Variation	A second limitation of the theory of constraints is its lack of consideration of variable factors. Constraints such as demand for a product might vary independently from any action taken by implementing the theory. If product demand is a temporary constraint and rises because of market dynamics, resources invested in increasing the demand might have been more beneficial in expanding production capacity. Other data, such as market studies, must be used to verify whether constraints targeted by the theory will remain fixed
Time Frame	The theory of constraints works effectively to address the current time frame for a business. It looks at actual situations and therefore limits itself to short-term effects. To overcome this limitation, one has to examine the long-term effects of the work on constraints that the theory identifies. If the short-term effect remains valid over a longer time frame, the strategy indicated by the theory may be valid. If the short-term effect does not last or causes long-term deterioration in other business variables, one has to identify other constraints that, if reduced, result in long-term benefits

CCPM is a TOC tool used for planning and project management. Resources in many projects have been used simultaneously in both the same program and many other project structures. CCPM is a new way of looking at project management based on the interaction of statistical fluctuation and dependent events. It works on the principle that tasks in a project do not consume the same time as originally estimated. Although project managers intend to make up the losses in one task with the gains in another, that never happens. As the execution of a task is dependent on the timely completion of the preceding tasks, the delays accumulate, but the gains do not materialize. For example, a train can never leave early from a station, but it has to make up late. The key benefits of the "TOC" philosophy and key principles of "CCPM" are summarized in Table 7.1.

Table 7.1 TOC and CCPM

Benefits of TOC	Key principles of CCPM
• Increased profit • Fast improvement • Improved capacity • Reduced lead time • Reduced inventory	• Parkinson's law: expanding work to fill work hours • Student's syndrome: people have started to work in a complete slowdown when the time is near • The Murphy act: what is wrong going wrong • Bad multitasking: the next task for bad multitasking cannot be started

7.2 Key Terminologies and Principles of TOC

Theory of constraints is a management philosophy first given by Eliyahu M. Goldratt in 1984 in his book called "The Goal."[2] Theory of constraints (TOC) focuses on eliminating or elevating the chain's weakest link, which is a bottleneck in the organization. TOC is a means to find out the weakest link in the chain, blocking the organization or the process chain to accomplish the desired target or attaining the standards set by the organization. TOC also aims at eliminating the constraints to help the organization reach the maximum efficiency and carry out the tasks properly and smoothly.

TOC is used to increase the performance of the organization. The principles of TOC come in handy for the managers of any organization. Managers use TOC to introspect the organization's structure to find any constraint preventing the organization from reaching its full potential. TOC comes in with applicability to almost anything. It applies to almost any sector, be it the manufacturing or service sector. TOC is also applicable to accounting, logistics, construction management, production, research and development, supply chain management, and other sectors. It aids managers to look and aim for any opportunity that gives them a chance to find out the constraint, exploit it, and provide the organization a chance to perform well and thus achieve high potential and profits and meet the objective and set standards.

Constraint:

The term constraint refers to anything that prevents a given operation from being done, hindering meeting an objective. A constraint can occur anywhere or at any stage. It may occur in the initial design stage or the planning stage. The constraint can also occur in the procurement stage or even when we are carrying out our operation. The constraint can also occur in bottlenecks where the whole line of operations may get stopped due to one bottleneck. Therefore, it becomes necessary for the organization to look out for the constraints at regular intervals and keep working on them to exploit them, remove them, and ensure that they do not have any further effect on the system. The constraint can also be in the form of company policy constraint, due date constraint, budget constraint, manpower constraint, or even natural constraint over which humans have next to zero control. A constraint may be physical (equipment inefficiency, raw material stockout, etc.) or non-physical (policy, terms, conditions, etc.).

TOC is governed by the idea that every system, whether having a single product or multiproduct, deals with at least one constraint. The constraint can be in any form. Now, having even a single constraint can have a significant impact on the system. Also, the ultimate aim of any organization is to provide quality service at minimum cost and achieve maximum profit. Given that a company has one or two constraints, it may have several impacts on the system. This gives managers of the company a chance to look out for the constraints and do whatever is necessary to remove the constraints and ensure the smooth functioning of the operations.

[2] Goldratt, Eliyahu M.; Cox, Jeff (1984). The Goal. Gower Publishing.

This will result in the successful completion of the task, increased efficiency, greater level of satisfaction among the workers, reduced cost, increased profit, and overall increased reputation of the organization in the very competitive environment as of today. A typical example of a constraint is presented in Fig. 7.1.

The TOC approach constantly looks for a continuous improvement process that enables the organization to focus on the constraints and work to eliminate/elevate the constraints. Three broad steps are considered in the theory of constraints; a system diagram is shown in Fig. 7.2.

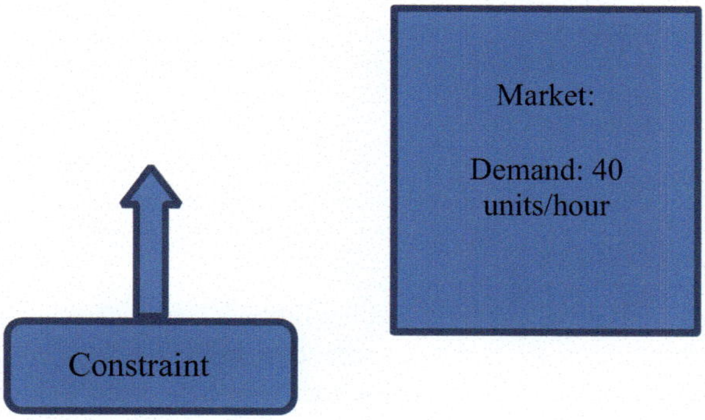

Fig. 7.1 Example of a constraint

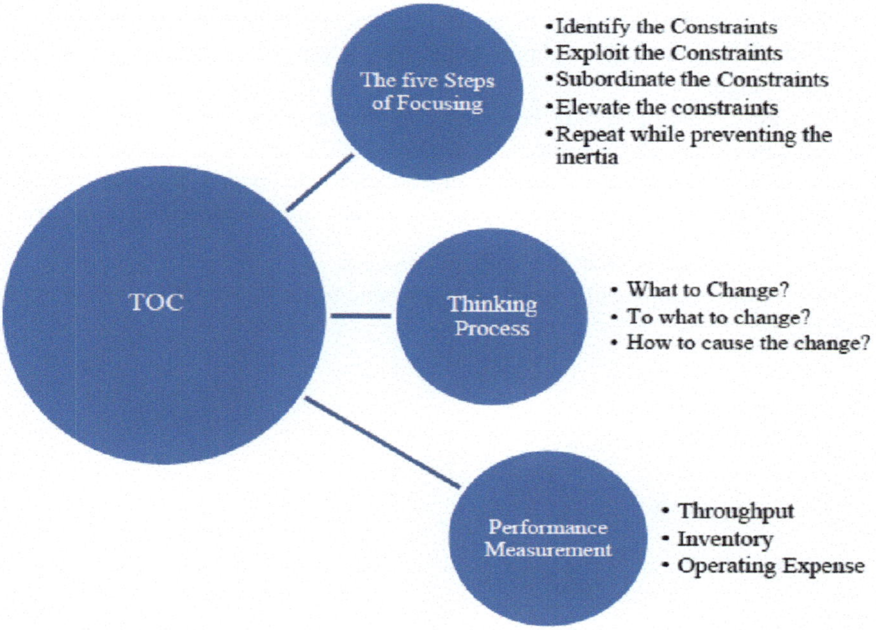

Fig. 7.2 Steps of TOC

1. The five steps of focusing:
   a. Identify the constraints
   b. Exploit the constraints
   c. Subordinate the constraints
   d. Elevate the constraints
   e. Repeat the process while preventing the inertia.
2. Thinking process:
   a. What to change?
   b. What to change to?
   c. How to cause the change?
3. Performance measurement:
   a. Throughput
   b. Inventory
   c. Operating expense

The five steps of TOC are summarized in Table 7.2.

The "TOC" thinking process is summarized in Table 7.3.

Performance measurement is the way to measure the system's efficiency after the constraints have been removed or modified. According to Eliyahu M. Goldratt, there may be three major performance measurement criteria, which are stated and defined below:

1. Throughput: The rate at which money is generated in the system through sales is termed throughput. Throughput excludes all the variable costs like raw materials cost, freight, etc.
2. Inventory: Inventory may be defined as the inventory kept in the stock in monetary terms awaiting sales.
3. Operating expense: This includes all the expense happened in giving a certain level of service. It is the money system spends to turn inventory into throughput.

Table 7.3 summarizes the TOC thinking process.

Some of the examples of constraints applicable to the construction industry can be stated as:

1. Economic constraints
2. Technical constraints
3. Environmental constraints
4. Social constraints
5. Legal constraints.

Table 7.2 Five steps TOC

Step 1: Identify the constraint	The first and foremost step is to define the constraints that are preventing the system to achieve its optimum target. Constraints may be one or many, and they may be acting independently or simultaneously. One's job is not only to identify the constraints but also to rank them according to the level of their importance in the system. For instance, a constraint may be of little or no impact on our ultimate aim, while the other may have a larger impact. So, it is better to cater to the larger constraint than waste time on the smaller constraint
Step 2: Exploit the constraints	This step tends to use the existing configuration to achieve the goal. Management must remove all non-necessary activities and steps from the constraints and try to maximize the use to achieve the organizational goal. Exploiting the constraints will involve changes in the organizational ways and policies rather than changes that will involve huge monetary expenditure. If the step is successful, we can skip stepping 5; otherwise, go to Step 3
Step 3: Subordinate the constraints	This step needs the focus of management to synchronize the behavior of non-constraining elements of the system to the strategy being utilized under the constraints. This may seem easy but is considered to be the most difficult step out of these five steps because this may include a paradigm shift of the manager's thinking process and ability
Step 4: Elevate the constraints	This step includes elevation or increasing the capacity of the constraining resource. The constraining resource capacity once increased may result in the elimination of that constraint, and thus, the operation would be carried out accordingly. While exploitation and subordination may not involve much change in the ongoing process, elevation may tend the organization for many out of the pocket expenditures, policy changes, increase in manpower, or increase in the equipment required
Step 5: Repeat while preventing the inertia	The above four steps will involve several changes in the ongoing process, and removal of one constraint does not guarantee that another constraint will not come into the system. Thus, it becomes extremely necessary that the system repeats the entire process in order to look out for some additional constraints Preventing the inertia simply means that while looking out for other constraints, the constraint that has just been overcome must not come back into the picture; otherwise, it would become a repetitive cycle, and the company cannot attain the intended goals

Table 7.3 TOC thinking process

What to change?	Identify what the things or links that are causing the conflicts, root causes of undesirable effects, and problems after understanding the current system patterns and provide a certain linkage between the cause and its effect that will help us relate better are
What to change to?	A solution is proposed and is examined for its validity. The proposed solution is again reviewed to identify whether there are any ill effects associated with this solution and again eliminating them to find proper corrections
How to cause the change?	Identify the hurdles and obstacles that are preventing the organization from achieving optimum goals. We plan a detailed step-by-step approach on how to tackle the problem, and we share the plan with concerned personnel so that they all work together for one goal

7.3 Principles of Critical Chain Project Management (CCPM)

Critical chain project management (CCPM) is considered a handy tool under the theory of constraints (TOC), which came into existence after Eliyahu M. Goldratt published another business novel called Critical Chain in 1997. This approach applies to single projects and multi-projects where resources are being used simultaneously during the project.

CCPM in a single project structure involves:

1. Elimination of that behavior and standards of organization that are otherwise harmful for the organization and the reason behind not achieving the specified objectives of the organization.
2. Planning of the project network and considering the relation between all tasks and resources for assigning the task time.
3. Inclusion of work schedule with the critical chain and all buffers. For multi-project structures, CCPM is inclusive of all the elements of the same tool used in a single project structure plus an additional tool that synchronizes the implementation of the project. The additional tool can be physical like a resource or even virtual like a policy or rule.

Critical chain project management requires constant updating of the task time used for buffer management and knowing whether it is even necessary to have any corrective measure. Critical project management (CPM) and project evaluation and review technique (PERT) were extensively used before introducing the critical chain concept.

CPM was usually the first choice of many organizations even after CCPM was introduced because:

(a) CPM identifies critical path.
(b) CPM aids in identifying total project duration.
(c) CPM also helps in knowing the earliest and latest start and finish times.

Even after all these benefits, CPM may have some drawbacks like.

(a) A project may have an unrealistic activity duration that combines project duration and redundant safety time duration, which aids in nothing but increases the overall project duration.
(b) CPM does not consider the resource constraints while planning and scheduling the project.
(c) CPM may give safety to all the activities, which in some cases may not even be needed.
(d) CPM considers the activity with total zero floats as the critical activity, which means that a project can have more than one critical path. This leads to confusion and multitasking.

CCPM can be considered more effective than CPM as CCPM obviates the redundant safety time from the activity duration and thus helps in producing a lean schedule.

CCPM also redefines the critical chain as the longest chain of dependent activities. A project is defined as any activity or any set of activities performed in a sequential manner to achieve a predefined objective within a certain time limit and with minimum cost.

It is also obvious that however small or large the steps of a particular project may be, the project must be finished within a specific time limit. Going beyond that time limit may yield losses in terms of cost and impressions, penalties, etc. Accelerating the projects may land the organization into fortune and give it a window for operation expansion, increased sales and profit, the opportunity for the newer products launch, and trust again in the globally competitive market. Therefore, it can be said that project management for companies involved in projects is necessary. Despite the wide range of project types, there are a few problems that are common with all types of projects:

a. Budget overruns and time overruns.
b. Frequent amendments and alterations.
c. Unavailability of resources.
d. Frequent changes in priorities.

All these problems have little or something to do with how the projects are managed in the companies. This gives a window for a universal, easy, and acceptable method that takes care of these common problems, eliminates them, and results in increased organizational efficiency.

The CCPM approach, derived from the theory of constraints, identifies three factors which are the cause behind the above-stated problems and many other problems such as those:

1. Bad multitasking.
2. The student's syndrome.
3. The Parkinson's law.

These factors are explained as below:

Multitasking:

Multitasking can be defined as the act of leaving one task before its completion and devoting time to other tasks, which may be perceived as more important than the current task. Each time the worker shifts from one task to another before completing the task, there is a considerable decrease in the worker's efficiency. The worker has to jot down all the previous efforts to end the initial task when he gets back to it, which is not good considering that the learning curve effect reduces each time the new task is assigned. The duration of a task is stretched due to bad multitasking (Fig. 7.3). A project is consisting of three tasks (A, B, C), each of which takes three days.

Considering half work is done initially (Fig. 7.3). It is visible that multitasking has increased the task duration unnecessarily, resulting in the project's delay (overrun). Many organizations have accepted that multitasking results in loss of efficiency, resulting in unwanted delay and overrun of the project.

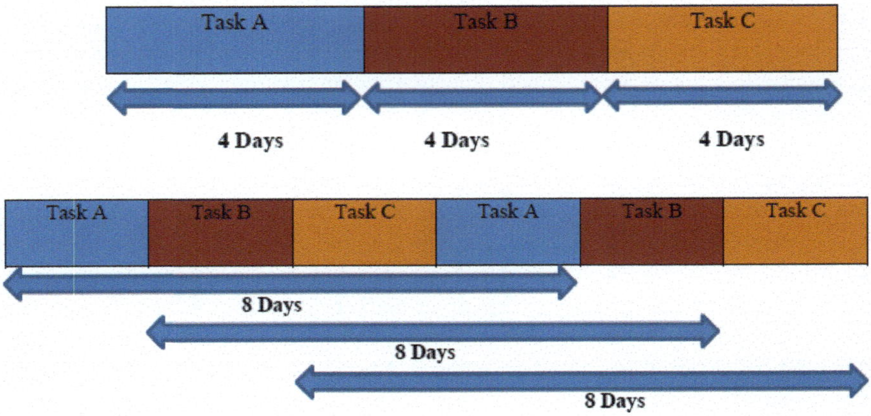

Fig. 7.3 Multitasking approach

Student's Syndrome:

The student's syndrome can be best understood by the term "procrastination." In a student's or worker's life, procrastination can be understood as students or workers working diligently on a specific project at the last minute, only to complete it right before the deadline. As soon as safety margin is incorporated in the time allotted for a particular project, the student's syndrome kicks in. When the worker assumes that the schedule is going to be busy, he requests the safety time. And as he gets the safety time, the urgency factor is removed, and the worker feels relieved as they feel that now they have got enough time to work on the project. But in reality, they only waste their safety time and only start to give sufficient time to the job only when the deadline is near. Student's syndrome is found to be responsible for delayed projects and overruns. The safe time that was induced in the beginning could have been used against unknown constraints arriving at later stages of the project. Still, since it gets wasted due to procrastination habits of the workers, it can no longer be availed, which results in delayed projects or overruns (Fig. 7.4).

Parkinson's Law:

There may be several tasks that are governed by Parkinson's law (see Fig. 7.5). It ensures that the task would not be reported even after completion of the safe time (buffer) has been assigned, but it did not come down to the use of the buffer, even if there are no constraints. The user may strive for extra effort on the task so that he gets recognition for it, or he may believe that if he reports that work, he might get a strict/tight deadline for the next task since he will think that they can be completed more quickly.

One task may take ten days on the first attempt, and in the next attempt, it may take 12 days because of uncertainty. That is why the worker may be concerned

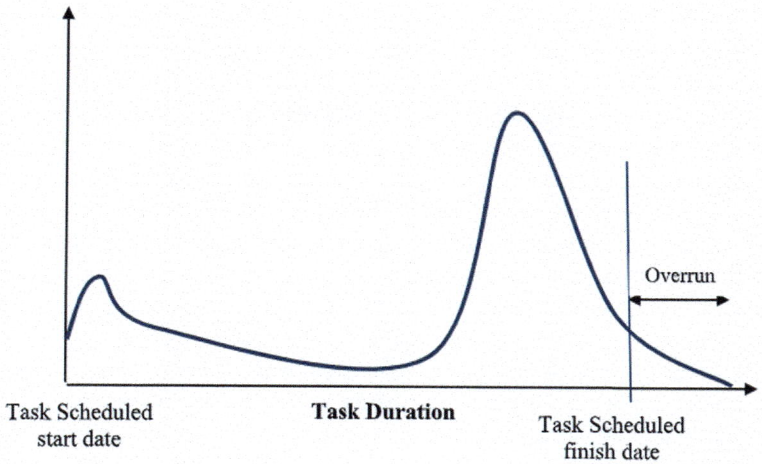

Fig. 7.4 Student's syndrome

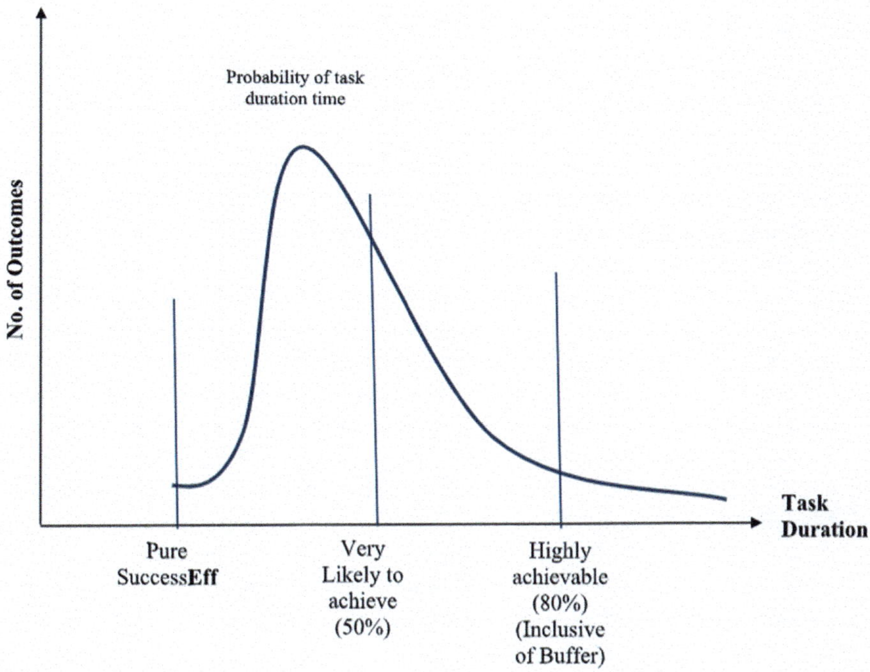

Fig. 7.5 Parkinson's law

about the buffer, which is why he may not report the project's early completion. Delays are generally passed on, while gains are not (Fig. 7.5). Out of the possible three outcomes, it is most likely the worker to choose 80% safety because the task would still be doable in case of any major uncertainty and can be finished within the deadline. But is that a feasible thing to do considering the student's syndrome?

The general approach of CCPM involves taking off all the safety from all the individual tasks and giving the project buffer at the end of the task. This approach is useful in keeping the worker on the edge of the seat always because when taken away half of the time from his estimated task time, we also take any opportunity of Parkinson's or student's syndrome occurrence. That means the worker will be sufficiently pressurized to complete his task within the deadline.

Even if there is some delay, the buffer is kept as a savior for emergencies. The buffers protect statistical variation. The buffer can be seen as a proactive warning mechanism. If activity variation consumes a buffer by a certain amount, the warning is raised to find a future course of action if the situation worsens. If the situation goes beyond a critical point, the necessary measures such as expediting, overtime, subcontracting, etc., must be put into effect.

The buffer is also classified broadly into three types:

1. Project buffer: It is used for maintaining project completion date to meet the target. CCPM shifts the safety associated with the critical chain activities to the end of the critical chain in the form of a project buffer. When all safety buffers are removed from individual tasks and added as project buffer, the project's duration is substantially cut. It is also imperative that when all safety buffers are pooled together, the project has better safety protection. Still, the project can also be completed in less time.
2. Feeding buffer: It is used for protecting the critical chain against disruptions from other chains. Critical chain computes the project buffer for the baseline schedule that does not yet contain the feeding buffer. During the execution of the project, the schedule keeps the provision for feeding buffers. The feeding buffers are placed whenever a non-critical chain activity joins the critical chain.
3. Resource buffer: It is used for notifying the scheduler of a new resource being employed on the critical chain. Resource buffers are placed whenever a resource has a job on the critical chain, and a different resource does previous critical chain activity. Resource buffers ensure that resources will be available when needed and critical chain tasks can start on time or early. Deciding on an appropriate buffer size is complicated. It is project-oriented and varies case by case (Fig. 7.6).

Aggregate buffer (project buffer in Fig. 7.6) can be smaller than the sum of individual safety.

In summary, the steps involved in CCPM can be summarized as below:

1. Clearly state the objective of the project and project path.
2. Determine the requirement and activities required to meet them.
3. Determine the logical relationship between activities and needs.

Task A: 4 days	Safety: 3 days	Task B: 6 days	Safety: 4 days	Task C: 4 days	Safety: 3 days

Task A: 2 days	Task B: 3 days	Task C: 2 days	Project Buffer (7 days)

Fig. 7.6 Basic approach of CCPM

4. Estimate the resource requirements, activity durations, and costs.
5. Calculate critical chain schedule and buffers.
6. Evaluate the schedule according to the project objectives.
7. Repeat the process if necessary.

Some of the problems that a project manager may have to encounter while implementing CCPM includes:

1. Poor communication throughout the hierarchy
2. Inadequate information
3. Inaccurate planning
4. Lack of training/education/motivation.

7.4 Differences in TOC and CCPM

The key differences in TOC and CCPM approaches are listed in Table 7.4.

7.5 Numerical Application of TOC

An automobile manufacturer produces two products X and Y that are processed in four departments, A, B, C, and D. The company is a leading supplier to the leading automobile manufacturers. The company has adopted the "theory of constraint" (TOC) approach to meet the manufacturers' quality and lead time expectations. The company uses four different types of resources—R1, R2, R3, and R4—to manufacture the products X (steering system) and Y (braking system). Product X requires three types of resources R1, R2, and R4. Product Y requires two types of resources R2 and R3. The company's production process is illustrated in Fig. 7.7.

Each department has 2000 min of available time per week. The weekly expense of the company is $25,000. Based on current demand, the organization can sell 75 units of product X and 50 units of product Y per week. The costs of the products are $500 for product X and $450 for product Y. All four materials are available in sufficient quantities. The availability of the needed workers is ensured.

Table 7.4 TOC versus CCPM

Theory of constraints (TOC)	Critical chain project management (CCPM)
The theory of constraints is an approach introduced in the mid-80 s of the twentieth century by Goldratt. The approach exploits the constraints and allows eliminating them for the proper functioning of the organization	A critical chain is a tool that is based on the principles of TOC. Basic steps and ideology of TOC are used for the CCPM approach
TOC aims at the removal of all the constraints for smooth functioning. TOC gives the user to identify the possible constraints and what are the things that can be done to make the best out of the situation	CCPM aims at finishing the project on time. CCPM allows forming a critical chain as the longest chain based on dependent activities
TOC helps the organization to manage, assess, and analyze the existing situation of the organization in line with the ultimate objective the organization	CCPM aims at finishing project within the given budget without compromising on service level
TOC helps an organization to formulate relevant steps or actions and execute a plan to effectively tackle the organizational problems	CCPM is adamant about avoiding multitasking. Multitasking may lead to unavoidable delays which should not occur in the projects
TOC allows to successfully implement the action plan that has been set up to achieve the targets and achieve desired organizational goals	CCPM aims at providing a buffer for the uncertainties that can occur any time in the system by aggregating all safety time at the end of the project
TOC allows certain performance measurement criteria to understand how they approach fairs after the implementation of an action plan	CCPM focuses on resource leveling and proper buffer utilization for better project execution
TOC makes sure that the inertia after tackling the constraints is maintained so that the same constraint does not occur again in the system and the new constraints coming into the system are also being taken care of	CCPM emphasizes focusing on start dates and exploitation of early finishes. It also focuses on aggressive completion of set targets of the tasks

The company is interested in investigating the following:

1. Find out the constraints in operation.
2. Find out the throughput per unit for each product.
3. Find out the throughput per minute of the constrained resource for each product.
4. Find out the number of units of X and Y that should be produced per week to maximize the throughput.
5. Find out the maximum net income of the organization per week.
6. If the company elevates the constraint by doubling it, what will happen to the system?

Solution:

1. Determining the constraints of the company

The total time required per week for each department is calculated in Table 7.5.

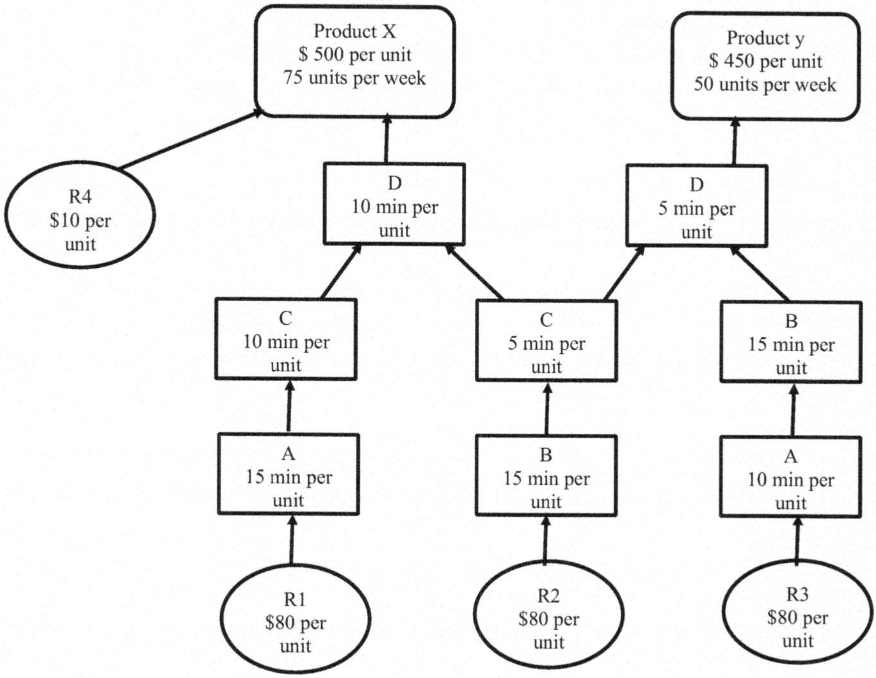

Fig. 7.7 Manufacturing process

Table 7.5 Total time required per week

Department	Product X	Product Y	Total time required per week
A	15 min * 75 units	10 min * 50 units	1625 min
B	15 min * 75 units	30 min * 50 units	2625 min
C	15 min * 75 units	5 min * 50 units	1375 min
D	15 min * 75 units	5 min * 50 units	1375 min

Table 7.6 Throughput per unit

Product	Material cost	Product cost	Throughput per unit
X	$80 + $80 + $10 = $170	$ 500	$500 – $170 = $330
Y	$80 + $80 = $160	$ 450	$450 – $160 = $ 290

Since each department has 2000 min required, Department B is the only department exceeding the time limit. Therefore, it is a constraint.

2. Determining the throughput per unit for each product (as given in Table 6).

Throughput per unit is defined as the difference between product cost and materials cost.

Table 7.7 Throughput per minute

Product	Minutes required for constraint B	Throughput per unit	Throughput per minute
X	15 min	\$ 330	\$ 22
Y	30 min	\$ 290	\$ 9.667

$$\text{Throughput per unit} = \text{Product Cost} - \text{Materials Costs}$$

3. Determining the throughput per minute of the constrained resource for each product.

Throughput per minute of the constrained resource can be calculated as given in Table 7.7.

$$\text{Throughput per minute} = \frac{\text{Throughput per unit}}{\text{Minutes required for constriant}}$$

4. Determining the number of products of X and Y for maximizing the throughput

To maximize the throughput, produce as many product units with the highest throughput per minute of the constrained resource. The company should produce 75 units of product X. This requires {75 units * 15 min} = 1125 min in the constraint Department B and leaves 2000 − 1125 = 875 min for the production of 30 units of product Y, i.e., 875 min ÷ 30 min per unit = 29.167 units of Z (30 units approximately).

5. Determining the maximum net income per week

Total throughput of the company can be calculated as:

$$\text{Throughput} = \sum \text{Optimum number of units of product} * \text{throughput of that product}$$

Therefore,

$$\text{Throughput} = 75 * 330 + 50 * 290 = \$39250$$

Net profit can be calculated as given in Table 7.8.

$$\text{Net Profit} = \text{Throughput} - \text{Expense of the company}$$

Table 7.8 Calculating net profit

Sales		
Product X	75 units * $ 500 = $ 37,500	
Product Y	50 units * $ 450 = $ 22,500	$60,000
Materials cost		
Product X	75 units * $ 170 = $ 12,750	
Product Y	50 units * $ 160 = $ 8000	$20,750
Throughput	Sales cost–materials cost	$39,250
Weekly operating expense		$30,000
Net profit		$ 9250

Therefore, the weekly net profit of the organization is $ 9250.

6. Elevate the current constraint by doubling the capacity

Suppose the capacity of Department B is doubled. In that case, that will allow the company to produce 60 units of product Y (approximately) and 75 units of product X. Since the demand for product Y is only 50 units, the company will have unused capacity in all departments of the organization. Therefore, external demand would become the constraint.

Project Management in Practice

Building a Multi-Specialty Hospital using TOC and CCPM.

In the era of cutthroat competition, a rapidly changing market, and a large number of competitors, the applications of the theory of constraints (TOC), and critical chain project management (CCPM) have received significant attention. There is hardly any sector left where TOC and CCPM have not found their use and implementation. TOC and CCPM are used in logistics, supply chain, IT, services sector, construction, etc. It is also being used in the healthcare industry. A hospital chain decided to build a multi-specialty hospital. With a focus on TOC and CCPM, a case study has been developed to understand the various problems and how to tackle them using TOC at various stages and how scheduling can be done following CCPM.

Design Stage

A team of capable officials is appointed for the initial design of the facility. The design team lays the initial foundation of steps that we need to follow for all successful infrastructural operations. Managers have to stick to deadlines to meet the targets without any delay.

The various types of constraints have to be addressed in the initial design phase following the principles of TOC. The constraints need to be identified, exploited, and if that does not resolve the issues, then subordinate other resources to the constraint. It is of equal importance to maintain the inertia and repeat the entire process over and over to look out for other constraints. Different designs have to be observed to improve the services to the patients because they may have to be shifted on stretchers to the top floor of the building, and lift may not be a good option for the patient and stairs do not seem viable. Therefore, a proper design should be kept in mind for such similar situations.

Manufacturing Stage

The manufacturing stage (construction stage) of the multi-specialty hospital is itself an arduous task. The various constraints in the manufacturing stage may include improper vendor and manufacturer relations for raw materials. The supplier may not be able to supply the required materials on time, or it may not have sufficient resources to fulfill all the demands. The site engineers and purchase department have to keep multiple vendors ready for such sort of situation. Another constraint may be natural reasons like weather, storm, etc., which could delay the project. Appropriate buffers should be used to be safe from such sudden constraints over which we have little or no control.

Implementation Stage

While implementation, various constraints may come into the picture, which we as practitioners must understand. A proper supply of skilled manpower is one prime example. If the manpower is not skilled, then the project will not be executed properly. For skilled manpower, we have to be sure that multitasking is prohibited. It is seen that multitasking leads to ill distributed concentration, which is not good for the company, and the worker is also less efficient. The managers also have to note student's syndrome, set time for activities, and allocate buffers in the system accordingly. Parkinson's law must also be considered, since even if the work gets done within the assigned time, the remaining time should not be wasted, and it should be utilized accordingly.

Equipment Setup

The procurement of advanced and high-quality equipment of international standards is a necessity of the healthcare center. It can be considered a major constraint that has to be appropriately handled by the system. The need for high-tech equipment triggers another constraint that is the requirement of skilled professionals to operate them. This is the essence of TOC that one constraint can give rise to another constraint, and it is essential to take care of both the constraints.

At times, there may arise the need for more than one piece of equipment when there is a higher requirement or significantly higher inflow of patients. The manager has to decide how much of each type of equipment is required.

Professional Team Requirement

A team of professional doctors, medicos, and other laboratory technicians and researchers are required for treatment, pharmacy, and other departments. It serves as a perfect critical chain scheduling problem where the number of professionals and their requirement and proper scheduling is done to make the most out of them. A considerable change in the organization will be seen once the steps and principles of TOC and CCPM are applied in the infrastructure project. Reduced work in progress, smooth functioning, and project completion within the due date is a few benefits. CCPM will also identify the critical chain that can lay out a plan for operations to be carried out with more attention. A typical list of activities included in constructing a multi-specialty hospital is given below, which can be used as an input for further analysis.

i. Raw materials arrangement (A1) = 10 weeks
ii. Design of the facility and other things (A2) = 15 weeks
iii. Arrangement of manpower (A3) = 10 weeks
iv. Construction of the facility (A4) = 25 weeks
v. Electrical connections (A5) = 7 weeks
vi. Equipment installation (A6) = 12 weeks
vii. Recruitment of all professionals (A7) = 15 weeks
viii. Quality control testing (A8) = 4 weeks.

Activity	Predecessor activity	Period
A1	–	10
A2	–	15
A3	A2	10
A4	A1, A2, A3	25
A5	A4	07
A6	A5	12
A7	A4, A5, A6	15
A8	A7	04

Questions for Class Discussion

1. What are the key advantages the hospital chain can derive, strengthening their revenue stream by adopting TOC approach?

2. What are the key challenges the case organization should face while implementing TOC?
3. How can a project manager integrate PERT, TOC, and CCPM to ensure effective project planning and execution?

Summary

Theory of constraints is a means of identifying a key objective factor (i.e., blocking) that leads to accomplishing the objective and limits that control system until it is a restricting factor.

TOC can be used to analyze business and identify possible weaknesses that might cause the company to perform below optimum levels.

The theory of constraints maintains that a few major constraints keep complex processes such as business functions from reaching optimized performance.

By focusing on single-unit renewable resource categories, a critical chain starts using multitasking resources for a particular job in a specific time frame. Goldratt (1998) defines it as “the biggest killer of the leading time.”

The CCPM approach, derived from theory of constraints, identifies three factors responsible for project delays. These are: (a) bad multitasking; (b) the student’s syndrome; and (c) the Parkinson’s law.

Questions for Discussion

1. What are the key terminology and principles in the theory of constraint (TOC)? Explain the importance and relevance of each one for a typical construction project.
2. What are the key terminology and principles in Critical Chain Project Management (CCPM)? Explain the importance and relevance of each one for a typical construction project.
3. What are the critical differences in TOC and CCPM approaches? Present your answer in a tabular format.

Web-based Exercise

Study the critical issues in a typical R&D project. Review the details given on the Internet on R&D projects executed by NASA and ISRO. Narrate the complexity involved in such kinds of technologically intensive and mission-critical projects. Evaluate the importance of each of the principles of TOC and CCPM of such types of projects. Develop critical recommendations for the research organizations.

Group Project

Undertake a TOC exercise in a group of three persons. Consider a project such as "Reducing vehicular pollution in Delhi." Think about possible alternatives. Evaluate each of the alternatives for the five steps of TOC.

Chapter 8
Project Time–Cost Trade-Off

Critical Questions

What is the importance of project time–cost trade-off?
What are the key concerns in project time–cost trade-off?
How do we evaluate the trade-off between direct and indirect cost?
What are the key insights a project manager can reveal through time–cost trade-off?
How do we employ a heuristics procedure for conducting time–cost trade-offs in a project?

8.1 Cost Considerations in Project

Project cost management is one of the most crucial areas of project management, particularly project planning or budgeting. This project management domain is extensively discussed and emphasized in the PMBOK Knowledge area, namely *Project Cost Management*. However, the importance of estimating all the different types of costs associated or likely to be associated with the project is not often appreciated by project managers while managing project costs. It has been frequently observed that project managers adopt a very optimistic cost estimation approach and fail to realize all the costs that might incur over the project's life. Such ignorance eventually leads to project failure or at least time and cost overrun. To have very realistic and reliable cost management of any project, listing down all the different costs associated with the project becomes paramount. The project cost is the total fund required to complete the project. In other words, it is the total cost that is likely to incur or estimated to be incurred over the entire lifecycle of the project. Broadly, the entire project cost can be divided into the following discussed groups:

J. J. Thakkar, *Project Management*, Management and Industrial Engineering,
https://doi.org/10.1007/978-981-15-3695-3_8

Direct or Variable Cost: Such costs are directly associated with the project and usually can't be avoided or minimized. They vary with output or scope and duration of the project. In simple terms, direct/variable costs are proportional to each output unit or, say, the scale of the project.

For example, material cost, labor cost, fuel for equipment, sometimes machine cost that performs an explicit function on every unit of output, specialized contractor fees, license fees, etc., are the examples of direct or variable cost.

Indirect or Fixed Cost: Such costs are not directly but indirectly related to the project and are fixed in nature, which means such costs do not vary with the length or unit output of the project. These costs are mostly calculated as lump sum amount or often some percentage of the direct cost of the project. For a given firm/company, these costs may be shared across several projects and are difficult to associate with a single project.

For example, office rentals, systems cost, software license cost, salaries of permanent employees, electricity consumption, stocking the coffee machine, cost of capital equipment, advertisement cost, etc., are examples of indirect or fixed cost.

General & Administrative Cost (G & A): Such costs are generally associated with administration. In some cases, these costs are not treated/listed separately but are usually included in overhead costs. In general, G & A costs are levied in terms of some percentage of direct cost.

For example, accounting, legal, human resource, computers and printers, official meetings, etc., are general and administrative costs.

Overhead Cost: Such costs are not usually related to any particular product or, say, class of products. These are also levied in terms of some fixed percentage of direct cost.

For example, health insurance cost, pension plans for employees, plant securities, operation and maintenance, cost of building, utility cost, etc., are overhead costs.

Sunk Cost: It is incurred by a firm, but the project has not delivered any value. These costs are gone costs, which means costs incurred because of an inappropriate decision-making. The adequate consideration and analysis of sunk costs are necessary to avoid such mistakes in future projects.

For example, a firm buys software to leverage its production output. Subsequently, the team realizes that the software is not fully compatible with their machines. In such a case, the cost incurred in buying the software is considered a sunk cost. Table 8.1 identifies the various cost components for construction projects and R&D projects.

8.2 Time–Cost Trade-Off in Project Management

In today's world of a market-driven economy, the success of any project manager is largely determined by his ability to evaluate the relationship between time and cost associated with the given project. Such a relationship between the associated time and

Table 8.1 Types of costs involved in a project

COSTS	Construction project (road development)	R & D project (software development)
Direct/variable	Land acquisition Planning and feasibility studies Site organization and leveling Soil stabilization Architectural and engineering design Materials, equipment, labor, fuel Subcontractor	Hardware and software Travel and training Effort cost Software engineers cost Subcontractor
Indirect/fixed	Field supervision Construction financing Permit and license Contracting Equipment and furnishings Inspection and testing	Acquiring license Background research Data collection Heating Lighting Office space Electricity Support staff Networking and communications Advertisements and sells
G & A	General office overhead Owner's other expenses Land and building rent Wages to permanent employees Pension plans	Accountants Administrators Cleaners Technicians System managers
Overhead	Insurances and taxes Utilities and facilities Workshops/stores/offices/warehouse Staffs and parking Periodic maintenance Operating staff Site security	Central facilities Pension and health insurance Social security Patent and copyright

cost of the project is nothing but what we call a time–cost trade-off in the language of project management. However, these decisions of making trade-offs are highly complex and require an appropriate approach to do so. The very primary aim of doing time–cost trade-off is to reduce the initially estimated project duration by deploying more resources to various activities to meet a particular deadline considering the least-cost approach.

In a real-world scenario, many times, a project manager faces a situation where he has to significantly reduce the project's initially scheduled completion time to meet some specific deadline. It should be noted that this can be done by deploying additional resources and also manpower to the project. However, these additional deployments of resources and manpower essentially increase the overall cost of the project. The decision regarding reducing the overall duration of a project largely depends on the trade-off between cost and time.

The moment we talk about reducing the scheduled completion time of any project, the concept of *Project Crashing* comes into play. Project crashing can be broadly defined as a widely used method for reducing the scheduled/initial duration of the project by shortening the duration of one or more activities on the critical path of the project network. Crashing any project is fundamentally done by deploying/allocating more resources and manpower to the critical activities to be crashed.

Project crashing is essential for several reasons. There can be multiple reasons for finishing the project in a particular period, such as recovering early costs and delays, freeing up significant resources, avoiding liquidated damages, avoiding adverse weather conditions, improving cash flow, or finishing the project before a specific date. Project crashing may have various implications on project completion, such as reduced scheduled project time, higher project cost than the initially planned, compromise on the quality of project deliverables. Sometimes, it is also possible that the project crashing is not an economical option and can be a recipe for disaster. It should be noted that as the project duration decreases, crashing cost increases. In addition, indirect costs also increase as the project duration increases.

Methods such as crashing enable a project manager to generate alternative cost and time scenarios and re-plan the entire project based on the time–cost trade-off. Consecutively, project crashing also allows the client to evaluate a range of feasible options and can accept a wide range of time-saving alternatives and strategies, and also their cost implications on the project. The key here is knowing the project's critical path. Once the critical path is known, the activities that need to be reduced for crashing can be identified. Apparently, by evaluating the associated costs of shortening the project's critical activities, the project manager would be able to select the most cost-effective crash sequence.

While doing project crashing or trade-offs, the project manager needs to consider various aspects of the project. A typical time–cost trade-off considering direct and indirect costs is presented in Fig. 8.1. The cost (direct/indirect) factor comes into play, but various other factors have to be taken into consideration. The critical path should be crashed one period at a time. Also, only the least expensive activity preferably crashed first. As a thumb rule, as long as the crashing costs of the project are less than the associated indirect costs, project crashing can be done to achieve the desired objectives. However, there is a limit to which crashing can be done. For technological reasons, it is not possible to reduce the duration below the crash limit even by deploying more and more resources. It will certainly increase the project cost without reducing its duration. Apart from these fundamental considerations, a project manager also needs to consider the client's expectations/requirements, availability of additional manpower and resources required for crashing, project priority, organizational goal and objectives, implication on other projects, will and support of upper management, the willingness of project team, and subcontractor's role. These considerations may vary over a range of projects, but the basic concept and prerequisite of project crashing will remain. If done properly and efficiently, project crashing can be a magical ingredient for project success and not to mention the success.

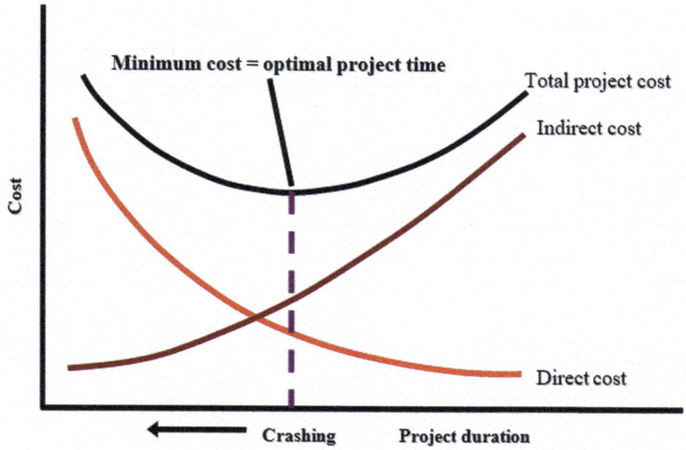

Fig. 8.1 Time–cost trade-off in project. **Direct Cost**: Raw material, labor, manufacturing or processing, travel, communication, transportation, etc. **Indirect Cost**: Managerial services, indirect supplies, equipment rentals, allocation of fixed expenses, site office maintenance, etc.

It is imperative from Fig. 8.1 that:

- Crashing costs increase as project duration decreases
- Indirect costs increase as project duration increases
- Reduce project length as long as crashing costs are less than indirect costs.

Construction projects: Construction projects are very unique in their way while discussing the concept of crashing. Often construction projects are huge, involving huge costs, and hence complex to manage when it comes to crashing. The risk associated with each activity that is being crashed is large, and therefore requires extra care and effort of the project manager. When we look at any construction project, we can indeed find that the cost of each activity and daily cost change due to many uncertainties such as natural events, economical stress, social stress, inflation, execution errors of contractors, and design errors. These uncertainties eventually result in high risk associated with each critical activity while deploying crashing. Hence, the project manager needs to set a certain level of acceptable risk associated with the crashing of each activity.

R&D projects: Unlike construction projects, R&D projects can be very unpredictable. Such projects are undertaken for the first time, and often project managers are not confident enough while doing the time–cost trade-off. Many times it has been observed that the project managers cannot identify all the implications of crashing the activities because of the very nature of the project chosen for crashing and hence fail to visualize the quality of the results. As a result of which quality of the final output is compromised. This is why firms often take utmost care while doing time–cost trade-off for R&D projects because not just the timing but quality of the output also matters equally and cannot be compromised at any cost. Unlike construction projects, the lack of similar project's crashing experience makes it complex for project managers to crash R&D projects. In such projects, time to market is often very crucial, and it may be very beneficial to accelerate the project activities.

Project Management in Practice

Why time–cost trade-offs are crucial for project managers?

The project manager is confronted with having to reduce the scheduled completion time of a project to meet a deadline. Project duration can often be reduced by assigning more labor to project activities, in the form of over time, and by assigning more resources, such as material, equipment, etc. However, the additional labor and resources increase the project cost. So, the decision to reduce the project duration must be based on an analysis of the trade-off between time and cost. Time–cost trade-off decisions are complex and require selection of appropriate construction method for each project task. Time–cost trade-off, in fact, is an important management tool for overcoming one of the critical path method limitations of being unable to bring the project schedule to a specified duration.

Cautions

- Most jobs can be reduced in duration if extra resources (men, machines, money and so on) are assigned to them.
- The cost for getting the job done may increase, but if other advantages outweigh this added cost, the job should be expedited, or crashed
- If there is no reason to shorten a particular job—if it has a generous amount of slack—the job should be done at its normal or most efficient pace, with a lesser assignment of resources
- There is no need to crash all the jobs to get a project done faster; only the critical jobs need be expedited

8.3 Step-By-Step Procedure of the Heuristic Used for Time–Cost Trade-Off

Time–cost trade-off or project crashing simply refers to the process of accelerating the project activities so that the project can be completed sooner than the earlier scheduled duration. It is known that the time required to complete any given project is usually determined by the time associated with its critical activities, so when we go for crashing, we focus on only critical activities along the critical path. The heuristic approach first determines the critical activities along the critical path, then finds out the normal completion time, followed by the crash time for each activity. The crash time is nothing but the shortest possible time under which a critical activity of the project can be completed successfully without compromising the quality. The heuristic approach works under the following assumptions:

1. The crash cost is higher than the normal cost for any activity which can be crashed.
2. Activity time and cost follow a linear relationship.
3. While crashing/shortening the activities, required resources are available.

The project manager identifies all the critical activities associated with the project and proceeds with shortening of activities in order of their lowest cost slopes. Eventually, overall project duration gets reduced as a result of the reduction in duration of activities on the critical path. Finally, a new path becomes critical with different sets of activities having a different cost and schedule. It should be noted that this approach gives a better schedule, but it is not guaranteed that the optimal schedule is achieved.

The following steps are used in the Heuristic solution procedure:

Step 1 Draw the project network

The project network, AOA/AON is drawn with all the activities based on its successor and predecessor relationship. The normal costs and durations for each given activity in the network are mentioned.

Step 2 Identify critical path using CPM calculations

Using the CPM approach, the project's critical path is typically determined and then all the critical activities on the critical path(s) are identified. This is done by considering the normal costs and durations for all the activities in the identified project network.

Step 3 Computation of cost slope

After identification of all the critical activities along the critical path, the cost slope for each of the critical activities is calculated using the following equation.

Step 4 Start shortening the activity duration

$$\text{Cost Slope} = \frac{\text{Crash Cost} - \text{Normal Cost}}{\text{Normal Duration} - \text{Crash Duration}}$$

As soon as the cost slope for each critical activity is calculated, we begin by reducing the duration of each critical activity on the critical path that has not yet been shortened to its total crash duration and has the least-cost slope.

Step 5 Continue shortening the activity duration

Continue reducing the critical path duration along the critical path with the least-cost slope until the original critical path changes or the desired crash duration is achieved. In other words, crash each of the critical activities (cheapest) until any other path than the current path becomes critical or until the crashed activity is fully crashed.

Step 6 Determine the most economical set of activities to be crashed or relaxed

In case of involvement of multiple critical paths, the activity or set of activities that is to be shortened is determined by comparing the activities' cost slope, which lies on all critical paths involved. By this, the most economical set of activities to be crashed or relaxed is determined to reduce the duration of all critical paths. Identification of an economic set of activities is the most crucial step. Each crashing will involve additional costs to the project. The entire concept of crashing is built on reducing the project time economically and acceptably.

Step 7 Adjust activity timings and floats

After shortening the original critical path, one should keep adjusting the activities' timings and floats in successive iteration.

Step 8 Calculate the cost increase

Whenever any activity is crashed, it results in a higher cost than normal. This increased cost can be calculated as the cost slope of that activity by time units shortened during crashing.

Step 9 Continue crashing

In successive iteration, continue until no further shortening is possible, and then the crash point is reached. Moreover, no crash is possible if at least one critical path cannot be further reduced.

Step 10 When two or more activities on any path are simultaneously crashed, some previously crashed activity on the path may be relaxed

Step 11 Graphical representation

The final result/critical path obtained can be represented graphically by plotting cumulative cost increase against project completion time. This is nothing but the depiction of project cost (direct) and time relationship or time–cost trade-off. By adding the indirect project cost to this curve, the resultant curve gives the optimum duration and the corresponding minimum cost. Trade-off analysis thus provides a look at the range of possibilities and assists in picking the best schedule.

8.4 Time–Cost Trade-Off: Illustrative Example 1

A construction company is executing a highway construction project between Bhubaneswar and Delhi. Mr. Deshpande (project in charge) decomposed the work package into a set of activities (P–V). The normal and crash duration and cost for each activity are determined in Table 8.2. All the cost numbers are in multiples of thousand.

Indirect cost varies as given in Table 8.3.

Suggest Mr. Deshpande the best trade-off between direct and indirect costs. If there is a penalty of Rs. 50,000 per day, would you advise Mr. Deshpande to expedite the project?

Solution

Based on the successor and predecessor relationship, the project network is derived in Fig. 8.2. The cost slope calculations are performed in Table 8.4.

Based on the network, the project can be completed through the following paths: PU, PSV, QV, RTV. The associated costs are given in Table 8.5.

Computations

Project Crashing

Using the heuristic solution procedure to determine the time–cost trade-off, the following results are obtained. (Note: Numbers in bold in each itteration indicate the duration of critical path.)

Iteration 1 (as given in Table 8.6*)*

The critical activity (on critical path PSV) with the least-cost slope is crashed first. In the next iteration, RTV becomes the critical path.

The successive iterations are performed as shown in Table 8.7.

Project Total Cost (as given in Table 8.8)

Table 8.2 Normal and crash time and costs

Activity	Immediate predecessor(s)	Time		Direct cost (Rs. in thousands)	
		Normal	Crash	Normal	Crash
P	–	5	2	80	140
Q	–	6	3	130	280
R	–	4	2	50	90
S	P	5	3	130	240
T	R	5	3	120	180
U	P	9	5	140	200
V	S, Q, T	6	4	100	180

Table 8.3 Indirect cost

Months	15	14	13	12	11	10	9	8	7	6
Cost (Rs.)	1200	900	500	450	400	350	200	180	90	60

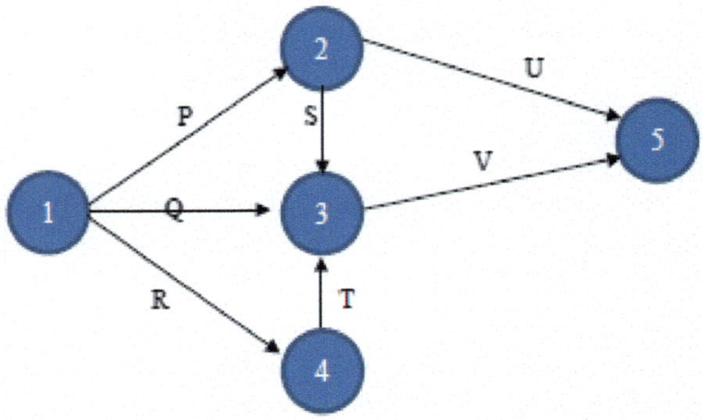

Fig. 8.2 Network representation of highway construction project

Table 8.4 Cost slope calculations

Activity	Normal (*t*)	Crash (*t*)	Normal (*c*)	Crash (*c*)	Cost slope
P	5	2	80	140	20
Q	6	3	130	280	50
R	4	2	50	90	20
S	5	3	130	240	55
T	5	3	120	180	30
U	9	5	140	200	15
V	6	4	100	180	40

Table 8.5 Project completion paths and associated costs

Paths	P	Q	R	S	T	U	V
PU	20					15	
PSV	20			55			40
QV		50					40
RTV			20		30		40

Table 8.6 Time–cost trade-off—iteration 1

PATHS	P	Q	R	S	T	U	V	1	2
PU	20					15		14	14
PSV	20			55			40	**16**	13
QV		50					40	12	12
RTV			20		30		40	15	15
1	3	3	2	2	2	4	2		
2	0	3	2	2	2	4	2		

Table 8.7 Time–cost trade-off for successive iterations (iterations 1–7)

PATHS	P	Q	R	S	T	U	V	1	2	3	4	5	6	7
PU	20					15		14	14	**14**	10	10	**10**	**10**
PSV	20			55			40	**16**	13	13	**13**	**11**	**10**	9
QV		50					40	12	12	12	12	10	**10**	9
RTV			20		30		40	15	**15**	13	**13**	**11**	**10**	9
1	3	3	2	2	2	4	2							
2	0	3	2	2	2	4	2							
3	0	3	0	2	2	4	2							
4	0	3	0	2	2	0	2							
5	0	3	0	2	2	0	0							
6	0	3	0	1	1	0	0							
7	0	2	0	0	0	0	0							

Table 8.8 Project total cost calculation

TIME (months)	Direct cost (in thousands)	Indirect cost (in thousands)	Total cost (in thousands)
16	750	Assuming 1200 (as not given in the question)	1950
15	810	1200	2010
14	850	900	1750
13	910	500	1410
12	950	450	1400
11	990	400	1390
10	1075	350	1425

The best possible trade-off for Mr. Deshpande is to complete the project in 11 months with a total cost of 1390 thousand. The trade-off is highlighted in the above table.

$$\text{Total saving} = 1{,}950{,}000 - 1{,}390{,}000 = 560{,}000$$

If there is a penalty of 50,000 per day, then the total penalty paid if the project could not be completed in the crashed time will be 750,000, i.e., (16–11) $\times$ 50,000 $\times$ 30 = 750,000.

This is significantly higher than the total saving. Hence, it would be a risky proposition to do the expedition and hence is not advisable.

Table 8.9 Mangalore infrastructure power plan project details

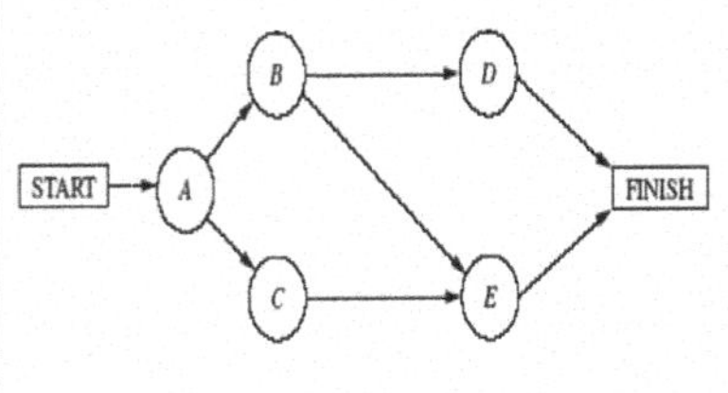

Activity	Normal time (in weeks)	Crash time (in weeks)	Normal cost (in Rs.)	Crash cost (in Rs.)
A	6	3	580,000	735,000
B	7	4	750,000	1,025,000
C	7	4	450,000	600,000
D	6	4	575,000	700,000
E	5	3	300,000	500,000

8.5 Time–Cost Trade-Off: Illustrative Example 2

Mangalore Infrastructure Ltd. is about to begin the power plant project. The company's President, Mr. Ghosh, is currently planning the schedule for this project. Mr. Ghosh has identified the five major activities (labeled A, B, …, E) that will need to be performed according to the following project network. He gathered the normal and crash, time and cost data as given in Table 8.9.

These costs reflect the company's direct costs for the material, equipment, and direct labor required to perform the activities. In addition, the company incurs indirect project costs such as supervision and other customary overhead costs, interest charges for capital tied up, and so forth. Mr. Ghosh estimates that these indirect costs run Rs. 100,000 per week.

He wants to minimize the overall cost of the project. Therefore, to save some of these indirect costs, Mr. Ghosh concludes that he should shorten the project by crashing to the extent that the crashing cost for each additional week saved is less than Rs. 100,000.

Task A: To prepare for analyzing the effect of crashing, find the earliest times, latest times, and slack for each activity when they are done in the normal way. Also, identify the corresponding critical path(s) and project duration.

Task B: Use marginal cost analysis to determine which activities should be crashed and how much to minimize the project's overall cost. Under this plan, what is the duration and cost of each activity? How much money is saved by doing this crashing?

Solution

Task A: Slack calculations for determining the critical path is given in Table 8.10.

Let,

ET Earliest Time
LT Latest Time

The critical path will be ABD with a project duration of 19 weeks.

***Task B: Cost slopes are calculated as given in* Table 8.11.**

Table 8.10 Slack calculations

Activity	ET	LT	Slack
A	0	0	0
B	6	6	0
C	13	14	1
D	13	13	0
E	19	19	0

Table 8.11 Cost slope calculations

Activity	Normal (*T*)	Crash (*T*)	Normal (*C*)	Crash (*C*)	Cost slope
A	6	3	580,000	735,000	51,667
B	7	4	750,000	1,025,000	91,667
C	7	4	450,000	600,000	50,000
D	6	4	575,000	700,000	62,500
E	5	3	300,000	500,000	100,000

Based on the network, the project can be completed through the following paths: ABD, ABE, ACE. The associated costs are given in Table 8.12.

Computations

Using marginal cost method—A step-by-step heuristic procedure explained in Sect. 8.3 is implemented and the various iterations for time–cost trade-offs are given in Table 8.13.

Activity E cannot be crashed. The crashing cost for each additional week saved should be less than 100,000 as assumed by Mr. Monideep.

Since the most desirable trade-off is with a 16-week time as shown in the next table. Hence, only activity A will be crashed completely. It can also be observed that when we crash the project for more than 15 weeks, the crashing cost per week becomes more than 1 lakh, which is not desirable.

Under this plan, the duration of activities A, B, C, D, and E will be 3, 7, 7, 6, and 5 respectively. Also, cost associated with each activity will be 735,000, 750,000, 450,000, 575,000, and 300,000, respectively.

Project Total Cost (as given in Table 8.14).

Table 8.12 Project completion paths and associated costs

Paths	A	B	C	D	E
ABD	51,667	91,667		62,500	
ABE	51,667	91,667			100,000
ACE	51,667		50,000		100,000

Table 8.13 Heuristics-based iterations for project time–cost trade-offs

Paths	A	B	C	D	E	1	2	3	4	5	6
ABD	51,667	91,667		62,500		**19**	**16**	14	**14**	**13**	11
ABE	51,667	91,667			100,000	18	15	14	**14**	**13**	**12**
ACE	51,667		50,000		100,000	18	15	**15**	**14**	**13**	**12**
1	3	3	3	2	2						
2	0	3	3	2	2						
3	0	2	3	1	2						
4	0	2	2	1	2						
5	0	1	1	1	2						
6	0	0	0	0	2						

Table 8.14 Total cost calculations

Time	Direct cost	Indirect cost	Total cost
19	2,655,000	1,900,000	4,555,000
16	2,810,001	1,600,000	4,410,001
15	2,935,001	1,500,000	4,435,001
14	Crash cost exceeds 1 lakh	1,400,000	Not acceptable
13	Crash cost exceeds 1 lakh	1,300,000	Not acceptable
12	Crash cost exceeds 1 lakh	1,200,000	Not acceptable

Initial project cost = 4,555,000
Crashed project cost = 4,410,001
Money saved = 4,555,000 – 4,410,001 = 144,999

Project Management in Practice

Time–Cost Trade-off in Road Development Project

A road construction agency is planning to develop a fourteen-lane expressway between New Delhi and Ahmedabad. The project has been sanctioned by the government with an estimated cost of 164,000 crores. The completion time for the project is estimated to be seven years. However, the government wants this project to be completed in a maximum of six years. Hence, the ministry issues notification to the agency management to complete the project in 6 years.

The Challenge

Because of the change order, the agency's management faces a considerable challenge to reduce the project completion time by at least one year. Time–cost trade-off has to be made for project accomplishment. The entire project is comprised of various activities, which require certain time and cost. Each activity will consume some amount of resources and time to complete. Initially, the project managers broke down the entire project into a list of activities and estimated the time and cost for each activity (presented in Exhibit 8.1). Based on this estimation, the initial budget of 164,000 crores was sanctioned for seven years. Now, to meet the new deadline, the project managers will have to make some sort of trade-off between time and cost. The project managers will have to reduce the planned execution time of activities using extra equipment and more manpower, which will undoubtedly add to the overall project cost. Hence, to expedite the execution of the project, an additional cost will have to be incurred. This will involve shortening the completion time of activities on the critical path network by incurring additional cost. The challenge for project managers is to find the most economical set of activities that can be crashed with minimum extra cost to complete the project within the given deadline period (see Exhibit 8.2).

The above 80,000 crores constitute the direct costs involved in the project, and apart from that, 12,000 crores per year are the estimated indirect costs for the project. The total project cost for seven years becomes 164,000 crores. Now the challenge for the project managers involved in the project is to complete the project in 6 years on the same budget by doing a time–cost trade-off. This will essentially require the project to crash, which will increase the direct cost, but the indirect cost will be reduced because the project is getting completed before one year of the scheduled time. So, delivering the project in 6 years will eventually save a direct cost of 12,000 crores. The project managers can use this saving to allocate extra resources to the critical activities.

Exhibit 8.1 Associated cost and time for various activities involved in the project

Activity	Code	Normal time (in months)	Normal cost (in crores)
Mobilization and setup	A	01	150
Clearing and grubbing	B	03	4000
Earthmoving	C	12	20,000
Drainage	D	06	6000
Noise barriers	E	02	1000
Granular base	F	08	1000
Sub-drain installation	G	03	2000
Electrical conduits	H	02	1000
High mast lighting	I	02	4800
Paving	J	19	28,000
Guardrails	K	04	3000
Landscaping	L	15	8000
Installation of signs	M	05	500
Lane marking	N	01	500
Checkout and Approval	O	01	50
Total		84 months (7 years)	80,000 crores

Project Crashing

In the construction industry, such as constructing huge infrastructure projects such as the one discussed above, project managers are often bound to the triple constraints of the project, i.e., scope, schedule, and cost. These three criteria are essentially interrelated as well as interdependent. So, it becomes imperative for any project manager to understand the projects' triple constraint properly and know how to handle these in balance with each other. As these constraints are nothing but a measure of the project's success, the project managers need to put some extra effort into maintaining these criteria. A successful project manager knows how to do a trade-off between these three criteria to deliver a quality project successfully. However, out of the three constraints, time or schedule is the most critical element because it would permanently affect the project's cost. This is essential because the longer the time taken by the project is, the higher the cost of the project will be.

If the scheduled project completion time is reduced by one year, the project managers cannot afford to incur any sort of project delays; they will have to resort to every option that they have in hand to deliver the project on time. To deliver the project before one year of the scheduled time, one of the project managers' options is crashing the project. In a nutshell, crashing is nothing but a method of strategically reducing the project's duration by deploying additional resources and manpower to a particular set of activities. No doubt that project crashing is essentially a handy tool to reduce the total project completion time.

However, it is not a simple task to do the project crashing, and it is crucial to know and understand how to handle and effectively manage the project crashing process. It is important to understand as a project manager that project crashing only works with the crashing of critical activities along the critical path; it will be worthless to crash any non-critical activity. The method of crashing intends to shorten the critical path. Any project manager should know the strategies to crash a project successfully.

The project managers responsible for delivering the project also need to consider the challenges that they may face while doing the project crashing. As per the very nature of the triple constraints, by shortening the project's duration by 1 year, the project's cost will be increased. This is very obvious as to complete the critical activities sooner than the initially assigned duration; more resources need to be assigned. This increase in one of the triple constraints will eventually affect the other constraints. This will overall affect the project quality. It is very much possible that depending on the project crashing, the project quality might get affected. It should also be noted that saving indirect costs by delivering the project earlier is the only money that needs to be utilized for crashing; the crashing cost should not exceed this 12,000 crores of saved indirect cost.

Heuristic Adopted for Crashing

The project manager identifies all the critical activities on the critical path and proceeds with shortening activities in order of their lowest cost slopes. Eventually, the overall project duration was reduced due to a reduction in the duration of activities on the critical path. Finally, a new path becomes critical with different sets of activities having different costs and schedules. It should be noted that this approach gives a better schedule, but it is not guaranteed that the optimal schedule is achieved. Crashing costs increase as project duration decreases. The project manager tries to reduce the project length as long as crashing costs are less than the indirect costs.

The project network for the predecessor–successor relationships (as given in Exhibit 8.2) is presented in Exhibit 8.3.

Exhibit 8.2 Normal and crash time–cost

Activity	Predecessor	Normal time	Crash time	Normal cost	Crash cost	Cost slope
A	–	01	01	150	150	Infinity
B	A	03	03	4000	4000	Infinity
C	B	12	10	20,000	24,000	2000
D	C	06	05	6000	8000	2000
E	D	02	02	1000	1000	Infinity
F	E	08	06	1000	1600	300
G	F	03	03	2000	2000	Infinity
H	–	02	01	1000	1200	200
I	H	02	01	4800	5000	200
J	I	19	12	28,000	35,000	1000
K	J	04	04	3000	3000	Infinity
L	K	15	13	8000	14,000	3000
M	L	05	03	500	800	150
N	M	01	01	500	500	Infinity
O	N	01	01	50	50	Infinity

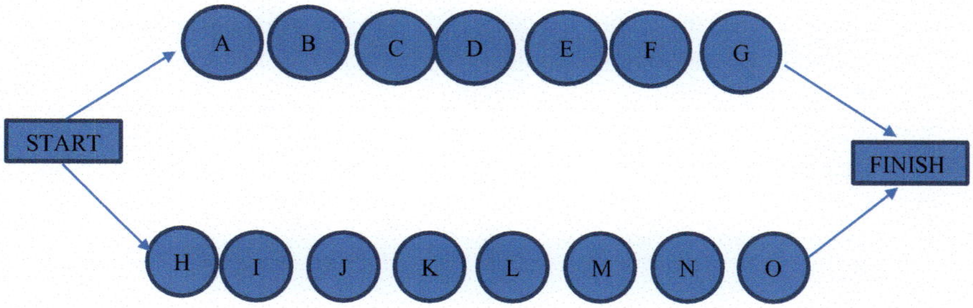

Exhibit 8.3 Project network

A tentative network with a preliminary understanding of activity relationships is developed as above. The activities are visualized as divided into two paths connecting source (start) and sink (finish). Activities A, B, C, D, E, F, G can be considered connected in one direction from A to G and set one path from source to sink. Similarly, H, I, J, K, L, M, N, O sets another path from source to sink. Real-life networks may involve more complex successor–predecessor relationships when considering the complete portfolio of activities and involvement of various agencies/subcontractors. The critical path for the project is H-I-JK-L-M-N-O, with a duration of 49 months. The project managers decided to crash the activities on this critical path to reduce the overall project time by 12 months. Also, the crash cost should not exceed 12,000 crores. There is a huge perceived benefit of delivering a project one year early if the crash cost does not exceed 12,000 crores.

In the critical path H-I-J-K-L-M-N-O, the activities K, N, and O cannot be crashed. Activity M has the least-cost slope so that it will be crashed first. Then, the activities H and I will be crashed. Then, J will be fully crashed, and finally, activity L will be crashed by one month.

This will give the total crash cost = 300 + 200 + 200 + 7000 + 3000 = **10,700 crores.**

It is expected that the crash cost should not exceed 12,000 crores. In this regard, Total saving = 12,000–10,700 = 1300 crores.

The project can be successfully crashed in 12 months with savings of 1300 crores in crash cost. Using a time–cost trade-off, the project managers planned a new schedule and budget to deliver the same project within six years without incurring any additional cost. They were even managed to save an estimated sum of 1300 crores if they go by the new project plan.

Key Insights

The main objective of time–cost trade-off or crashing of the project is to reach a maximum possible decrease in the scheduled completion time of the project for a minimum additional cost. The various methods to accomplish this include early procurement, extra supervision, working longer hours, deploying machinery/equipment, hiring more men, changing the design or specification, fast-tracking (in case of overlapping activities), adopting different techniques/technologies, etc. Many of these strategies adopted for project crashing will eventually lead to incurring additional costs to the project, which may be called cost uncertainty. In project crashing, the number of tasks/activities needed to be executed remains the same; they are in a way just compressed into a shorter duration, hence are likely to require additional manpower and resources. In addition to such complexities, there can be other complexities such as purchasing costs might go higher because of time pressure, incomplete information, and also the complexity of managing the interfaces between various elements.

When doing project crashing, particularly for a large infrastructure project involving huge costs, several risks are attached to it. While resources are primarily oriented toward the project's critical activities, still possibilities are there that the non-critical activities/paths would also get affected. A critical path crash should never compromise compliance, quality, or safety. Apart from that, the project managers often ignore another risk because new resources may not be as productive as existing resources because they may be unfamiliar with the project, the program, and the tasks at hand. This points to a critical aspect of project crashing; the crashing becomes difficult and risky when performed in the middle of the project. In the discussed case study of the construction of the Delhi–Ahmedabad Expressway, the crashing was done very effectively with additional cost-saving because it was performed before the project was started. The project crashing is like a magical tool; if done correctly, it can do wonders, as evident from the case study.

Question for Class Discussion

1. How do you appreciate the importance of the time–cost trade-off for the case organization?
2. Do you think that information considered by the case organization for the Time–Cost trade-off is adequate? If not, what is the additional information that can help the case organization plan the project?
3. What are the key managerial implications for the case organization if they utilize the time–cost trade-off in project scheduling?

Summary

Importance of estimation of all the different types of costs associated or likely to be associated with the project is not often appreciated by project managers while managing project costs.

It has been often observed that project managers adopt a very optimistic cost estimation approach and fail to realize all the costs that might incur over the life of the project. Such ignorance eventually leads to project failure or at least time and cost overrun.

Methods such as crashing enable a project manager to generate alternative cost and time scenarios, and re-plan the entire project based on the time–cost trade-off.

Project crashing allows the client to evaluate a range of feasible options and can accept a wide range of time-saving alternatives and strategies, and also their cost implications on the project.

Questions for Discussion

1. What are the different types of cost (direct/indirect) in the project? Give at least five examples of each one. Consider two projects—Construction Project and R&D project—identify the various types of costs involved. Classify them properly.
2. What is a time–cost trade-off in project management, or What is Project Crashing? Why is it important? What implications may it have on project completion? What are the key considerations in the time–cost trade-off in project management? Explain the severity of such trade-offs for construction and R&D projects.
3. Explain the step-by-step procedure of the heuristic used for the time–cost trade-off.
4. A construction company is involved in building 5 BHK Luxury Apartments in Baroda. Mr. Himanshu is a project CEO with 30 years of experience in the construction industry. He had broken down the work package into activities (A–G). The normal and crash duration and cost for each activity are determined as below. All the cost numbers are in multiples of thousand.

Activity	Immediate predecessor(s)	Time		Direct cost (Rs. in thousands)	
		Normal	Crash	Normal	Crash
A	–	6	3	180	190
B	–	7	5	170	380
B	–	5	3	80	120
D	P	5	3	230	350
E	R	6	4	320	280
F	P	10	6	140	200
G	S, Q, T	6	4	200	380

Indirect Cost Varies as Follows

Months	15	14	13	12	11	10	9	8	7	6
Cost (Rs.)	1400	1000	600	550	600	450	500	280	190	90

Suggest Mr. Himanshu the best trade-off between direct and indirect costs. If there is a penalty of Rs. 90,000 per day. Would you advise Mr. Himanshu to expedite the project?

Web-Based Exercise

Study the project execution issues in the Central Public Works Department in India. Enlist the various project implementation challenges from a planning and organizational point of view. How do you appreciate the use of time–cost trade-offs in the timely execution and implementation of various projects by such an organization?

Chapter 9
Project Resource Allocation

Critical Questions
What is the importance of resource in project?
What is the different between resource allocation and leveling?
What is the limited resource problem?
What are the heuristics used in allocation of resources?
What is the well-established Weist's procedure for resource leveling?
How do we allocate resources for tasks with splitting permitted and tasks with splitting not permitted?

9.1 Importance of Resource Allocation and Leveling in Project Management

In project management, the scheduling of activities and the resources required by them, considering their availability, project duration, and the organization's strategic goals, is defined as resource allocation. It allocates fixed resources to all the available activities to determine the actions that can be delayed in case of resource unavailability. It helps to determine the shortest critical path based upon the available resources.

Project Management in Practice
What are the real-life issues with project scheduling?

J. J. Thakkar, *Project Management*, Management and Industrial Engineering,
https://doi.org/10.1007/978-981-15-3695-3_9

- A shortcoming of the scheduling procedures is that they do not address resource utilization and availability issues. The focus is on time rather than physical resources.
- It will not be sufficient to refer to resource usage simply as "costs." Instead, we must refer to individual types of labor, specific facilities, kinds of materials, individual pieces of equipment, and other discrete inputs that are relevant to an individual project but are limited in availability.
- We must not forget that time itself is always a critical resource in project management, one that is unique because it can neither be inventoried nor renewed.

Caution: *No system should be loaded beyond 85% for a long !*

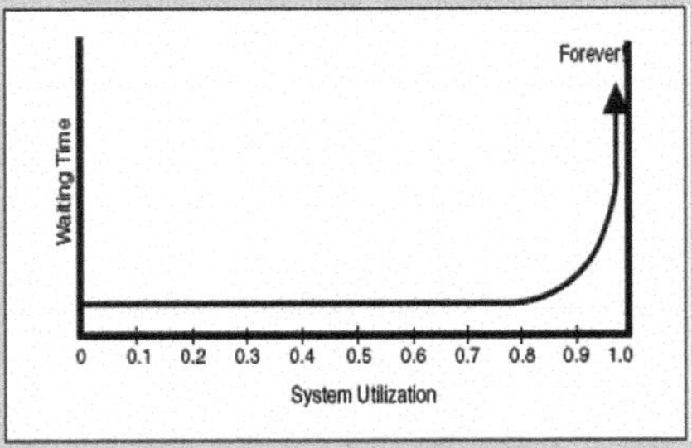

The advantages of resource allocation are:

- It gives a clear picture of the amount of work to be done to complete the project.
- It helps in the better organization of the project.
- It increases the effective use of resources available across the company to maximize their utility.
- It helps in better tracking and rescheduling of the resources.

The following important issue after the allocation of resources is resource leveling. It is defined as the smoothening of the peaks and valleys of the resource allocation chart to ensure a relatively constant level of resource throughout the project. Resource leveling helps in the identification and prevention of imbalance in resource allocation and resource over time. It is essential to evaluate a trade-off between the project completion time and the effective utilization of resources.

The idea behind resource leveling is to reduce wastage of resources, i.e., to stop over-allocation of resources. The project manager will identify unused time by a resource and take measures to prevent it or make an advantage out of it. The benefits of resource leveling are:

- It prevents delay because of inadequate allocation of resources.
- It makes the project manager aware of the slack available and helps him plan the utilization of this slack.

- It ensures that there is no over-allocation of resources when there is a limited resource allocation.
- The number of disruptions harming the schedule is reduced.

Resource allocation and resource leveling have severe implications for the execution of the project.

- Proper resource allocation and leveling help the organization in promising a realistic completion date of the project.
- Non-compliance to resource allocation and leveling may lead to tarnishing of the image of the company.
- It ensures optimal utilization and minimal wastage of resources.
- If resource allocation and leveling are not done adequately, then the project has a high probability of going over budget because of the uncertainty associated with the cost of resources.

Let us consider the case of a construction company. The importance of resource allocation can be appreciated as follows:

- The growing number of projects: Contractors are taking on more and more projects due to increased residential and non-residential construction projects. It leads to severe resource allocation and management problems.
- Shortage of skilled labor: The significant rise in demand for skilled labor creates a shortage of such resources.
- Problem due to poor scheduling: Overscheduling of workers leads to high overtime costs for the company, and the efficiency of the workers can go down because of increased physical and psychological stresses.
- Materials not delivered on time: Proper resource planning is necessary to ensure that the right resources are made available at the right time. Typically, this is essential for the context of India as most of the construction sites do not have a good storage area. Thus, an early delivery leads to high storage cost, and a late delivery leads to a delay in project completion (Table 9.1).

A typical construction company encountered the resource leveling issue in the ongoing project. The project network of this company is presented in Fig. 9.1.

Table 9.1 "Resource loading" v/s "resource leveling"

Resource loading	Resource leveling
• It identifies project resource requirements over the project life cycle • The objective of resource loading is to minimize the duration of the project being scheduled given the stated constraints on available resources	• The leveling process ensures that demand for resources does not exceed available resources at a particular time and is almost constant throughout the project's lifetime • Resource leveling may lead to a trade-off between timely completion of the project and the smooth use of resources throughout the lifetime of the project

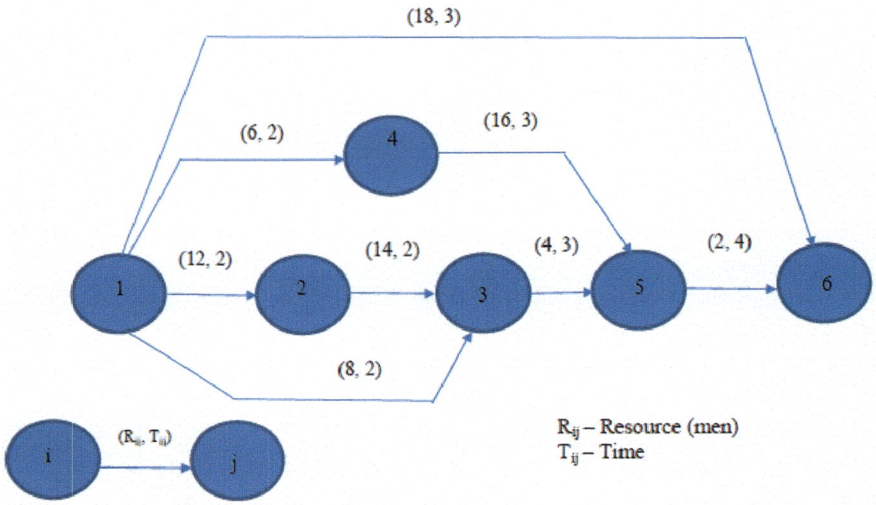

Fig. 9.1 Project network

The numbers in the bracket indicate "resource (men)" (first number) required for executing an activity and "time required" (second number) to complete the activity. A flowchart showing the order of activities and the total resources required per day is shown in Fig. 9.2. As seen in Fig. 9.2, there is an uneven distribution of resources. The resource allocation is followed by resource leveling, as shown in Fig. 9.3. A uniform resource consumption profile can be obtained by rescheduling the activities without compromising with the predecessor–successor relationship. This helps the

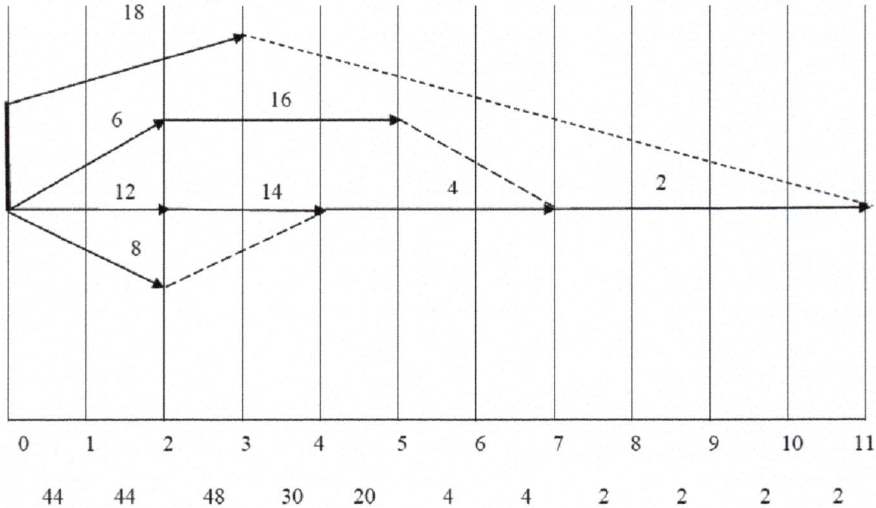

Fig. 9.2 Resource profile

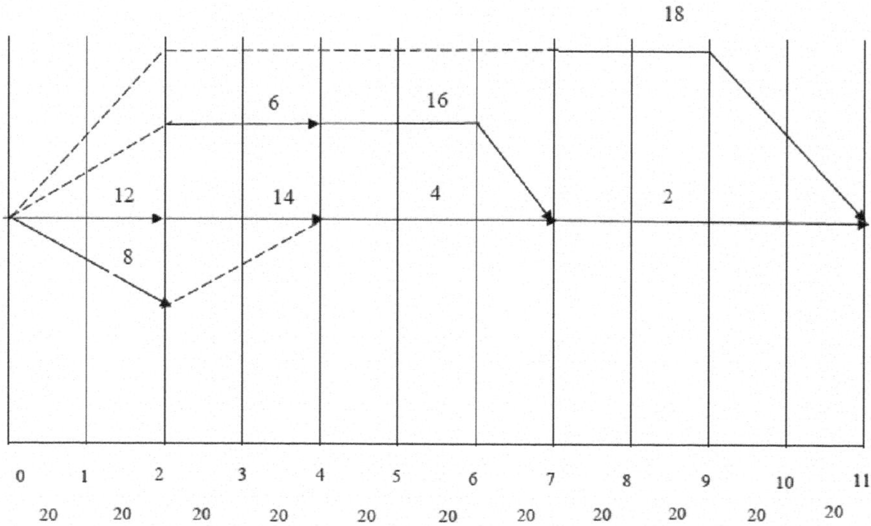

Fig. 9.3 Resource leveling

company retain skilled workers for a longer time and avoids the hassle of hiring and firing.

Project Management in Practice

What are the issues in network scheduling with limited resources?

- Project managers face the problem of relatively fixed manpower availability, a certain number of machines or other pieces of equipment, and considering money as a resource—a limited budget.
- Jobs that occur on parallel paths through the network may compete for the same resources, and even though precedence constraints would not prevent they're being scheduled simultaneously, a limited supply of resources might force them to be organized sequentially.
- Job start times are constrained not only by precedence relationships but also by resource availability.
- How does one decide which subset of jobs to schedule on a given day when the whole set of jobs that could start, as far as technological constraints are concerned, would require more than the available resources?
- Problems of resource scheduling vary in kind and severity, depending upon the nature of the project and its organizational setting.
- In some cases, there may be just one essential resource—perhaps a large crane or a test facility—that is the bottleneck in scheduling a project.

- Activities must be scheduled so that no two of them requiring the same facility occur at the same time.
- Scheduling projects with limited resources is a type of problem that mathematicians refer to as a significant combinatorial problem.

Project Scheduling Approaches

Ordinary PERT and CPM	No explicit consideration of resources is made
Resource scheduling program	The program calculates a total resource requirements profile for each resource by summing, period by period, requirements for that resource of all activities as they occur in an early start (or any other) schedule. No resource leveling is involved
Resource leveling program	It uses the resource requirements profile of the early start schedule. The program attempts to reduce peak requirements by shifting slack jobs to non-peak periods. Resource limits are not specified, but peak requirements are leveled as much as possible without delaying the selected due date
Resource allocation program	According to specific scheduling heuristics, fixed amounts of resources are allocated to available jobs that determine which jobs will be postponed if total requirements for a given period exceed the resources available. The completion date of the project may be pushed ahead to keep within specified resource limits

Time-Limited v/s Resource-Limited Tradeoff

- **Time-Limited**: The project must be finished by a particular time, using as few resources as possible. But, it is time, not resource usage, that is critical
- **Resource-Limited**: The project must be finished as soon as possible, but without exceeding some specific level of resource usage or some general resource constraint.

9.2 Types of Resources in Project

There are **two** different types of project resources:

1. **Non-Renewable Resources**

The total resource consumption over the entire project is of concern. As shown in Fig. 9.4, the total amount of resources required for the period "*t*" are the typical "non-renewable resources" needed for the project. The availability of these resources dictates if the project can even be started.

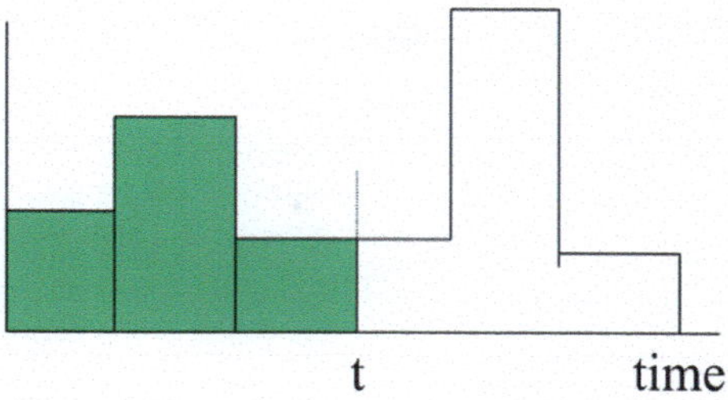

Fig. 9.4 Non-renewable resources

Examples of Non-Renewable Resources

Money: The total capital cost for the entire project is of concern as it is an essential indicator of the project's feasibility.

Energy: The energy requirements and the technology to provide that amount of energy are of concern.

Raw Materials: Proper and timely availability of inputs directly impacts the costs of a project.

2. **Renewable Resources**

The total resource usage at each point of time during the project is of concern. As shown in Fig. 9.5, the number of resources required at period "*t*" is the project's typical "renewable resources." This type of resource can be changed according to the demand.

Examples of Renewable Resources

Manpower: Manpower can be increased or decreased anytime according to the needs, or they can also be outsourced.

Machines: Machinery can be increased or decreased according to the order, demand, or nature of the project at hand.

The need for various resources for a typical infrastructure project like "construction of a building" is summarized in Table 9.2.

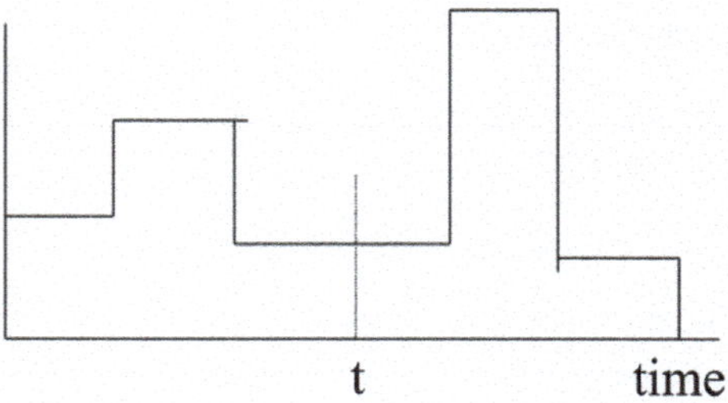

Fig. 9.5 Renewable resources

Table 9.2 Infrastructure project: construction of a building

S. No.	Resource	Significance
1	Money	The capital required to construct the building is essential for the feasibility of the project. If the cost escalation is too high, it is better to drop the project and invest it elsewhere
2	Energy	The building area should have adequate availability of energy supply to ensure timely completion of the project
3	Raw materials	The raw materials required to construct the building should be available readily or at least should be made available through a robust supply chain. This ensures timely completion of a project and meeting the stakeholders' requirements
4	Man power	The necessary skilled and unskilled manpower is required for the effective utilization of capital and material
5	Machines	The use of efficient technology reduces costs, power consumption and improves productivity

9.3 Heuristics Used in Resource Leveling

As Soon as Possible

The purpose of this heuristic is to start the activities as early as possible to avoid delay because of any uncertainty or unforeseen events in the later stage. The objective is to achieve the most effective utilization of time, people, quality, and other resources. This is used as a default rule for project scheduling to provide a general solution for critical path and time.

As Late as Possible

This heuristic starts the activities as late as possible and helps to defer the cash flows. It places the completion time of the task as close to the end of the project as possible.

Shortest Task First

It accommodates the task shortest in duration first. It minimizes the average amount of time each process has to wait until its execution is complete. However, it can cause starvation for the processes that require a long time to complete.

Most Resources First

The task which requires the most resources is completed first to ensure there is no slack available in the later stages.

Least Resources First

The task which requires the least amount of resources is completed first so that there is no time wasted on waiting for materials to continue with the next operation.

9.4 Weist's Heuristics for Resource Leveling

- The resource usage profile shows peaks and valleys as resource consumption varies throughout the project execution.
- The resource leveling aims to move the peaks by dragging the jobs with available slack to a non-peak period and intend to level or evenly allocate the resources throughout the project.
- A change in the project completion deadline makes the resources leveling process difficult.
- Under resource constraint situations, a project manager needs to extend the project completion deadline while leveling the resources.

An illustrative example is presented to appreciate the application of Weist's heuristic for resource leveling. A network considered for this purpose is given in Fig. 9.6. Figures 9.7, 9.8, 9.9, and 9.10 represent the network and resource usage profile with leveling and without leveling.

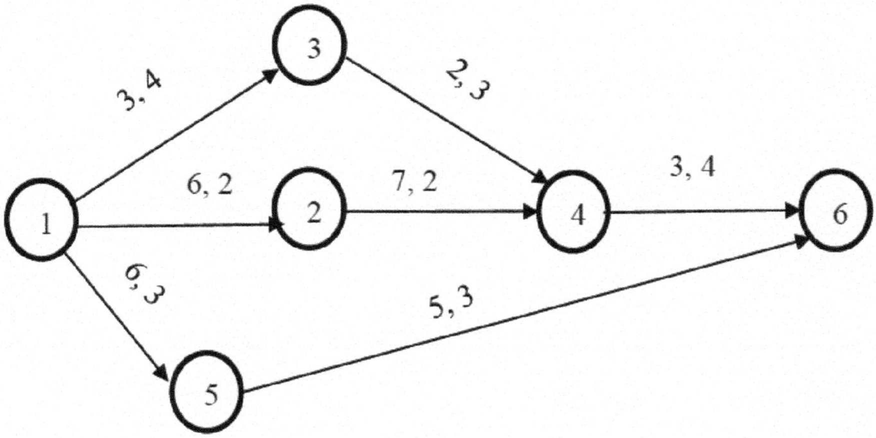

Fig. 9.6 Project network for resource leveling

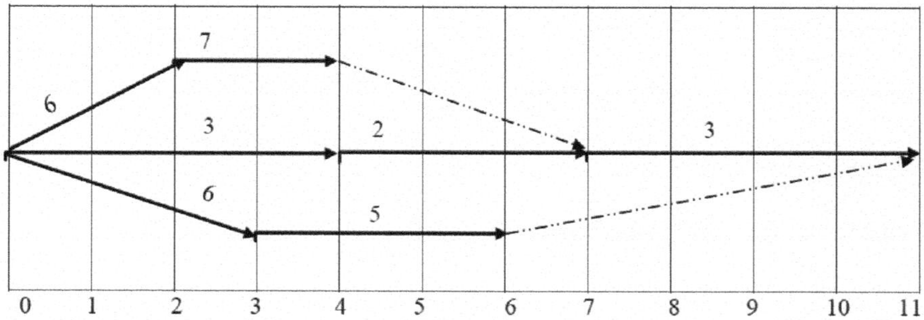

Fig. 9.7 Project network without leveling

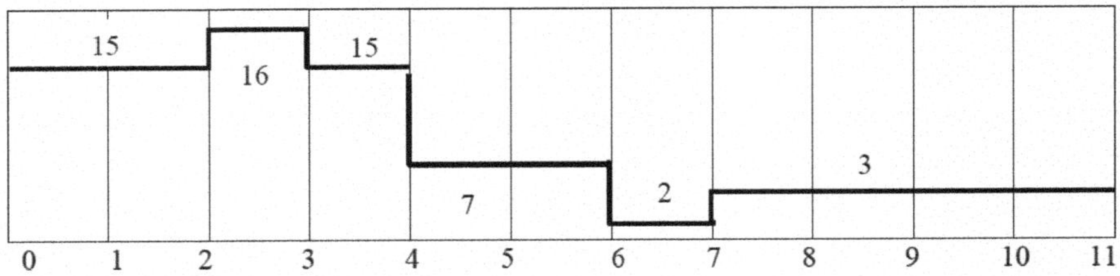

Fig. 9.8 Resource usage without leveling

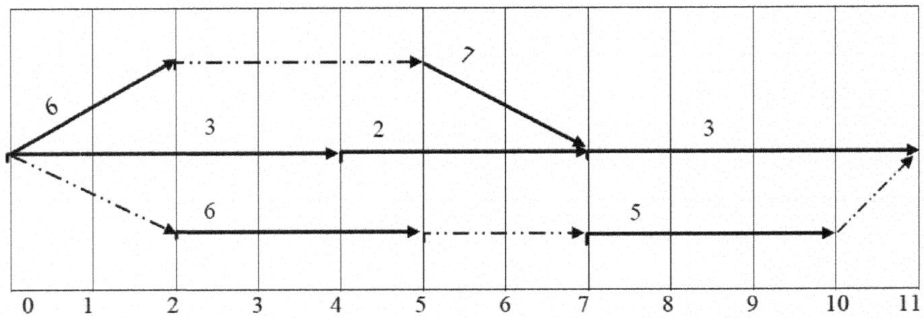

Fig. 9.9 Project network with leveling

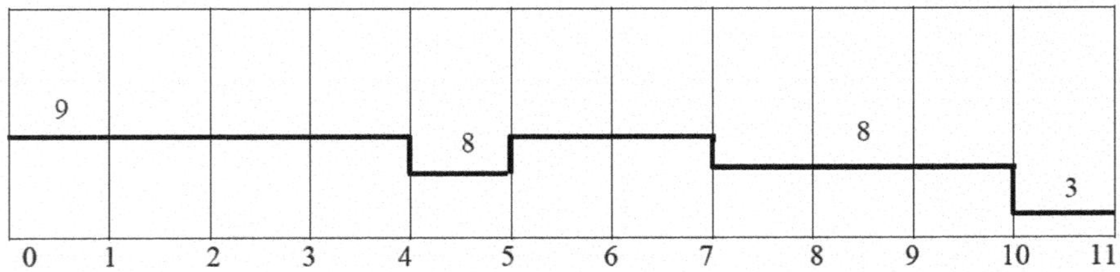

Fig. 9.10 Resource usage with leveling

Here, 6, 2 denote 6 (resource) and 2 (time taken).
The dotted line in Figs. 9.7 and 9.9 represents the available slack.

Network (Without Leveling)

Resource Usage (Without Leveling)

Network (After Leveling)

Resource Usage (After Leveling)

9.5 Resource Leveling: Illustrative Example

A company is executing a typical construction project of building a railway station. The project is described in the table below in terms of activities, timing, and manpower needed.

Job	Predecessors	Duration (days)	Manpower needed
A	–	6	9
B	A	8	8
C	A	3	6
D	B	5	7
E	C	7	5
F	D, E	6	10
G	C	9	6

Your Tasks

Task A: What is the minimum time to complete assuming an unlimited supply of manpower?

Task B: If only 15 men were available, what is the minimum bound on the duration? Using Wiest's heuristics for resource allocation, determine the minimum duration schedule for the project.

Task C: Formulate the resource leveling problem for this project as an integer linear program assuming (a) no job splitting and (b) job splitting permissible.

Task A

The critical path (the longest path) of the project is A-B-D-F. Assuming unlimited resources, we can complete the project in (6 + 8 + 5 + 6) 25 days minimum (Fig. 9.11).

Task B

If we have to satisfy the resource (number of men available) constraint (which is 15 in this case), we need to reschedule the activities and carry out resource leveling (Fig. 9.12).

Day 1: Activity "A" is going on, which requires nine resources. As the resource requirement is nine which is less than the available resources 15, there is no violation of resource constraint. Here, $9 < 15$, and hence, an activity can be accommodated.

Day 7: Activity "B" and "C" are going on, requiring $8 + 6 = 14$ resources. This is less than the resources available, i.e., 15, and hence, both the activities can be scheduled. Similarly, the analysis for the other activities is carried out, and the decisions regarding scheduling or postponing the activity are necessary.

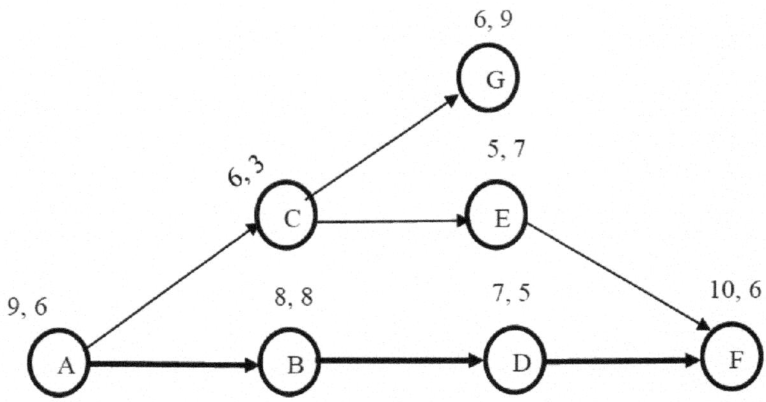

Fig. 9.11 Project network with time and resource constraints

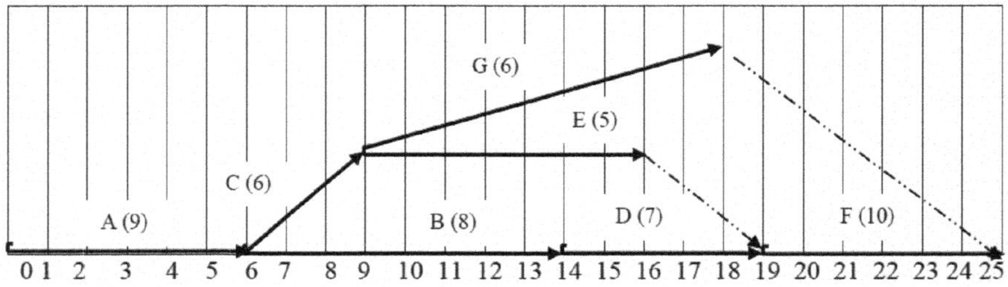

Fig. 9.12 Project network without resource leveling

Day 9: B 8 + E 5(3) + G 6(7) > 15. The explanation of this expression is as follows. The activity "B" is going on with resource requirement 8. The activity "E" can be accommodated, requiring five resources and three days of slack. This means the activity "E" can be delayed by three days. Similarly, in the current schedule (without resource constraint), the activity "G" is going on, which needs six resources and has seven days of slack available. Therefore, postponing G has more slack by one day.

Day 10: B 8 + E 5(3) + G 6(6) > 15 Therefore, postponing G has more slack by 1 day.

Day 15: D 7 + E 5(3) + G 6(2) > 15.

There is a need to accommodate activity "D," which has zero slack and is on the critical path. Adjusting this activity "D" with the ongoing activities "E" and "G" will exceed the availability of the resources (in this case, 15 men). Hence, we postpone the activity "D" by two days. It will delay the activity "F" by two days and therefore increase the project completion time.

Day 17: D 7 + G 6(2) < 15 which satisfies the condition.

Day 22: F 10 + G 6 > 15.

Therefore, postponing F by 2 days.

Day 24: F 10 < 15. This satisfies the resource constraint.

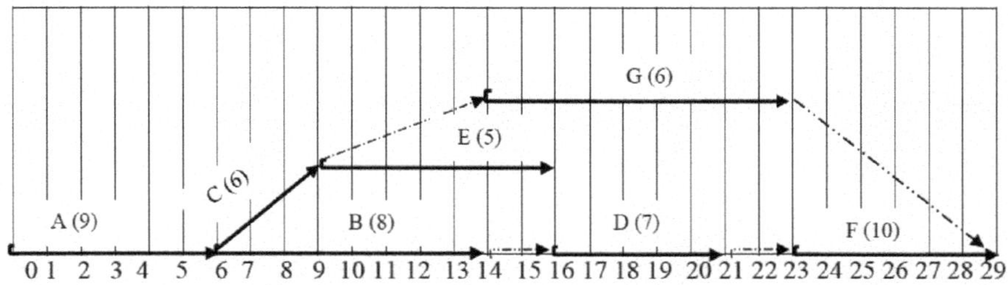

Fig. 9.13 Project network with resource leveling

The total project time has extended to 29 days. It can be observed that there is a trade-off between project completion time and leveling resource consumption profile. In the present case, we are constrained by the resource availability and to satisfy this condition, we have to accept the increase in project completion time by four days (i.e., 25 days (without resource constraint) to 29 days (with resource constraint)). It is necessary to appreciate that the heuristic used here does not permit "activity or task splitting." As soon as the activity or task has been scheduled, it must be completed. Otherwise, the entire activity or task can be delayed depending upon the slack available, resulting in an extension of the project deadline (Fig. 9.13).

Task C

No Job Splitting

Total number of days: 25.

So job A can be started from 1st day to 20th day.

$$X_{A1} \ldots X_{A20}$$

Similarly

$$\begin{array}{ll} X_{B1} \ldots X_{B18} & X_{F1} \ldots X_{F20} \\ X_{C1} \ldots X_{C23} & X_{G1} \ldots X_{G17} \\ X_{D1} \ldots X_{D21} & X_{E1} \ldots X_{E19} \end{array}$$

Each activity should be done only once so,

$$X_{A1} + \cdots + X_{A20} = 1$$

So on,

$$X_{G1} + \cdots + X_{G17} = 1$$

Let us take resource available for each job as

$$r_A \ldots r_G$$

Maximum resource available at a particular period = R.
For Day 1,

$$r_A X_{A1} + r_B X_{B1} + r_C X_{C1} + \cdots + r_G X_{G1} \leq R$$

For Day 2,

$$r_A(X_{A1} + X_{A2}) + r_B(X_{B1} + X_{C1}) + r_C(X_{C1} + \cdots + r_G(X_{G1} + X_{G2}) \leq R$$

In the similar manner, the expressions can be derived for all 25 days. For Day 25,

$$r_A X_{A20} + r_B X_{B18} + r_C X_{C23} + \cdots + r_G X_{G17} \leq R$$

We have totally 138 X_{ij} variables.

- If job i starts at time $(t - 1)$ it would end at $t - 1 + d$
- For precedence,

$$\sum (t-1) X_{jt} - \sum t X_{jt} \geq d_i$$

Now precedence parameters,

$$\text{A} < \text{B}, \text{B} < \text{D}, \text{D} < \text{F}, \text{A} < \text{C}, \text{C} < \text{F}, \text{E} < \text{F}, \text{C} < \text{G}$$

ABDF is the critical path, which is fixed, and start and end times cannot be varied.
C, E, H can be changed.
C can start in 7th, 8th, 9th positions ($X_{\text{C7}}, X_{\text{C8}}, X_{\text{C9}}$).
E can start in 10th, 11th, 12th positions ($X_{\text{E10}}, X_{\text{E11}}, X_{\text{E12}}$).
G can start in 10th, … 16th positions ($X_{\text{G10}}, \ldots X_{\text{G16}}$).
For A < C.
$6X_{\text{C7}} + 7X_{\text{C8}} + 8X_{\text{C9}} \geq 6$ (finish time of A).
For E < F.
$9X_{\text{E10}} + 10X_{\text{E11}} + 11X_{\text{E12}} + 7 \leq$ Start time of F.

With Job Splitting

$$X_{it} = (1, 0)$$

d_i is duration.
r_i is resource.
N is the total no of jobs for the project.
T is maximum of d_i.

We have a variable for every task,

$$X_{A1}, \ldots X_{A9}$$
$$X_{B1}, \ldots X_{B9}$$

So on,

$$X_{G1}, \ldots X_{G9}$$

Let us consider the time constraint for the project,

$$\sum (j = 1 \text{ to } 9) X_{Aj} = 6$$
$$\sum (j = 1 \text{ to } 9) X_{Bj} = 8$$

So on,

$$\sum (j = 1 \text{ to } 9) X_{Gj} = 6$$

Now let us consider the final resource constraint,

$$\sum (i = 1 \text{ to } 7) \sum (j = 1 \text{ to } 9)\, r_i X_{ij} \leq R$$

Our objective is to minimize R.
We have $9 * 7 + 1 = 64$ variables and 16 constraints.

Project Management in Practice

Resource allocation and leveling for an infrastructure project.

An urgent requirement to meet the ever-growing traffic around a railway station has made the state government act by approving the tender winner XYZ Limited (a hypothetical company) to construct a flyover from the station to decongest the area. The order for constructing the flyover has been sent to the Municipal Corporation, which will evaluate the environment clearances, height approval, etc. The Civil Engineering Department and Architecture Department of an eminent engineering research institute was roped into conducting the soil tests, designing the flyover, etc. The project manager of XYZ Limited was expected to ensure the availability of necessary resources timely, coordinate with the sub-contractors, maintain schedule, and ensure that human capital is insufficient supply. Let us consider an infrastructure project of construction of a flyover bridge in front of the railway station to stretch 5 km.

The various stakeholders are:

(1) Municipal Corporation—**Owner**
(2) Civil Engineering Department and Architecture Department—**Consultant**
(3) XYZ Limited—**Contractor**.

The project involves the following activities.

Preliminary Activities

(a) Setting up of site office
(b) Site and Topographic Survey
(c) Geotechnical Investigation
(d) Setting of batching plant
(e) Setting up of fabrication yard
(f) Setting of labor camp.

After the initial setup, the construction will start with the foundation.

Foundation

(a) Pile Boring
(b) Pile Capping.

After the foundation, the initial support pillars will be put up.

Substructure

(a) Pier
(b) Pier caps
(c) Bearing.

Superstructure

(a) Cast in situ PSC Box Girder
(b) Construction of Deck Slab
(c) Diaphragm.

The final few touches, which will provide basic features to the bridge, will be constructed.

Finishing Items

(a) Anti-crash barrier and railing
(b) Drainage system
(c) Wearing coat and expansion joints
(d) Painting
(e) Street lighting and electrical works.

Cost Analysis

The budget allocated for the entire project is **Rs. 5 crores**.

Machineries used:

Transit mixer	4
R.M.C. plant	1
Concrete pump	2
Tire mounted crane	1
Hydraulic pile rig	2
Equivalent excavator	1
Hydraulic tipper and dumper	6
J.C.B.	2
Prestressing jacks	2

As it is a massive project with a long timeline that requires many resources, various complexities arise because of resource allocation and resource leveling. This includes:

- The safety of the road users has to be maintained while exploring alternative paths to the one where the bridge is getting constructed, and proper permission has to be taken.
- The construction of a bridge is a more linear and repetitive task compared to the construction of buildings. This can ease resource leveling as an almost similar amount of resources are needed to build the bridge in parts.
- Resource allocation is an essential step as there will be no inventory to store resources in case of over-allocation. Thus, care has to be taken to prevent over-allocation that may cause increased storage cost and over-allocation of resources than required.
- Resource leveling may lead to elongation in the project's finish date, which causes problems mentioned in the previous points.
- There could be scope for expediting the project as the laborers are skilled, and the materials are available.

If all these measures are taken appropriately, we can expect the project to be completed without much delay and cost escalation.

Questions for Class Discussion

1. What are the critical project resources in infrastructure projects?
2. What are the key trade-offs a project manager needs to evaluate in allocating resources in infrastructure projects?
3. Based on your understanding of the case study, what are key recommendations you would like to extend to utilize the resources to the case organization effectively?

Summary

In project management, the scheduling of activities and the resources required by them, taking into account their availability, project duration and the organization's strategic goals is defined as resource allocation.

Resource leveling determines the activities that can be delayed in case of unavailability of resources.

Resource leveling utilizes the slack available with non-critical activities to level the resource consumption.

Resource allocation can be done by considering job with splitting and job without splitting depending upon the nature of the project.

Questions for Discussion

1. What is the importance of resource allocation and leveling in project management? What profound implications may this have on the execution of project management? Consider a case of a typical construction project and explain what the severity of the resource allocation issue in project management is?
2. What is the difference between "Resource Loading" and "Resource Leveling?" Explain each one in detail by considering a network diagram with activity timing and resources. Show the necessary calculation systematically.
3. What are the different types of resources in the Project? Classify them appropriately? Explain each resource under another category critically. Explain the significance of each of the resources for a typical infrastructure project in a Tabular format.
4. Explain various heuristics useful in resource leveling for a project. Demonstrate the utility of each resource for a typical example specific to a project environment.
5. Explain steps involved in Wiest's heuristics for solving a resource leveling problem for a project.
6. DIMHILL is developing an ERP application for a leading automobile manufacturer. This IT project is described in the table below regarding activities, timing, and manpower needed.

Job	Predecessors	Duration (days)	Manpower needed
A	–	8	12
B	A	10	9
C	A	3	6
D	B	5	7
E	C	7	5
F	D, E	8	12
G	C	10	8

Your Tasks

(a) What is the minimum time in which it can be completed assuming an unlimited supply of manpower?

(b) If only 12 men were available, what is the minimum bound on the duration? Using Wiest's heuristics for resource allocation, determine the minimum duration schedule for the project.

(c) Formulate the resource leveling problem for this project as an integer linear program assuming (a) no job splitting and (b) job splitting permissible.

Group Project

In a group of four persons, visit the actual site of the ongoing project in your neighborhood and identify the critical resources necessary for meeting the various deadlines of this project based on your discussion with the project manager, contractors, and sub-contractors. Prioritize the resources and map them with the requirements of the project. Evaluate the consequences of the non-availability the resources on project completion time.

Chapter 10
Project Monitoring and Control

Critical Questions

What is the importance of project monitoring and control?
How IT helps to enhance the performance of the project?
What is Earned Value Analysis (EVA)?
What is the role of IT in project monitoring and control?
What are the key indicators for monitoring the progress of project?

Project Management in Practice

Rail project for linking State Capitals of North East in India

- The Indian government had planned to connect all State Capitals of North East States with the rail network. However, during the previous five years, the Central Government has emphasized the timely completion of infrastructure and safety projects, and funding for infrastructure projects on Indian Railways has increased significantly.
- The completion of new line projects for capital connectivity has been hampered in various states, primarily due to land acquisition delays and law and order difficulties. In addition, due to the rugged topography of the Himalayas, all of these capital connection projects entail a high number of tunnels and considerable bridges in a geologically demanding setting.
- Some organizations claim that railway connectivity will result in the influx of outsiders, which has sparked local opposition to the project in Meghalaya. However, after the entire site is handed over to railways, the project is expected to be completed by 2023.

J. J. Thakkar, *Project Management*, Management and Industrial Engineering,
https://doi.org/10.1007/978-981-15-3695-3_10

Sources:

*(a) Capitals of NE states to be connected to rail network by 2023: Railway Board, Mint, 18th July 2020 (*https://www.livemint.com/news/india/capitals-of-ne-states-to-be-connected-to-rail-network-by-2023-railway-board-11595029292137.html*), accessed on 17th March 2021.*

(b) N-E state capitals to be connected via *rail network by 2023: Railway Board, Business Standard, 18th July 2020 (*https://www.business-standard.com/article/current-affairs/n-e-state-capitals-to-be-connected-via-rail-network-by-2023-railway-board-120071800115_1.html*), accessed on 17th March 2021.*

10.1 Importance of Project Monitoring and Control

> It is the process of keeping the project on target and as close to plan as feasibly possible. This means you must have a way of detecting when a project is off the target.

Project monitoring and control involves tracking the actual project performance with the planned project management activities. It can mainly be looked at as a control function at all stages of a project. Monitoring and controlling projects can make it easier for organizations to achieve the following:

(a) There is a need for monitoring and control systems that the management can use to ensure timely completion of the project within the allocated budget.
(b) Monitoring and control are required to assess the need for any mid-course corrections needed for the project.
(c) To assess the need for diverting resources across various projects.
(d) To get a timely update about the activity status.
(e) It helps in cost control as the cost of activity during the execution phase might be higher or lower than the cost estimated during the planning phase.

Several aspects need to be monitored and controlled in a project. Some of them are as follows:

(a) Schedule/Time/Progress: It establishes targets or milestones (for comparison) regarding the quality of work done and the cost of work done. It is monitored by generating timely review reports and circulating them to the team members.
(b) Cost: It is an indicative parameter used to measure the rate of the project's progress.
(c) Quality
(d) Safety
(e) Inventory

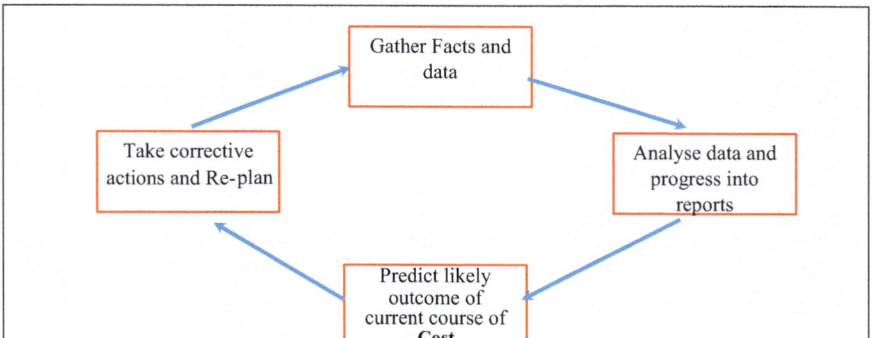

Areas of Concern: Time-Cost-Performance		
Time	**Cost**	**Performance**
• Difficulties took along to solve • Initial estimates were optimistic • Sequencing was incorrect • Resources unavailable • Preceding tasks were incomplete • Change orders • Governmental regulations were altered	• Difficulties may need more resources • Scope may increase • Bid was too low • Reporting was poor • Budget was inadequate • Correction not in time • Input price changed	• Technical problems • Technical difficulties • Quality problems • Client wants changes • Inter-functional complications • Technological breakthroughs • Intra-team conflict • Market changes

Fig. 10.1 Basic monitoring and control process

The primary monitoring and control process is illustrated in Fig. 10.1.

The monitoring and control process aims at regular monitoring of planned vs. actual achievement. It attempts to discover any deviations at the earliest so that corrective mechanisms can be applied on time.

Project performance, time, and cost are the three significant watchwords in planning and control. These can be evaluated and used to compute the time and cost overruns in a project at a specific milestone.

The consequences of project monitoring and control on the performance of the project are as follows:

(a) If the project is being well monitored and controlled, it leads to the project's well-synchronized behavior. Here, synchronization means all the resources are timely utilized and applied in the project to achieve enhanced performance
(b) Better project monitoring and control will cater to the fluctuations in the market. Therefore, timely monitoring and controlling the project by various means will be a very effective way to amalgamate the market fluctuations and thus maintain the project's health.
(c) Often, intra-team conflicts in a project occur between two groups or between two departments, which significantly hamper the project's growth. Timely monitoring and control by conducting joint meetings or social meetings will be the panacea for getting rid of such problems.

(d) Often, the requirements of customers change and, if not addressed carefully at the right time, may cause a significant loss for an organization.

The consequences of project monitoring and control on the cost of the project are as follows:

(a) A disciplined approach toward project monitoring and control is an excellent guide for the top executives to allocate and release the budget.
(b) The price of raw materials and various inputs required for the project keeps fluctuating because of so many factors. The increased cost can be analyzed timely, and appropriate steps can be taken.
(c) The cost of the project is directly related to its scope. Therefore, the project's scope may change during the execution of the project and disturbs the budget requirements. Project monitoring and control is an essential tool that can cater to these changes.
(d) The complete cost analysis of the project depends upon the timely and accurate reporting of the cost details and expenditures incurred. Any lacuna in this can lead to imbalance in the expenditure and budget allocation. Hence, proper project monitoring and control are indispensable for the project's health.

The consequences of project monitoring and control on the time of the project are as follows:

(a) Government regulations may change with time. For example, regulations over land acquisition, the minimum PF for the labor class, daily change in the price of diesel/petrol, regulations to reduce greenhouse gases emission, etc. The changes in regulations directly affect the project and thus the deadline of the project.
(b) Project monitoring and control gives a much better idea about the availability of resources at a given point in time and thus provides a direct indication of the time required to complete the project.
(c) Timely monitoring and control will help the project manager solve the new difficulties and technical problems well before the time and thus the timely completion of the project.
(d) A project is the aggregation of successive tasks. In a project, it may happen that some tasks were left behind because of unforeseen circumstances. Proper monitoring and control will cater for these incomplete preceding tasks, and thus, timely action could be taken to complete the project well in time.

10.2 Earned Value Analysis (EVA)

Earned Value Analysis is an aggregate performance measure for measuring the overall performance of the project. It is a well-known project management tool that uses information on cost, schedule, and work performance to monitor the project's current status. It can forecast the project's completion time and final cost and also provide schedule and budget variances along the way. It offers consistent numerical indicators with which one can evaluate and compare projects. For example, if the earned value chart shows a cost overrun or scope underrun, the project manager must figure out what to do to get the system back on target. He might borrow resources from activities performing better than expected or hold a meeting of project team members or notify the client that the project may be late or overbudgeted. The essential terms involved in Earned Value Analysis are explained below. A typically earned value chart presented below indicates these measures to project the cost and schedule variances. A typical earned value understanding is illustrated in Fig. 10.2.

(a) **Budgeted Cost of Work Scheduled (BCWS) or Planned Value (PV)**: The planned cost of the total amount of work is scheduled to be performed by the milestone date.

(b) **Actual Cost of Work Performed (ACWP) or Actual Cost (AC)**: It is the total of direct and indirect costs incurred to accomplish the work that has been done to date.

(c) **Budgeted Cost of Work Performed (BCWP) or Earned Value (EV)**: The earned value of work performed for tasks in progress is found by multiplying the estimated percentage of physical completion of work by the planned cost of each task those tasks.

(a) **Budget at Completion (BAC)**: It is the baseline budget for completing the project.

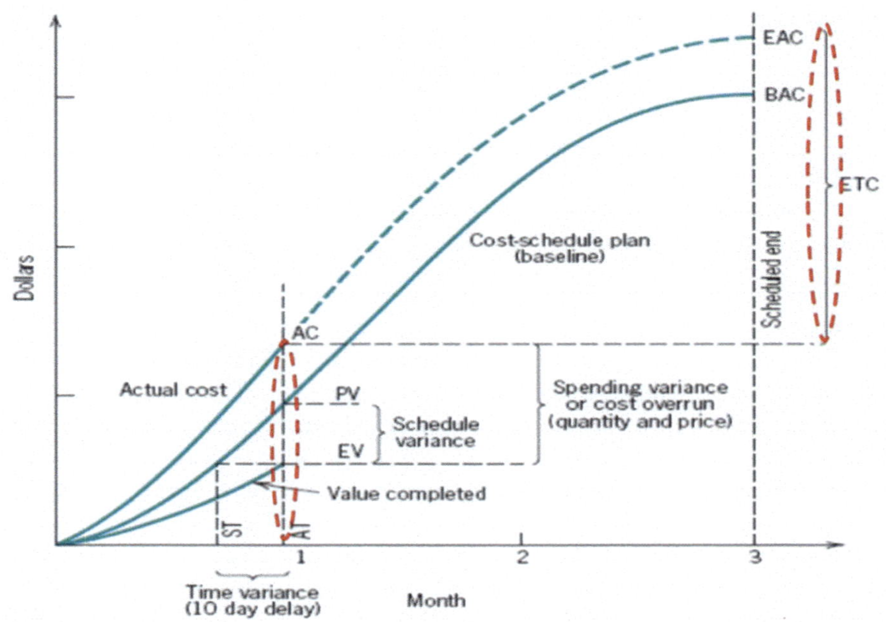

Fig. 10.2 Earned value chart (*Source* Project Management: A Managerial Approach by Meredith and Mentel, 2009 John Wiley & Sons, Inc., Seventh Edition)

(b) **Estimated cost at Completion (EAC)**: It is the actual cost to date for completing the project.

(c) **Cost Variance (CV)**: It is the difference between earned value and actual cost. An unfavorable variance is not acceptable. A negative variance means the project is over budget.

$$\mathrm{CV} = \mathrm{EV} - \mathrm{AC}$$

(d) **Schedule Variance (SV):** It is the difference between earned value and the planned value. A negative variance means the project is behind schedule.

$$\mathrm{SV} = \mathrm{EV} - \mathrm{PV}$$

(e) **Time Variance (TV):** It is the difference between the time scheduled for the work that has been performed (ST) and the actual time used to perform it (AT). TV, the delay is negative.

$$\mathrm{TV} = \mathrm{ST} - \mathrm{AT}$$

(f) **Cost Performance Index (CPI):** CPI = EV/AC (Ratio of EV and AC)

(g) **Schedule Performance Index (SPI) :** SPI = EV/PV (Ratio of EV and AC)

(h) **Time Performance Index (TPI) :** TPI = ST/AT (Ratio of ST and AT)

(i) **Cost Performance Index (CSI) :** CSI = CPI*SPI

$$\begin{aligned} &= (\mathrm{EV}/\mathrm{AC}) * (\mathrm{EV}/\mathrm{PV}) \\ &= \mathrm{EV}^2/(\mathrm{AC} * \mathrm{PV}) \end{aligned}$$

If CSI < 1, then it is an indication of a problem, and there will be a less likely chance of recovery of project.

(j) **Estimated cost to complete (ETC)**: ETC = (BAC + EV)/CPI

(k) **Estimated cost at completion (EAC)**: EAC = ETC + AC

A topology presented in Fig. 10.3 can be used for evaluating the performance of the project.

Example: A construction company is executing a metro rail project connecting Ahmedabad and Gandhinagar in Gujarat. The company uses Earned Value Analysis to monitor the progress of the project continually and initiate the necessary corrective actions at the right time. Suppose that operations on a project were expected to cost rupees 15 thousand lakhs for the total amount of work scheduled to be performed at the planned date (i.e., today).

Initially, the work was to be completed by today (per plan). At this point, Rs. 13.50 thousand lakhs have been spent, and we estimate that two-thirds of the work has been completed. The project manager calculates the various indicators of earned value analysis as follows.

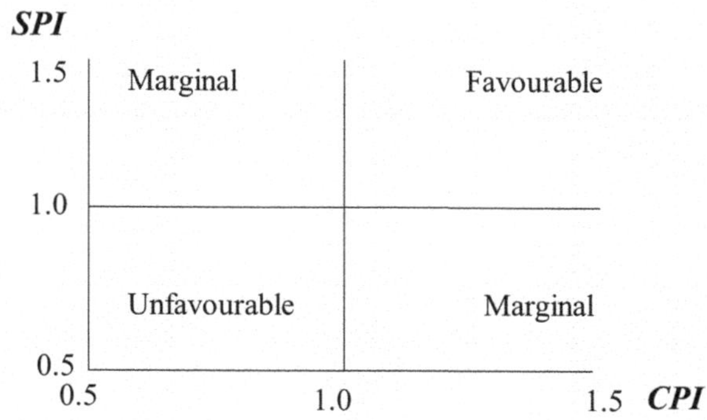

Fig. 10.3 Combined performance index for project performance

Cost Variance, CV = EV − AC
= (15*2/3–13.50) thousand lakhs
= −3.50 thousand lakhs

Schedule Variance, SV = EV − PV
= (15*2/3–15) thousand lakhs
= −5 thousand lakhs

CPI = EV/AC
= (15*2/3)/13.50
= 0.740

SPI = EV/PV
= (15 * 2/3)/15
= 0.666

CSI = CPI * SPI
= 0.740 * 0.666
= 0.492

The analysis indicates that CSI is less than one, and it is indicative of a severe problem. Therefore, the project manager and team need to look into the critical issues behind time and cost overrun. If required, top management should be consulted for necessary assistance and advice.

10.3 Role of IT in Project Monitoring and Control

Information technology (IT) plays a vital role in project monitoring and control. In this present era, where projects are vast, complex, and comprise many Work Breakdown Schedules, proper use of IT is inevitable for continuous monitoring of the project. If a project manager is aware of the project's status, only he can control the pace and progress of the project. IT enables a project manager to ensure that the roadmap laid down for executing a project is being followed accurately. By using suitable tools, the project manager can always keep track of where the project is at any given moment, whether it is running behind schedule, ahead of schedule, or on schedule.

Numerous techniques are available that help a project manager to get a good insight into the project performance. Some methods compare actual project performance with ideal project performance in terms of cost or budget, while some give the same comparison relative to the percentage of job completion. The available tools provide a visual indication of the performance that a project manager can easily understand. Also, the manager has access to all the past performance figures, which helps him understand the project better. With the various methods available, it is straightforward to detect a problem early, which further helps take appropriate actions beforehand, thus avoiding unnecessary delays and waste of money. A well-designed IT system ensures effective project monitoring and control.

Technology plays a significant role in nearly every aspect of business, and project management is no exception. Technology can significantly enhance the performance of projects. The following are the applications of information technology that can improve project management.

- *Communication is instant:* The idea of having team members in the same location was once considered accessible. Recent technological advancements have changed this setting. Members of a team on different projects do not necessarily have to be from the same company or location. Technology has made communication more accessible and instantaneous. Despite being across the world, many team members can make themselves available almost any time of the day or night.
- *Management of data:* You can analyze and measure the data collected on a project to improve your process from now on. Tools are developed to collect and analyze project data and suggest ways to improve the current project. While data analysis is sometimes complex and leaves you wondering how it all makes sense, several management tools simplify the process and streamline data breakdown.
- *Keep up with deadlines with a project management platform:* Project management platforms have made it easier for teams to meet deadlines. Technology and software can help a lot in keeping deadlines.
- *Automating workflows:* Automating workflows is a great way to automate repetitive and routine tasks in your organization, so your team can better focus on the project advancement. With the proliferation of workflow automation tools, there are a variety of choices available. Take the time to research the tools that suit the particular needs of your team.

Information technology has become a vital part of many organizations' future, and the demand for good project managers has increased. Although project management has existed for many years, managing information technology projects requires ideas and information that go beyond standard practice. A lack of executive support, weak user involvement, and unclear business objectives cause many IT projects to fail.

Enterprise Resource Planning (ERP) is an advanced manufacturing system that integrates transaction-oriented data and business functions throughout a company.

Table 10.1 Comparison of key approaches used for project monitoring and control

S. No.	S curve	Network cost accounting	Earned value analysis
1	It is a plot of cumulative cost, hours of labors, and other quantities concerning the time	The project forms the basis of cost measure and control	Project is measured in monetary terms
2	It allows visual tracking of project progress over time and helps in determining slippage, if any	Project tracking is done based on a comparison between actual and budgeted costs	It gives the amount of work achieved relative to the cost incurred
3	A historical accounting is possible with S curve that allows a project manager to see past growth of the project	Past growth can be observed in terms of cost	Historical accounting is also possible with EVA
4	Simple to visualize project performance from the graph	It needs simple calculations, and graphical presentation provides an intuitive understanding	The method gives good insight as to how the project is performing. However, it involves complex calculations to get the necessary performance data
5	Less agile	A method is agile and can evolve with changing of plans	It is heavily dependent on initial planning. Any change in plans at a later stage drastically leads to the failure of this method. Hence, it is not suitable for agile projects

An ERP system can significantly enhance organizational and project performance and establish a competitive advantage. The ERP implementation process can help a team reduce implementation costs. The ERP implementation process can be divided into five phases: initiating, planning, executing, controlling, and closing.

Three systems can be used to monitor and control the project: S curve-based project management, network cost accounting, and Earned Value Analysis. The salient features of these approaches for project monitoring and control are presented in Table 10.1.

10.4 Earned Value Analysis: Illustrative Example

An IT firm located in Bangalore is executing an ERP development project for a banking and financial services company. First, the project with the information given in Table 10.2 needs to be implemented.

Table 10.2 ERP development project

Activity	Duration (months)	Cost (Thousand Rs.)
P (1,2)	8	08
Q (1,3)	4	10
R (2,3)	3	12
S (2,4)	4	12
T (3,4)	5	16
U (3,5)	5	10
V (4,5)	4	14

Your Tasks

Task 1: Compare the pattern of expenditure for both an early start and a late start schedule.

Activity P: 100% *complete*

Activity Q: 70% *completed*

Other activities: *not yet started*

Total spending to date: *Rs.* 48,000.

Task 2: Estimate time and phasing of expenditure from now onward if the project has been completed eight months. The late start schedule is chosen for implementation, and the progress is monitored after eight months.

Task 3: What are time and cost overrun? What are the remedial measures to control time and cost overrun in a typical project?

Task 4: Compute the cost and schedule performance indices and show the project's performance on a consolidated front.

Task 1:

The cost calculations for early start and late start schedules are listed in Tables 10.3, 10.4, and 10.5 and Figs. 10.4, 10.5, and 10.6.

A comparison between early and late start scheduling can be viewed from the curve given in Fig. 10.5.

Cost overrun: Cost overrun is also known as budget overrun in a project. Due to mistakes in estimating the actual cost during the budgeting phase, unexpected costs may arise during the project's execution phase, termed cost overrun. Cost overrun and time overrun are the typical problems associated with a megaproject like a construction project and a technical project.

The following are the remedial measures to control the cost overrun in a typical project:

Correct estimates: Inaccuracy in the cost estimates during the initial phase is the main reason for cost overrun. To begin the project's planning phase, highly effective

Table 10.3 Basic scheduling of ERP project

Activity	Duration (in months)	Early start	Late start	Total cost (in thousands)	Cost per month (Rs./month)
P(1,2)	8	0	0	8	1
Q(1,3)	4	0	7	10	2.5
R(2,3)	3	8	8	12	4
S(2,4)	4	8	12	12	3
T(3,4)	5	11	11	16	3.2
U(3,5)	5	11	15	10	2
V(4,5)	4	16	16	14	3.5

bidding procedures and market surveys should be conducted. Moreover, the planning committee should also be updating itself about the changing market conditions.

(a) **Error-free design**: Error in the design stage is an essential factor that leads to the wrong application of techniques and thus finally leads to cost overrun. The prominent reasons for the design errors are insufficient field investigation and frequent changes in design. By proper and effective communication and thorough field investigation, the planning and monitoring committee can reduce these design errors and thus can minimize cost overrun.
(b) Proactive approach: Any change in the project's scope during the execution phase leads to the entire review of the project. A proactive approach could be adopted to incorporate the stakeholders' needs through the project life cycle to minimize the change in scope.
(c) Appropriate and adequate procurement: The main reason behind the inappropriate and inadequate procurement is the faulty contractual system with specific clauses. This ambiguity leads to disputes between the contractors and managers and thus leads to cost overrun. It demands approaching the most deserving contractor through an ethical tender system.
(d) Reduction in complexity: The complexity in the project is also one of the reasons for cost overrun. Adequate planning using Work Breakdown Structures (WBS) is helpful.

Time overrun: Time overrun is defined as the unexpected delay in the project's completion time beyond the planned one or delay in the final delivery of the product to the customer. The following are the remedial measures to control cost overrun in a typical project:

(a) Excellent cash flow plan: Improper and untimely cash flow is an essential reason for time overrun. An appropriate cash flow plan minimizes financial issues with contractors.
(b) Effective decision-making: Delay in the decision-making of various tasks in a project leads to time overrun. So, effective leadership is required so that the right decision can be taken at the right time.

Table 10.4 Cumulative cost (early start)

	1	2	3	4	5	6	7	8	9	10	11	12	13	14	15	16	17	18	19	20
Cost (in thousands)	3.5	3.5	3.5	3.5	1	1	1	1	7	7	7	8.2	5.2	5.2	5.2	5.2	3.5	3.5	3.5	3.5
Cumulative cost (in thousands)	3.5	7	10.5	14	15	16	17	18	25	32	39	47.2	52.4	57.6	62.8	68	71.5	75	78.5	82

Table 10.5 Cumulative cost (late start)

	1	2	3	4	5	6	7	8	9	10	11	12	13	14	15	16	17	18	19	20
Cost (in thousands)	1	1	1	1	1	1	1	3.5	6.5	6.5	6.5	3.2	6.2	6.2	6.2	8.2	5.5	5.5	5.5	5.5
Cumulative cost (in thousands)	1	2	3	4	5	6	7	10.5	17	23.5	30	33.2	39.4	45.6	51.8	60	65.5	71	76.5	82

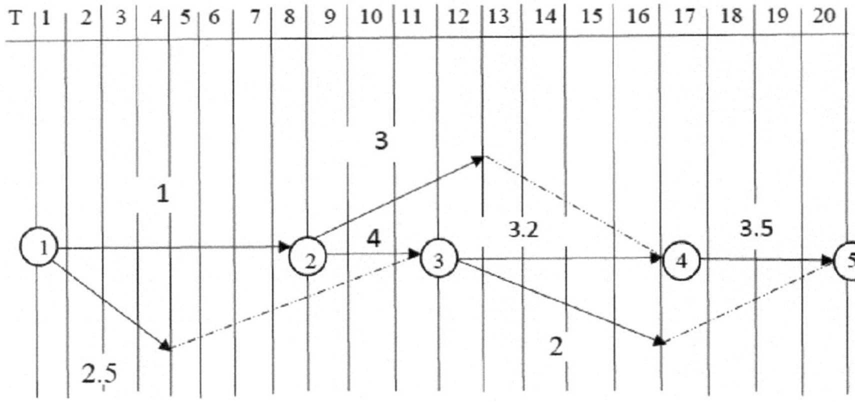

Fig. 10.4 Time-scaled network (early start)

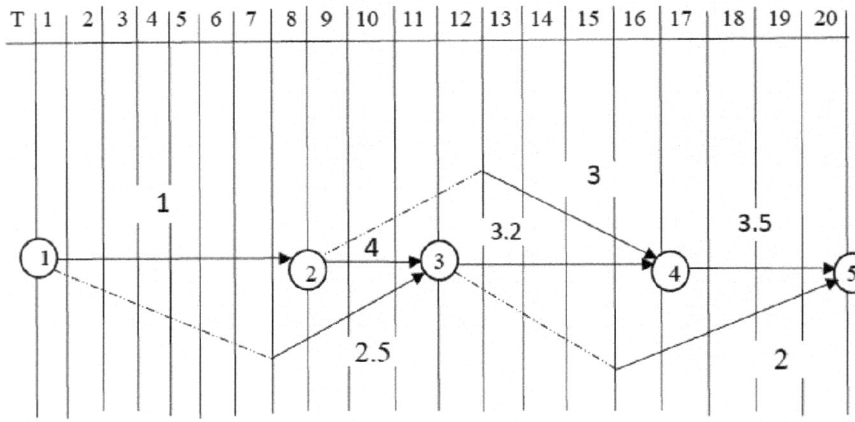

Fig. 10.5 Time-scaled network (late start)

(c) Availability of site workers: Many times, especially in construction projects, the time overrun is due to a shortage of workers. Contractors and sub-contractors handle the workers, and any dispute between them may lead to time overrun.

(d) Documentation and proper scheduling: Proper documentation and schedules are vital for a project's health. The timely completion of the project requires systematic documentation of the schedule and the updates. Moreover, it also helps extensively in managing financial resources.

(e) Hiring efficient and skilled manpower with sound managerial capabilities can significantly help in reducing time overrun.

Let us consider P (1, 2)—100% complete and Q (1, 3)—70% completed
Now after 8 months, we have

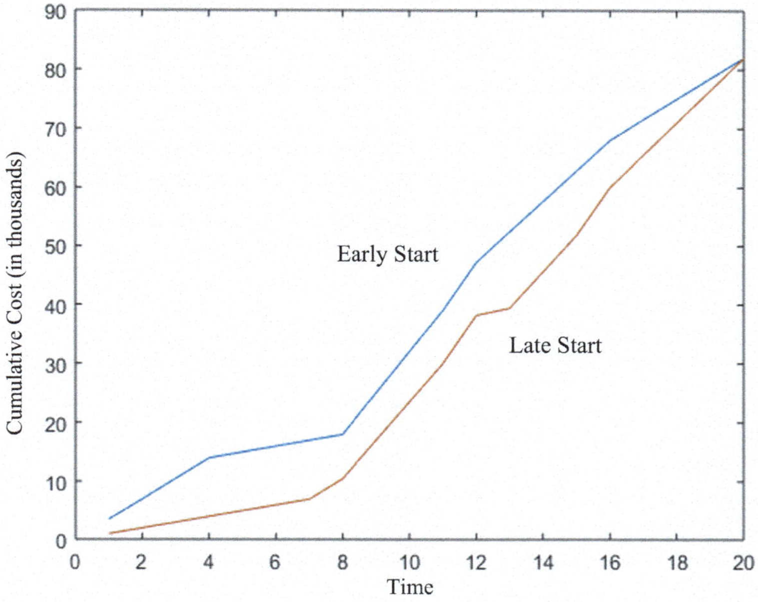

Fig. 10.6 Early start versus late start schedule

A: Budgeted cost = 10,500 (cumulative cost in eighth month in late start schedule table)

B: Actual money spent = 48,000

C: Value of work completed = 15,000 (100% of cost P + 70% of cost of Q)

Then,

(a) Cost Overrun = (A − C)/c

Cost overrun = (48,000 – 15,000)/15,000

Cost overrun = 220%

(b) Time Overrun = Review month (8th month in the present case)—Budgeted cost/Actual money spent = 8 − 10,500/48,000 = 8 − 0.22 = 7.78.

Time overrun = 7.78 months **(time overrun is positive, i.e., a project is running 7.78 months late)** There is a precaution to practice if the time overrun becomes negative (i.e., the project is running ahead of schedule). This kind of situation when time overrun is negative may not always indicate a sign of good progress. The reasons need to be investigated for moving ahead in the schedule.

The negative time overrun may be because of inflated time estimates provided by the project team, and hence, this excessive safety results in moving ahead of schedule.

(c) Cost Performance Index (CPI) = BCWP (Value of work completed)/ACWP (Actual money spent)

CPI = 15,000/48,000

CPI = 0.3125

(d) Schedule Performance Index (SPI) = BCWP (Value of work completed)/BCWS (Budgeted Cost)

SPI = 15,000/10,500

SPI = 1.43

From the values of CPI and SPI obtained above, we can calculate the cost schedule index (CSI) to check whether the project suffers from any problem. CSI is given as

CSI = CPI * SPI

For the above example, we get

CPI = 0.3125 * 1.43

CPI = 0.44 < 1

Based on the above calculation and assuming that the duration and cost of the activities remain unaltered, it is evident the CPI is significantly less than 1. It indicates an inferior performance of the project at the end of the eight months. Since the project duration is 20 months and we are evaluating the progress at the end of the eighth month, there is still a scope to apply tighter control over cost and time overrun and see that project time and cost overrun can be controlled. It is evident from the S curve that there is always initial inertia in the inception phase of the project, and the project may run out of schedule and cost in the beginning. However, suppose the situation persists beyond a point. In that case, the project manager must inform the top management about the project's performance. Taking timely decisions regarding technological interventions or abandoning a project (accepting the sunk cost) can improve organizational efficiency.

Project Management in Practice

Key Issues in Information Systems Project Management: Freight Operations Information Systems of Indian Railways as IS Megaproject

by

Mr. Sharad Sharma, Doctoral Candidate, IS Area, IIM, Ahmedabad

Information systems (IS) projects are some of the most complex and risky projects and typically have low success rates. The complexity arises not only because of the rapidly changing technologies but also because all IS projects deal with representations and abstractions of real-world phenomena and do not have a physical manifestation as in, for example, in the construction or transportation sector. This often leads to a lack of realization of benefits, scope creep, and poor estimation of resources required to manage the project. The Chaos Report[1], which has consistently tracked IS projects for more than 20 years across various domains and countries, has identified the success rate (considered in a very basic way as completion on time, within budget and scope) as only 31%. This number has largely remained unchanged over the years.

---

[1] *The Standish Group is a primary research advisory organization that focuses on software development performance. It brings out an annual CHAOS Report is brought out by the Standish Group and is considered an authority on the success rates and causes of IT/IS Projects.*

IS megaprojects (ISMPs) have features of megaprojects like complexity, large-scale, high investment, use of new technology, long duration and multiple stakeholders, and the traits of IS projects. ISMPs are typically implemented in/by government organizations with a focus on more efficient public/government service delivery. The government context makes it more challenging to achieve success in ISMPs. Low incentives for IT adoption, including resistance from internal users and process and role changes consequent to the project, create further barriers to the design, implementation, and acceptance of ISMPs. In India, the government's push for Digital India[2] has resulted in several ISMPs such as Aadhaar and GSTN. Freight Operations Information System of Indian Railways, although earlier not a part of Digital India, is an ISMP.

Given their large-scale and magnitude, high cost, and involvement of the government, ISMPs are subjected to intense scrutiny from the public, media, and political parties, over and above the mandatory regulatory and statutory audits, which are associated with large-scale government projects. Poor institutional and technology infrastructure in emerging economies makes the implementation more difficult. As a result, ISMPs in an emerging economy face formidable hurdles all of which may not be anticipated. Implementing ISMPs thus requires a very strong commitment from the key stakeholders, in the absence of which, these face premature termination, cancelation, time and cost overrun, and scope creep.

FOIS, implemented in the Indian Railways (IR), provides scope for better understanding of project management of ISMPS. IR is one of the largest and busiest railway networks globally divided into 17 zones which are further sub-divided into 68 divisions. FOIS covers 8,479 goods trains per day, 293,077 wagons of 25 different types (which differ from each other due to their tare weight, carrying capacity, suitability for commodity, and speed), and 2400 stations across different levels of management hierarchy. FOIS covered the Rakes Management System implemented in 246 locations (completed in 2002) in Phase 1, the Terminal Management System (TMS) implemented in 523 locations (completed in 2008) in Phase 2, and extension of the same to 1317 locations in Phase 3, completed in 2013. In the ongoing Phase 4, the TMS has been provided at 225 more locations, implementation of the Yard Management System and various e-interfaces, for both internal and external customers (based on the information collected by the author).

The challenges in implementing ISMPs are highlighted because FOIS is not yet complete, even after more than 30 years. During this time, it had to consider both technological and organizational policy changes. Technological changes included the availability of GPS as a positioning system for tracking freight assets, web interfaces, hand-held devices, etc. An example of an organizational policy change in managing freight assets was IR moving from managing wagons to managing only rakes. This led to redesign of the application architecture, as focus shifted to tracking and managing rakes rather than wagons. Such changes are typical in an ISMP. Even when there is flexibility in the application architecture, it may sometimes not be able to incorporate

---

[2] *Digital India is a programme of the Government of India with the aim of transforming the country into a digitally empowered society and knowledge economy.*

such drastic policy changes. Therefore, a buffer for resource and effort requirements for such changes needs to be provided in the design phase itself.

FOIS has multiple external and internal stakeholders, as is usual in any IS megaproject. Managing their expectations and seeking acceptance is often a long drawn-out process involving several levels of bureacracy. The internal stakeholders also tend to protect and serve their departmental interests, more than the organizational goals.

Recognizing the greater contribution of profitability and revenue for freight at approximately 65%, compared to passenger revenue at 28%, FOIS was the obvious choice for prioritization in computerization, when IR began its computerization at scale. However, the Passenger Reservation System was given priority and development started in 1984. This was clearly a case of leveraging the high visibility for political gains in the 1984 general elections. IS projects/applications in IR. FOIS implementation started only in 1989 at a cost of nearly 1000 cr (1986 prices).

The choice of platform, whether to import an already implemented system (TOPs by Canadian Rail) or to develop the system indigenously were key decisions, often involving inter-ministerial committees, foreign and Indian experts leading to long decision-making cycles. In the early days of FOIS, there were no large system implementations in any part of the government and there were few Indian organizations that had the expertise to develop large systems. After almost deciding to go for the imported system, IR decided to develop the system indigenously. A new organization, i.e., Centre for Railway Information System (CRIS), registered as a society under the Ministry of Railways, was established in 1986 to design and implement FOIS. Over time, the scope of CRIS has increased to cover several other applications for IR.

Today, FOIS has evolved as a complete freight operations management application, including the route-finder, on-line freight calculation, and generating and processing the billing. With an aim toward greater integration with stakeholders, FOIS plans to include electronic data interchange with Customs' database for import–export freight. CRIS is working toward a robust disaster recovery and data warehousing and developing IR's own cloud system.

FOIS has brought in sea changes in the way freight operations are conducted on IR. It brought in standardization of practices across the various zones and divisions. The streamlined freight operations became easier to analyze, compare, and monitor the performance across the divisions and zones and improved customer satisfaction. It has also enabled better monitoring and hence improved utilization of the freight assets and brought in transparency in data and accounting of freight operations. It has improved the work environment of supervisors and managers, saving them from manual collection and management of data.

The implementation of e-payment, electronic transaction of railway receipts (e-TRR), and electronic registration of demand (e-RD) modules, further provided customer service and enabled integration with customer's IT systems. Such visibility also allowed the government to showcase the success more easily and use it as an exemplar for the other ministries to emulate. FOIS resulted in better business outcomes as IR launched various freight incentivizing schemes based on the analysis of the data and information yielded by FOIS. IR's policy of only booking rakes has enabled the emergence of freight forwarders, who were basically aggregators and acted as an interface between the freight customers and IR. Going forward, integrating FOIS with crew management and scheduling freight trains (which currently do not follow any schedule) would allow IR to give reliability of freight arrival as it competes with private freight carriers who are able to track and predict delivery times precisely.

The success of ISMPs may be evaluated by assessing user acceptance and satisfaction, achieving the goals and objectives and fulfilling the time, cost, and quality requirements. Technological advancements and evolving business needs led to increased expectations from IR as well as users, managers, and freight customers. Therefore, up-gradation for future-readiness, customer orientation, benefit, usefulness, indispensability, and ease of use are the other subjective criteria.

Summary

Project monitoring and control is one of the most important project management activities. It involves tracking the actual project performance with the planned project management activities.

It can mainly be looked as a control function that takes place at all stages of a project.

Earned Value Analysis is an approach for measuring how much work has been completed in a project at a given point of time and performance. This analysis can be done by calculating how much time the work has taken and the resources it has utilized.

Earned Value Analysis (EVA) helps the project managers to keep a track on the progress of the project and initiate appropriate actions at the right time.

Questions for Discussion

1. Why project monitoring and control is essential? What are its consequences for executing the project with time, cost, and performance trade-offs?
2. What is "Earned Value Analysis"? Explain each term involved in Earned Value Analysis with suitable numerical calculation and graphs?
3. What is the role of IT in project monitoring and control? Explain and compare (in Tabular format) the critical feature of at least three computerized systems widely used for project monitoring and control.
4. A construction company located in Kolkata is executing a flyover construction project. The project with the following data has to be implemented.

Activity	Duration (months)	Cost (Thousand Rs.)
P (1,2)	10	12
Q (1,3)	6	8
R (2,3)	4	10
S (2,4)	4	12
T (3,4)	5	16
U (3,5)	7	12
V (4,5)	4	14

Your Tasks

Task 1: Compare the pattern of expenditure for both an early start and a late start schedule.

Task 2: Estimate time and phasing of expenditure from now onward if the project has completed five months. The early start schedule is chosen for implementation, and the progress is monitored after five months. The following status is observed:

Activity P: *60% complete*

Activity Q: *80% complete*

Other activities: *not yet started*

Total spending to date: *Rs.* 67,000.

Task 3: What are time and cost overrun? What are the remedial measures to control time and cost overrun in a typical infrastructure project?

Task 4: Compute the cost and schedule performance indices and show the project's performance on a consolidated front.

Task 5: Assuming that the duration and costs of the project remain unaltered, what is your projection of the project completion?

Chapter 11
Project Risk Control and Management

Critical Questions
What is risk in project management?
Why risk control and management is crucial?
What are the approaches to manage risk in the project?
What is the role and importance of communication and coordination within a project team and with stake holders? How an effective communication helps to minimize the risks?

11.1 Importance of Project Risk Management

> It intends to minimize potential problems that may negatively impact the performance of the project. Risk is any unexpected event that might affect the people, processes, technology, and resources involved in a project.

Any undesirable event or condition that may negatively impact people, processes, technology and resources is termed "risk." Social, political, and cultural are the few types of risk found in a project. Risk management is an efficient way to control the risks. Explore, identify, analyze, and mitigate; these are the basic steps in doing risk management. Risk management is an essential aspect of management, and if done efficiently, the project will triumph. An action plan of risk management will consist of various steps to ensure that the risk is eliminated. In case of an uncontrollable risk, an action plan should focus on minimizing the effect of these risks.

J. J. Thakkar, *Project Management*, Management and Industrial Engineering,
https://doi.org/10.1007/978-981-15-3695-3_11

> The alternative to proactive management is reactive management, also called crisis management. This requires significantly more resources and takes longer for problems to surface.

Managing projects of any size requires a multifaceted approach. It consists of planning, designing, manpower management, legal and regulatory works, risk assessment, and other decisions like choosing the right stakeholder, identifying areas of investment like financial institutions and public funding, and the mode of operation like a private–public partnership (PPP), built, operate and transfer (BOT), fully private funded, etc.

Successful project managers can understand that risk management is essential because achieving a specific project's goals depends on planning, preparation, results, and evaluation that contribute to achieving strategic goals. Risk management contributes to the success of projects by determining the external, internal risks and predicting undetected risks that arise out of taking impromptu decisions during the project's life. It is less expensive to mitigate risks or be proactive than deal with issues later that appear or be reactive. The risk factor is synonymous with project management, especially since the scope is vast and not limited to only a few aspects. Risk is an unforeseen and unpredictable occurrence and can manifest in multiple ways. It is predominantly associated with the failure of the project. Hence, it becomes imperative to assess various possible risks during the early planning stages of the project. Risk assessment involves identifying the risks and predicting the impact it will have on the project in terms of the degree of disruption and the additional costs, if any, associated with it. It is followed by probing into different scenarios and the number of times these risks are likely to occur. Risks can also be beneficial. Such positive risks are called opportunities. They have a different set of risk responses than the usual negative risks because opportunities tend to get maximized.

For example, in a transport infrastructure project like building a road tunnel or a metro corridor, the main risks will include the core operational risks, usually construction-related or technical. Depending on the scale or scope, these projects will also attract relevant stakeholders like the general public, other government institutions, and political parties. Consequently, accountability becomes critical, which is a socio-environmental risk. For such projects, there is another kind of risk which is the economic market/market risk. This relates to the demand and supply aspects of the project. In this case, the actual viability of the project, including cost overruns, the direct beneficiaries, or commuters who will use the new road or metro, all fall under the category of market risk. The degree of impact of these risks varies for different kinds of projects. A lot of risk management strategies are available. However, the most common one for mega infrastructure projects is assessing the risks and developing mitigation strategies to reduce the impact or eliminate them. Since such projects cannot be undone, certain risks have to be embraced. For example, the social risk of vacating residents for road tunneling cannot be entirely avoided. Instead, adequate remedial measures can be taken. Another frequent challenge that project managers might face is the Ricochet Effect which means that any modifications made in one area or section of the project will lead to unplanned changes in other areas. This is

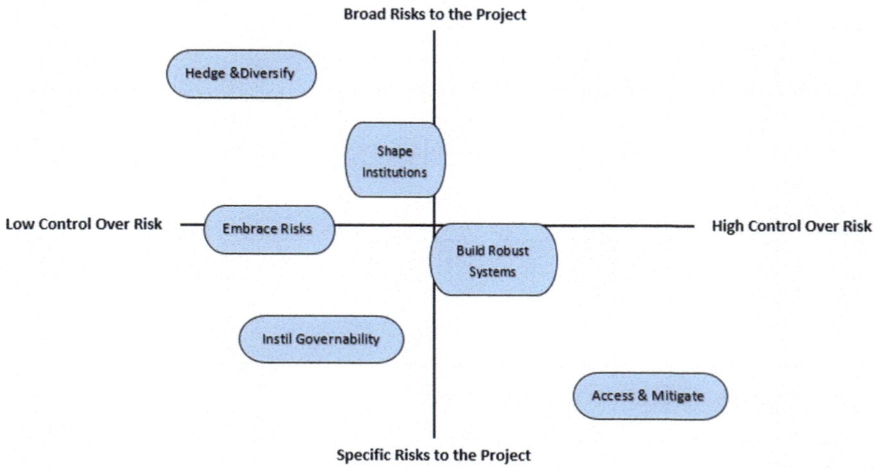

Fig. 11.1 Risk management strategies

quite common and unavoidable in mega projects and can also be considered a form of risk that cannot be easily predicted and prone to being overlooked. The various risk management strategies are illustrated in Fig. 11.1.

The key examples of risks arising in various project areas are given below.

- Integration: Inadequate planning, poor resource allocation, poor integration management, lack of post-project review
- Scope: Poor definition of scope or work packages, the incomplete definition of quality requirements, inadequate scope control
- Cost: Estimating errors, inadequate productivity, cost, change or contingency control, poor maintenance
- Time: Errors in estimating time or resource availability, poor allocation, and management of float
- Quality: Poor work culture, substandard design, material, workmanship, poor implementation of quality assurance programs
- Human resources: Absence of leadership and support of top management, poor conflict management, the inadequate definition of roles and responsibilities
- Communication: Poor planning and communicating, inadequate involvement of stakeholders
- Procurement: Adversarial relationships with suppliers, issues in contracts

Significance of Risk Management

- Project objectives can be adequately accomplished with the assistance of risk administration.
- The negative impact associated with risks can be avoided, and the cost of addressing the risk at an early stage is less.
- It gives control over the possibility and impact of risks.

- Early risk identification helps to explore and plan alternate options proactively.
- It gives the power to steer the project implementation as the risks are known and risk responses are planned.
- It helps the project managers to come up with better strategic decisions.
- The negative effect of litigation, etc., which are the results of project failure, can be avoided.
- Shareholder value can be protected.
- It provides a competitive edge over competitors.

Risk management focuses on the future. Risk and information are inversely related.

11.2 Phases of Risk Management

The phases of risk management include:

1. Defining the context
2. Identification of risks
3. Assessment of risk
4. Risk control
5. Effective communication
6. Prior planning
7. Supervision
8. Review of the process

A risk management cycle operates systematically, as presented in Fig. 11.2. This demands the managers identify risks exercising significant impact on the system, qualitative and quantitative analysis of the identified risks, development of strategies and determining the capabilities to accommodate or counter the prioritized risks in the system, action plan, and monitoring and control.

***Phase 1:* Defining the Context**

The consideration of external and internal factors is necessary while developing a risk management plan. External factors include external stakeholders, local, national, and international market environment. Internal factors include internal stakeholders, contractual relationships, and the capabilities of the organization.

***Phase 2:* Identification of the risk**

It is a process that involves finding, recognizing, and describing the risks that might hinder the process of attaining the objectives. It is accustomed to determine the possible sources of risk before the occurrence of the events and circumstances.

Historical data, theoretical processing, expert advice, and stakeholder input can be used to determine the risks that an organization might face. As per PMI, there are three types of reasons. This includes: a) External (unpredictable, predictable); b) internal (non-technical, technical); and c) legal.

This helps to identify the potential risks in the early stage. It minimizes the possibilities of time and cost overruns. Additionally, risk recognition encourages one to make a detailed understanding that can impact partners and make better venture choices.

***Phase 3:* Assessment of risk**

The assessment of the identified risk is usually done based on the following two factors.

- Chances of the event to occur.
- Active and passive consequences of the event.

The strategies to deal with the risks of different severity (i.e., the amount of damage a hazard could create) are explained in Table 11.1. It is also essential to judge risk using the probability scales as given in Tables 11.2 and 11.3.

The risk rating can be defined as probability (also called likelihood) $\times$ consequences $=$ risk rating.

For instance, if the probability of a fire mishap is "improbable" (a score of 2), however, if it happened, the outcomes are "serious" (a score of 4). Consequently, fire has a risk rating of 8 (i.e., $2 \times 4 = 8$). The risk rating table (as given in Table 11.4) can be used to understand the level (severe, high, moderate, and low) of the risk.

***Phase 4:* Risk Control**

Risk control employs a set of methods for evaluating potential losses and undertakes necessary actions to reduce or eliminate such threats. It utilizes the findings of risk

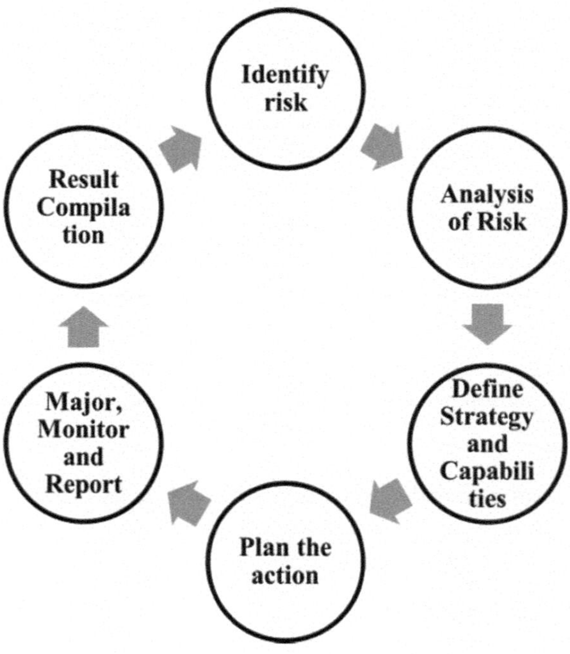

Fig. 11.2 Phases of risk management

Table 11.1 Severity of risks

Operating conditions responsible: Human error, design deficiencies, element, environment, subsystem or component failure, or procedural deficiencies	
Catastrophic	Operating conditions may commonly cause death or major system loss. This demands an immediate termination of the hazardous activity or operation
Critical	Operating conditions may lead to severe injury or illness or major system damage, thereby requiring immediate corrective action
Marginal	Operating conditions may cause minor injury or illness or minor systems damage. This can be controlled without severe injury, illness, or major system damage
Negligible	Operating conditions will result in no or less than minor illness, injury, or system damage

Table 11.2 Risk probability scale for likelihood

Level	Likelihood	Description
4	Very likely	More than once a year, this happens in the industry
3	Likely	Just once every year, this occurs in the business
2	Unlikely	Every ten years or more, this happens in the industry
1	Very unlikely	Happens only once in the industry

Table 11.3 Risk probability scale for a consequence

Level	Consequence	Description
4	Severe	Losses financially greater than $50,000
3	High	Financial losses between $10,000 and $50,000
2	Moderate	Financial losses between $1000 and $10,000
1	Low	Losses financially less than $1000

Table 11.4 Risk rating and actions

Risk rating	Description	Action
12–16	Severe	Immediate corrective action required
8–12	High	Corrective action can be done within one month
4–8	Moderate	Corrective action can be done within three months
1–4	Low	Does not require corrective action currently

assessment and aims to implement necessary steps to minimize the impact of risks. This helps an organization to control damages and limit losses. The strategies used for risk control include:

- Transfer of risk
- Exclusion of risk
- Reduction of risk
- Acceptance of risk (totally or mostly)

The implementation of a risk control mechanism (transfer, exclusion, reduction, and acceptance) depends upon the situation and condition of the organization at a given point in time. The selection of risk control mechanism is also governed by both short-term and long-term financial and operational losses. The few risks that cannot be exchanged or avoided are acknowledged. The acknowledgment is specifically beneficial when the risk has:

- The probability of the occurrence is low.
- Consequences have little relevance.
- Benefits are extraordinary if fruitful.

***Phase 5* : Effective Communication**

Communication of risk is a highly effective strategy in the process of managing risk. A risk management report documents the following details:

- Profile
- Matrix
- Risk treatment
- Control planning

The above details should be timely shared with the project team members. Risk communication is an essential tool for spreading data and understanding risk management decisions. This understanding of the data should enable the stakeholders to appreciate how the choices can impact their interests and values. Effective communication should involve:

1. Identification of stakeholders
2. Create and analyze stakeholder profiles (prioritizing stakeholders in terms of their relative influence, interest, and perspective toward the project)
3. Define the purpose (keeps everyone informed of developments)
4. Timely delivery (timing is critical, and communication often fails because it is done too late. People always need time to adjust to the change and accept)
5. Measure results (communication works best once it is two-way. Therefore, ensure there are mechanisms in place to capture measure and answer feedback).

***Phase 6:* Prior Planning**

This defines the risk prevention methods, which are:

- Analyzing and storing incoming information for the control process
- Suitable level and localization for the decisions and activities
- Preventive instruments to be used
- Interpreting and capturing outgoing data for the control process

***Phase 7:* Supervision**

It is essential to decide the periodicity of checking and supervision. This result ought to be constantly reported, assessed, and recorded. One plan is never enough in the hazard administration process, and hence, continuous monitoring and updating are essential to the success of the risk management process.

***Phase 8:* Review of the process**

The risk management process is dynamic, and continuous checking is essential. The purpose of this review should include:

- Possible advancement that worries any period of the procedure ought to be assessed
- Efficiency and adequacy of the embraced risk management plan ought to be assessed
- Checking and overseeing results ought to be assessed

It is noteworthy to appreciate that risk management is a systematic process, and any ad hoc approach leaves the project and team members in an awkward situation. A complete idea of project risk management with its key components, necessary inputs, and risk mitigation strategy is presented in Fig. 11.3.

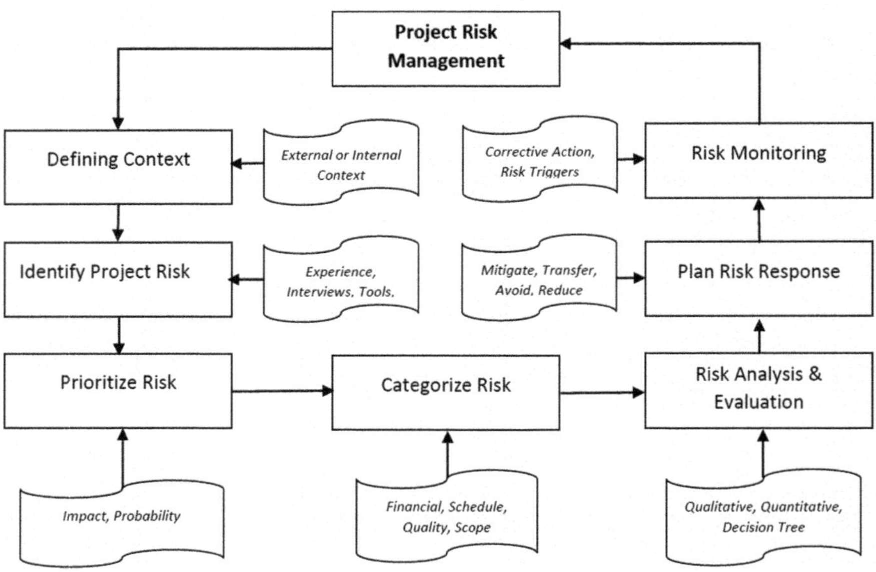

Fig. 11.3 Project risk management

Exhibit 11.1 RACI matrix

Deliverable	Executive sponsor	Project sponsor	Agency CIO	Project manager	Development team	Project stakeholders
Complete system	I	A	A	R	I	I
System documentation	I	A	A	R	I	I
Implementation notice	I	I	A	R	I	I
Readiness document	I	I	A	R	I	I
Version description document	I	I	A	R	I	I
Post-implementation review Report	I	A	A	R	I	C
Standard operating procedures	I	A	A	R	I	I

Project Management in Practice

Why Do Construction Projects Experience Delays?

Doloi et al.'s (2012) study of delays of Indian construction projects identified the following seven critical factors in descending order of importance:

(1) Lack of commitment by the client, contractor, and vendors leading to accidents, improper or obsolete construction methods, delay in material delivery

(2) Inefficient site management due to ambiguous specifications, unskilled labor, ineffective supervision, the inadequate experience of the contractor, lack of control over subcontractors

(3) Poor site coordination due to lack of coordination between site and design office, non-availability of drawings/designs on time, unrealistic schedule built in the contract

(4) Improper planning in ignoring extreme weather conditions, which lead to low labor productivity and therefore lead to errors in time estimation; improper planning for recruitment of skilled operators for specialized equipment; improper planning for the requirement of equipment and their utilization

(5) Lack of clarity of project scope resulting in rework or scope creep due to misunderstanding by the contractor or project manager

(6) Lack of communication with local authorities resulting in delays in permissions; lack of communication between contractor and client results in delay in the approval of stages substandard contract, selection of contractor with inadequate experience or skill sets, optimistic cost, and time duration built in the contract

Source: Doloi, H, Sawhney, A, Iyer, KC & Rentala, S 2012, Analysing factors affecting delays in Indian construction projects, International Journal of Project Management, vol. 30, 479–489.

11.3 Types of Risks

If a company can identify its internal risks and external risks, then the adverse impact can be minimized.

11.3.1 Internal Risk Factors

An organization faces inside dangers that may emerge from the routine tasks of the organization. The organization has a decent possibility of decreasing internal risks as these dangers can be anticipated easily. The internal risk factors are three types: Human factors, technological factors, and physical factors.

1. Human factor risk can be:
   - Union strikes
   - Dishonesty by employees
   - Ineffective management or leadership
   - External producers or suppliers may fail
   - Clients and customers fail to pay, etc.
2. Technological risk arises because of unexpected events like information security incidents, cyberattacks, password theft, service outages, sudden machine failures, etc.
3. Physical risk occurs when there is a loss or damage to the assets of a company.

An organization can devise suitable strategies to hedge against internal risks. This may include having a backup supplier, improving employee and staff awareness for handling emergency situations arising because of natural disaster, pandemics, security features, storage, and backup for IT systems.

Throughout an organization, there are internal risks that arise during regular operations. Internal risks may be forecasted and can be avoided or mitigated. They are caused by one (or a combination) of human, technological, or physical factors. Project risk drivers reside with the project or with the association's reams, like the financial efficiency of the overseeing body, the ability to deal with human resources, etc. Figure 11.4 summarizes the internal risks in the construction industry.

The key terms associated with the identification of internal risk drivers include:

1. Cause: Empathetic action can be taken before preventing or reducing the probability of effectiveness.
2. Event: Knowing the event allows you to identify properties and determine when it occurs, so no plan to take or unplanned requires action.
3. Time window: This allows the project team to focus more on their endeavors.
4. Impact: This gives an estimate of the effect and the potential size.

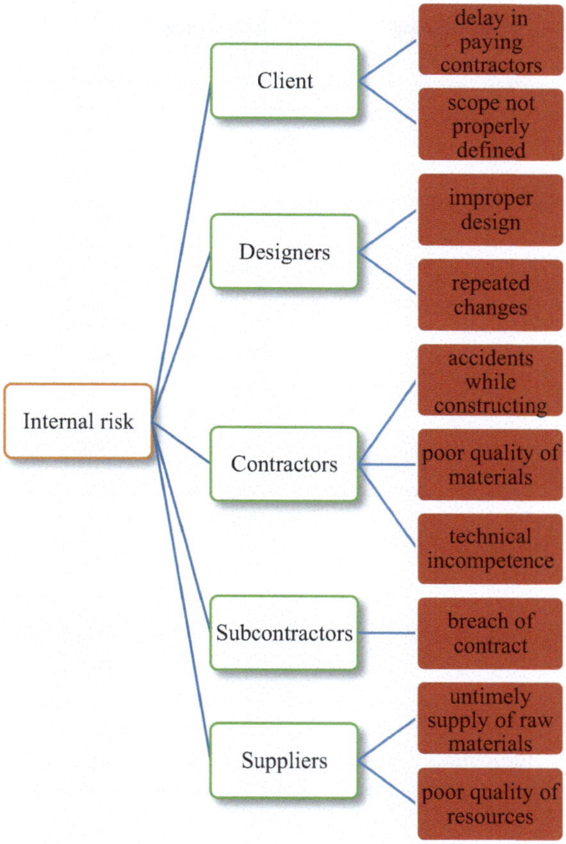

Fig. 11.4 Internal risks in construction industry

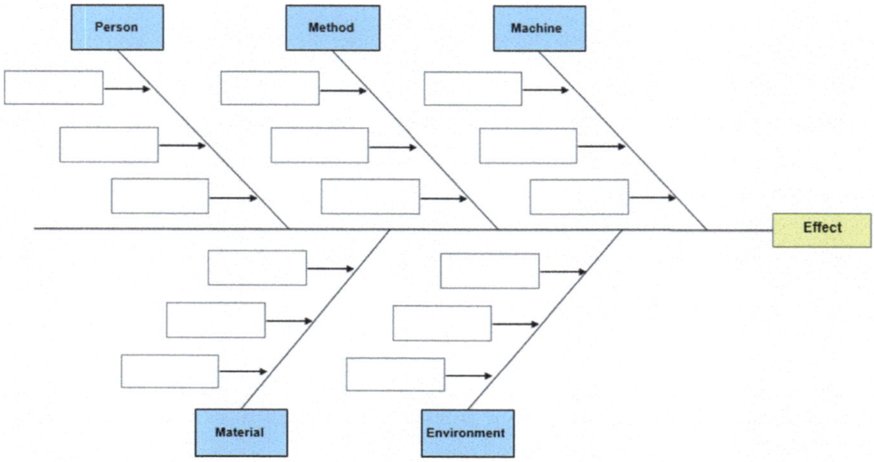

Fig. 11.5 Fishbone (cause and effect) diagram for risk assessment

Table 11.5 5W2H method for identification of internal risk

	5W and 2H	Response
5W	What is the issue? We need to depict it in a solitary sentence, with the goal that others will have the capacity to comprehend what we mean	The issue is…
	For what reason is it an issue? What is the torment?	This is an issue because…
	Where do we experience the issue?	We experience the issue at (Location) (Time) when (Specific situation)…
	Who is affected?	This affects: (Staff) by…, (Patients) by…, (Other suppliers) by … (others) by…
	At the point when did we first experience the issue?	We originally experienced this issue…
2H	How could we know there was an issue?	The indications of this issue are…
	How regularly do we experience this issue?	We experience this issue (x) times, and each experience is (this huge). The issue is getting (better/more regrettable)

The "fishbone" diagram (Fig. 11.5) for identifying internal risk shows the potential risk factors affecting the problem graphically. The layout identifies the causes and possible reasons. It is used in the analytical phase. We need to put the impact or issue proclamation on the right side of the paper, mostly down, and afterward draw an even line with an arrow indicating the impact or issue articulation. The fishbone diagram helps to identify the causes (factors responsible for an effect) under five categories such as:

- Person
- Method
- Machine
- Materials
- Environment.

5W2H method for identifying internal risk: This method helps to apply critical thinking for identifying various internal risks. The details of 5W and 2H are summarized in Table 11.5.

11.3.2 External Risk Factors

External risks are beyond the organizational boundaries. This may include economic crisis, natural and political disturbances, etc. It is difficult to estimate and control such risks. External risks are of three types:

1. **Economic risk:** A financial downturn may cause a sudden, unforeseen loss of income.
2. **Natural risk:** The events that influence routine project activities. For example, a seismic tremor may disturb the supply chain activities; COVID-19 lockdowns have created a crisis for businesses worldwide in the year 2020.
3. **Political risk:** Any change in government policies or structure impacts the project execution. This may cause war, strikes, reallocation, exchange restrictions, and changes in import/export structure.

Table 11.6 explains various risk categories and their possible impact on the project.

Table 11.6 Risk category and impact on project

Risk category	Factors	Specific risk (Examples)	Impact on Project
Internal risk	Human	Union strike	• The project schedule will be altered; if the project needs to be completed on time budget may increase • Other workers' involvement in the project will drop, which will bring the quality of the project down • May cause panic among stakeholders if not dealt with in time
		Dishonest Employee	• May take off the competitive edge in the market by sharing the project details • May corrupt co-workers
		Ineffective management	• The project team may move out of control • This risk can lead to other internal risks like strikes, supplier failure, etc. • Stakeholders may start to doubt the management and project success
		Suppliers fail	• Tough to keep up the schedule • Finding another supplier may be time and cost consuming
		Customer fails to pay	• The time value of the project will increase, challenging for a company to move to another project • Directly affect stakeholder
	Technological	Manufacture problem	• Project completion will be delayed • Solving the problem will be time and cost consuming
		Distribution Problem	• This may lead to an unhappy client • Can bring the reputation of the company down
	Physical	Asset Damage	• This will increase the cost of the project • Replacing or repairing the asset will be time and cost consuming

(continued)

Table 11.6 (continued)

Risk category	Factors	Specific risk (Examples)	Impact on Project
External	Economic	Economic downturn	• Funding for the project will be affected • Stakeholder may cancel or withdraw from the project • Demand for the project outcome might change
	Natural	Natural disasters	• Damage or loss of assets/lives • Alters supply and demand • Drop-in sale
	Political	Change in law, taxes, etc.	• Getting approvals might get complicated • The entire project or part of the project might need to be changed

11.4 Risk Monitoring

Step 1 : Create a Management Summary

A summary indicates whether the project is on target to meet its important delivery dates and budget objectives. If not, mention the key reason. Also, forecast any significant risk that might be in danger of occurring or have occurred. Additionally, to make the project successful, mention what is expected of management and the client in the short term. This section should not be more than half to one page, regardless of the size of the project. It can even be just one sentence if everything is going according to plan.

Step 2: Tabulate key project milestones

ID	Title	Plan completion date	Forecast completion date as reported two months ago	Forecast completion date as reported last month	Current forecast completion date	Actual completion date
1	Description	dd-mmm-yy	dd-mmm-yy	dd-mmm-yy	dd-mmm-yy	dd-mmm-yy
2	Description					
	**					
	**					
	**					
N	Description					

Step 3 : Report the activity progress and deviations from the plan

Achievements showing progress toward the key milestones should be noted. This section should not be a meager commentary on what was done last month.

Table 11.7 Risk register

ID	Description	Mitigation plan (what is being done to prevent the risk)	A contingency plan (what will be done if the risk occurs)	Likelihood of occurring	Potential impact (dollar/schedule/quality, etc.)

Step 4: Create a Risk Register

Based on the project progress table and other risk analyses, make a risk register. Make sure that this section contains only the critical project risks. To document the risk, one can use the format given in Table 11.7.

Project Management in Practice

Importance of safety management system to improve the effectiveness of project

by

Dr. O. B. Krishna

Ex Senior executive of Tata Steel & Current Professor, IIT Kharagpur

Project management is intended to complete the task chosen in time and within the cost. It has to look into the life cycle stages of the task like design, construction, operation/maintenance, and decommissioning for the above two aspects. The project managers used to adopt the reactive mode of interventions or deal with the consequences when safety issues arise, resulting in more cost, disruption in the schedules, and a huge reputation of the organization. This is to happen till recently, even with the project management in reputed organizations. The main reason for this thinking is the lack of knowledge in safety engineering and management and its consequence by the project management team.

It has been proved, now, beyond any doubt, that the effort and resources spent to look into safety by the project managers had improved not only the schedules and cost but many other domains. To cite some examples, the pedestrian paths, roads at the Greenfield sites provided initially have improved the movement of materials greatly and reduced the incidents. Provision of the electrically operated transmission (EOT) cranes in the shop floor was traditionally the last activity. This used to involve mobile cranes and huge manpower in the installation of equipment. These EOT cranes were installed on the shop floor before the equipment was unloaded and erected as a safety requirement. Tata Steel had seen huge gains at their Jamshedpur site to expand their integrated steel plant from 6 to 10 MT and 6 MT plant at their Tata Steel Kalinga Plant. Fire protection systems used to be ignored due to cost, resulting in the loss of humans and equipment at the commissioning and operation stage. Energy isolations were never thought to be necessary at the construction activities to the mobile equipment and fixed equipment. Adopting these so-called safety measures at the beginning while planning and during execution has resulted in gains in all the project management objectives. In fact, this has given birth to the concept of "Prevention through Design (PtD)," which envisages shifting safety to the upstream side of the project activities. Its cost of implementation is reduced, and the ease of implantation is improved.

Being a person involved in the design, manufacture, commissioning of steel plant equipment on a turnkey basis and in the maintenance of integrated steel plant, I have seen the benefits of thinking about safety engineering and management at the project stage, as it improves the effectiveness of the project management immensely.

11.5 Stakeholders in Project Team

A stakeholder can either be an individual, group, or organization that the project's outcome will impact. The project's success will be their prime interest, and they can be within or outside the organization that is sponsoring the project. Stakeholders can have either positive or negative influences on the project. The key project stakeholders are indicated in Fig. 11.6.

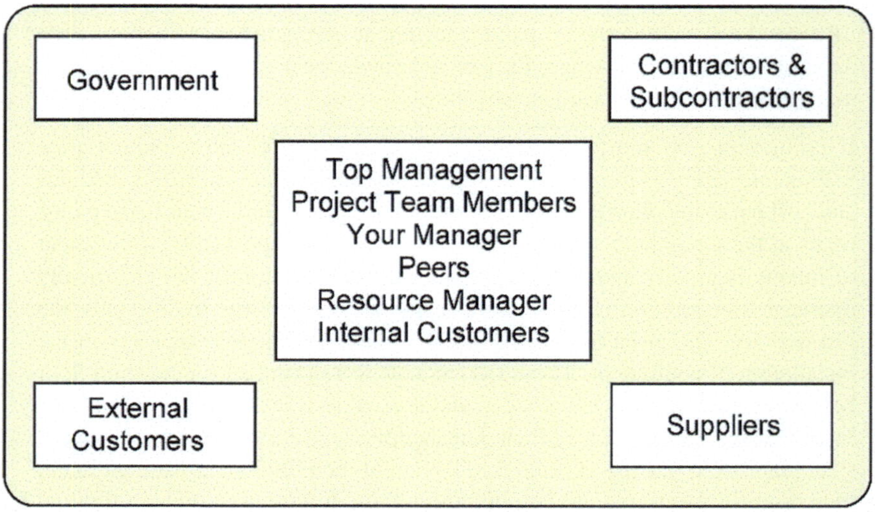

Fig. 11.6 Project stakeholders

Project Stakeholders

Top Management

The company's president, vice-presidents, directors, division managers, the corporate operating committee, and others constitute the top management. They direct the company's strategy and development. The top management support will make it easier to recruit the most effective employees and secure the required materials and resources. This visibility enhances a project manager's professional standing within the organization. On the negative aspect, failure will be quite dramatic and visible to all, and if the project is massive and costly (most are), the price of loss will be a lot more substantial than for a smaller, less visible project.

Few suggestions to deal with top management are:

- Develop in-depth plans and significant milestones that have to be approved by high management throughout the planning and design phases of the project.
- Ask top management related to your project for their data coverage needs and frequency.
- Develop a standing coverage methodology to be distributed regularly.
- Keep them aware of project risks and potential impacts at all times.

The Project Team

The project team primarily consists of people dedicated to the project or borrowed on a part-time basis. As a project manager, you need to provide leadership, direction, and especially support to team members as they accomplish their tasks. Operating closely with the team to resolve issues will assist you in learning from the group and build rapport. Showing your support for the project team and every member can help you get their support and cooperation.

Some difficulties in dealing with project team members are:

- As they are borrowed, their priorities may be different as they do not report to you.
- They may be dealing with many projects at a time to have difficulty meeting the deadlines.
- Personality conflicts may arise.

Managing project team members requires interpersonal skills. The following suggestions can help the project manager.

- Team members can be involved in project planning.
- Arrange for meeting privately and informally with each team member at several points in the project.
- Always be available and listen to team members' concerns at any time.
- Encourage team members to help each other at times of need.
- Develop a complete project performance review for the project team members.

Resource Managers

As project managers are borrowing resources, other managers are there to control these resources. So, the relationships with these resource managers are essential. If the connection is good, it will help consistently acquire the best possible resources (staff and equipment) for the projects. If the relationship is not favorable, it might be challenging to avail necessary resources timely.

Internal Customers

Internal customers are individuals among the organization who are customers for projects that meet the requirements of internal demands. The internal customer holds the facility to accept or reject your work. Early in the relationship, the project manager needs to talk terms, clarify, and document project specifications and deliverables. Once the project begins, the project manager should keep tuned in to the customer's considerations and problems and keep the customer enlightened about the project.

Common stumbling blocks that might be faced while dealing with internal customers are:

- A lack of clarity about customer wants
- A lack of documentation of the requirements
- A lack of understanding about the organization of the customer and its operating characteristics
- Requests by the customer can be unrealistic deadlines, budgets, or specifications
- The hesitancy of the internal customer to sign off the project or accept responsibility for the decision.
- Changes in the project scope.

The following strategies can be adopted to overcome the above-mentioned bottlenecks.

- Develop knowledge about the client's organization, culture, and business
- Discuss and clarify all project requirements and specifications in the written agreement
- Specify a change procedure if needed
- Make the project manager the center of communications in the project organization

External Customer

External customers are the customers who come into the picture when the project enters the market. For example, in an infrastructure project like the construction of a COVID-19 hospital, the people infected by coronavirus are the external customers.

Government

Project managers faced with stringent regulatory norms (e.g., industries in pharmaceutical, banking, or defense sectors) must strictly adhere to the government regulations and compliances.

Contractors, Subcontractors, and Suppliers

When the organization does not have the expertise or adequate resources in-house, then the work is outsourced to contractors or subcontractors. The problems that can arise with contractors or subcontractors are:

- Quality of the work
- Cost overruns
- Schedule slippage
- Scope change

More than half of the project manager's time is spent on managing contractor and supplier relationships, depending upon the nature and complexity of the project. It is purely intuitive; it also involves a skill set that includes managing conflicts, negotiating, and other interpersonal skills.

Chapter 12
Project Procurement and Contracts Management

Critical Questions

Why project procurement is an important issue?

What are the key considerations in project procurement management?

What are the various contracts in project management?

What are the approaches to evaluate procurement and contract decisions in project management?

What is opportunism in project procurement? How does it impact the performance of the project?

12.1 Issues in Project Procurement Management

Procurement is an indispensable activity during the project life cycle. It deals with obtaining goods, supplies, and/or services. It ensures a smooth execution of the project by obtaining all materials and services essential during the various phases of the project. Project procurement management aims to ensure the availability of goods or services on time and within budget. In addition to its cost-saving benefits, it can also provide a competitive advantage by improving customer satisfaction, time to market, and organizational efficiency. The importance of procurement management in greater detail can be explained as follows:

1. **Cost Savings**: Improving cost savings can be the most critical function of procurement management. It involves analyzing the product from various suppliers spread across vast geographies and identifying the best option to balance the cost and quality as required. It is the biggest reason why large retailers are able to maintain highly competitive prices despite massive infrastructure and costs overheads.

J. J. Thakkar, *Project Management*, Management and Industrial Engineering,
https://doi.org/10.1007/978-981-15-3695-3_12

2. **Ensuring Compliance**: Even though procurement may seem concerned only with purchasing activities, it ensures that the organization complies with all regulations across the global locations. An organization needs to make sure that it is aware of the plethora of rules and regulations governing the corporate sector globally and adhere to these norms to prevent paying hefty fines and damaging its reputation.
3. **Enhancing Supplier Relationship Management**: Developing a long-term relationship with the suppliers to get the best product at the best price is the primary objective of the procurement team. Procurement Management is about going one step further and striving to integrate the suppliers as partners to assist in the value creation process.
4. **Increasing Organizational Efficiency**: Procurement management ensures the best match to needs with goods and services, but it also helps maintain the quality of goods and ensure timely delivery. A dynamic procurement practice can help an organization develop resistance to market volatility like labor shortage, price changes, economic volatility, and financial problems.
5. **Supporting Strategic Vision**: Procurement management helps understand the tactical needs and requirements to maximize its return on investment and its long-term strategic planning guides every organization. Procurement management assists an organization in achieving its strategic goals. For instance, an organization might have a plan to create the best-in-class product. Procurement teams can develop ideal product and part specifications and help identify suppliers who can provide them with high-quality materials. Bad procurement management practices can have very critical short-term and long-term consequences on any organization. Short-term issues include things like confusion, redundant spending, and misinterpretation of contracts and needs. These all result in a company being unable to obtain the products it needs, at the right quality to deliver its goods and services. A project management office must execute the procurement plan as per defined procedure and schedules. The various elements involved in project procurement management are presented in Fig. 12.1. This begins with the procurement planning, followed by its conduction, which decides the set of policies for supplier selection, decision to make or buy items, etc., control the procurement, and finally closure of the procurement process.

We may hypothesize an example, say Mr. Lanka owns a club and needs to order lighting for swimming pool, garden, and guest areas. Consider the number of ways a light bulb can be described: Regular light bulb; 50-W light bulb; 50-W energy-saving light bulb; 50-W white light bulb; 50-W frosted, white light bulb, 20-W CFL light bulb, halogen light bulb, LED light bulb, etc. Also, consider the number of suppliers from whom it can be ordered each of whom may be offering light bulbs of the same type at a different price range.

The point is, if one is asked to order a light bulb, numerous types of bulbs can be procured. Whenever a lot of people are involved in the decision-making process, a product can be described in a variety of ways. And every light bulb person X buys that

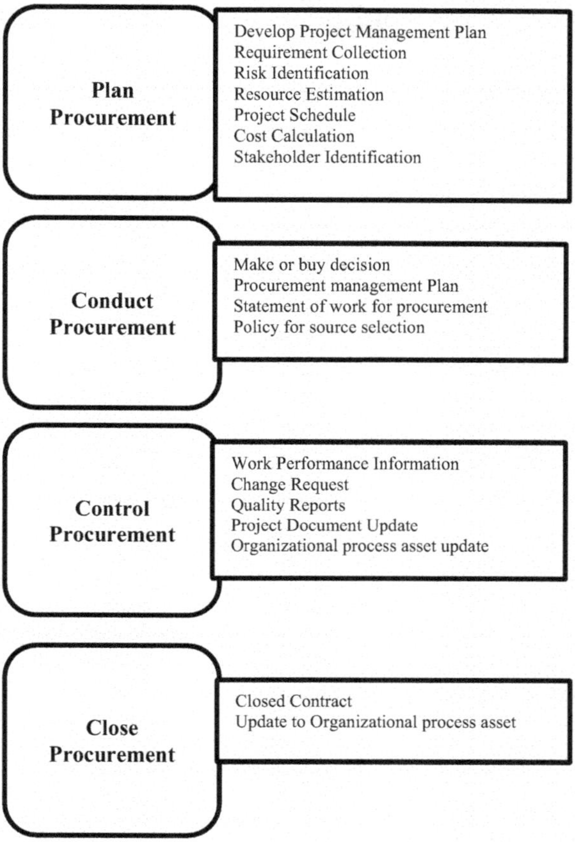

Fig. 12.1 Project procurement management

he cannot use is wasted money and time. Hence, poor procurement management can mean no proper analysis of suppliers available for the same product domain, resulting in lower quality purchases or purchases that do not meet the needs of a company. The repeated problems with procurement can lead to some long-term issues, including money loss, diminishing profitability, bad reputation, and negative publicity. Having poor procurement strategies means making redundant purchases, paying too much for transportation, suffering from a lack of quality control, or developing problems with contract scope. For example, a family purchases a car that does not satisfy their requirements regarding the number of persons to be accommodated, safety for children and elders, fuel economy, etc.

12.2 Phases in Project Procurement Management

As per the PMBOK guidelines, the project procurement process involves four phases as summarized in Table 12.1. The application of each of the four phases of the procurement process is illustrated for two projects: Construction projects and R&D projects in Table 12.2.

Table 12.1 Phases of project procurement

Phase I	Plan procurement management	• Identifying which materials, products, or services should be procured from outside the organization, products that should be manufacturacted at the plant or site itself, identifying potential suppliers who can meet our needs • It deals with the identification of procurement needs. Procurement documents usually consist of a request for proposal, invitation to tender, or the like, with the statement of work included as an attachment • A statement of work is composed for each external vendor requirement, which can be named in many ways (terms of references, scope statement, etc.), but serves as a written statement of what work will be performed by the contractor • Additionally, the procedures used to seek proposals and bids and the decision-making criteria are finalized at this stage. Potential sellers are identified, and the preferred vendors might be contacted
Phase II	Conduct procurements	• After identifying the procurement needs of an organization and finalizing the procurement procedure, the procurements are executed • This is done by sending out requests for proposal (RFPs) or invitations to tender and analyzing the responses • Selection criteria are determined in advance. This is followed by signing the agreements and updating the project management plan with the new cost and schedule information obtained from the vendor
Phase III	Control procurement	• Once the procurements have been finalized, it is indispensable that both the buyer and the seller conduct the control of these procurements to ensure that the other party meets its contractual obligations • It is often seen that poor contractor management leads to the budgets and schedules going out of control and derailing the project as a whole • During regular project status intervals, the project manager must review the subcontractor agreements, progress updates, and work performance information to ensure that subcontractors are on pace to meet their budget and schedule commitments • It is not enough to assume that subcontractors are "good at what they do" and meet their contractual obligations. That is simply negligence on the part of the project manager
Phase IV	Close procurement	• Upon completion of work or termination of the contract, procurements are closed • Early termination may result from a mutual decision or default on the contractors or the buyer's part • The work that has to be done during these phases may include product verification, negotiation of settlement, financial closure, procurement audit, updating of records, final contract performance reporting, recording the lessons learned, and procurement file generation

Table 12.2 Relevance of project procurement phases for "construction" and "R&D" projects

Phase	Construction project	R&D project
Plan procurement management	• The quantity and quality of raw materials required, the cost of procurement, the time of procurement as per the construction schedule, etc., are finalized based on the scope and scale of the project. Since the project's scope is clear and well-defined, these things do not require extra precautions and contingencies to be put in place as such Fixed-price contracts are generally preferred • If some work is being outsourced to an external contractor, his scope of work is defined using the statement of work, which is used to invite tenders	• It is a significant challenge in any R&D project as the scope is not defined. So it becomes a challenge to balance the objectives of the work with the cost limitations • One should not be too ambitious if the cost is a significant concern • The scope should be defined as closely as possible, even if it is not fixed. Since most R&D projects are pilot projects, the material requirement is small, and the purchase of equipments are usually one-time events • The contracts signed for such projects are always cost—reimbursable type, and most of the risk lies with the client • The contractors are generally large firms with the knowledge, experience, and ability to provide state-of-the-art equipment and suitable quality materials • Quick delivery of the materials is the key to any R&D project. Time is a more important consideration than cost when designing contracts
Conduct procurements	• The invitations to tender are sent, and the most economical bid is accepted • The focus is on the economy and the ability of the contractor to carry out the procurements	• It is essential to ensure the availability of suppliers of the equipment for a research project • Good brands are generally preferred for the contract • The supplier has a higher bargaining power here • Finding a contractor willing to deliver the materials as per the contract terms is a major challenge, let alone finding one that promises timely delivery of equipment at an affordable price
Control procurements	• It is ensured that all the deliveries made by the contractor comply with the quality norms and provides timely delivery at an agreed cost • Poor execution at this stage can lead to the delay of the project	• It is often seen in R&D projects that the delivery gets delayed, sometimes enormously. The vendor was unable to complete the job due to the item's developmental status, despite the agreed delivery schedule. At the beginning, it was not possible to estimate the completion time precisely • In an R&D setup, there is a lack of adequate information on sources, likely price, and a justified delivery schedule • A specification often remains a variable until just before the placement of order

(continued)

Table 12.2 (continued)

Phase	Construction project	R&D project
Close procurement	• After the project is complete, the contract is terminated unless it has to be discontinued for some other reasons • The contractor's performance is reported along with a financial audit to see if a long-term relationship can be established with the contractor to benefit future projects	• Regardless of whether the R&D effort was successful or not, the contract must be terminated once the project is completed • The learnings from the procurement do not generally apply to other R&D projects as they are unique • Even trying to establish a long-term relationship with the contractor might not be a concern

12.3 Importance of Contracts in Project Management

Implementation of concrete project management is crucial to the successful completion of the project. Contracts are the legal terms signed between the client and contractor to complete the project as per the quality standard within the stipulated time duration.

It helps manage the client's requirements effectively, saving time and effort for both the client and the contractor. Both parties can benefit from a contract that complies with the needs and properly allocates risks. The accomplishment of a project mainly depends on the exhaustiveness and the character of the project contract. The contract describes how conveyance and acceptance of the product and service will be done.

Contracts are vital as they command the critical business strategy and associations. According to the stipulated requirements, it is a general practice for the contractual parties to spend a substantial amount of their money and resources finishing the contracts.

Due to the lack of proper supervision and monitoring, contractual obligations may not be fulfilled even after the agreement is established between parties and services are procured as required. Failure on the part of the client or contractor can lead to loss in profits, penalty on contractor, legal battles between different parties, loss of goodwill, and broken associations. All these lead to reduced public benefits and loss of public money in typical government or PPP projects.

PPP project contract defines payment terms, negotiation for risk allocation, the flow of work, and likely service levels of the completed project. Effective project management ensures a good relationship between the different contractual parties involved; in addition, it helps ensure that the work is carried out in compliance with the contract terms, eliminating risks for all stakeholders involved in the project. Projects carried out by infrastructure builders on behalf of the government require effective implementation of project management. Usually, government projects are of large scale since critical infrastructure requirements of the country (i.e., catering to the needs of its population) are usually met by government projects. Due to the lengthy period of projects and significant funds involved, the situation often becomes complicated. Thus, proper implementation of contracts and management are the only ways of ensuring low risk and uncertainty in the project.

Experienced professionals from the industry select contractors. This is often done by evaluating the contractor's experience with similar projects in the past and financial stability. The contractor's approach for safe execution and arrangements made to accomplish the anticipated results are also examined. There is generally no rigid fixed approach in the selection process, and it changes from project to project. The following factors need to be considered when selecting a contractor:

(1) Financial status includes financial firmness, turnover, obligations, profit, amounts due, and available monetary funds.
(2) Technical capabilities of the contractor, such as knowledge, plant and machinery, and qualification of employees along with their relevant experience.

(3) Management competencies, such as performance on similar projects, quality management system, quality control policy, project management system, and management knowledge.
(4) Examining profiles of the senior management, including knowledge, occupancy with firm, and division of duties.
(5) Existing projects/backlog with size, and location of projects, percent of capacity being utilized, status and anticipated completion, previous failures in finished projects, number of years in building, past customer associations, and collaboration with contractors.

12.4 Types of Contracts

Risks are always associated with projects. It might be described as the probability of danger, harm, hazard, mishap, or other adverse incidents caused by external or internal vulnerabilities. Some of the critical risks faced by contractors are summarized in Table 12.3.

Table 12.3 Contractors' risks

Damage of equipment	Many heavy, high-tech, and costly equipment are involved in modern construction to maintain the quality of work and to construct in large numbers. Their maintenance is very crucial for the contractor because his livelihood is based on these machines. Therefore, workers must be properly trained to use this equipment in order to reduce wear and tear and extend the life of the equipment. In case of any damage, property protection arrangements and insurance can be very useful and act as lifelines
Accident on site	Ensuring safety for workers and other people visiting the site should be the utmost priority for the contractor. Workers must be equipped with protective measures like helmets, jackets, boots, etc. The location of construction projects are full of probable hazards. Despite having taken all necessary precautions, even experienced construction workers can get hurt. Hence, everyone who works at the site or visits should be familiar with the safety standards. Everyone working on site should have health insurance that covers medical expenses in case of an accident
Defective work	According to the contract between client and contractor, it is evident that the contractor is responsible for any complete project which is not structurally sound or unfollowed provisions written on code or for any other type of defect. If the contractor does not follow the local or state bylaws regarding construction, then his work may be demolished, and he may be required to do the job again or to reimburse the clients for the losses. Strict compliance with the government guidelines is the best way to avoid financial losses and damage to reputation
Missed deadline	Since most projects are very large and take a long time to complete, it is hard to predict every unforeseen event which can delay the project. However, any number of unanticipated snags in the infrastructure development business can keep a project from being finished on time. Building materials can be procured late, equipment and machine can unexpectedly break down, and plans might have to be redrawn altogether on such events. To limit the effect of missed due dates, the contractor should keep the lines of correspondence open with the client

The different types of contracts in project management are:

1. **Firm-Fixed-Price Contract (FFP)**: In a firm-fixed-price contract, the contractor has to quote a fixed price depending on the preliminary investigation for completing the project. If given the contract, the contractor has to complete the work below the quoted price to gain profit. In the event that the cost of work increases due to some unforeseen circumstances, the contractor will suffer losses. For maximum profit, he must achieve the work at the lowest possible cost without compromising on quality. Here, the risk lies with the contractor. The client can give incentives to the contractor for completing the project on time. In this arrangement, the client has minimum risk allocated. When the scope of work is clearly defined, and all the subtasks and activities required to complete the project are identified in advance, this contract can be used.

   A firm-fixed-price contract is appropriate when the clients can bargain with the suppliers/contractors to get the work done for a project whose events are clearly defined. The following criteria should be met when using a firm-fixed-price contract:

   - The number of contractors should be enough so that there is no monopoly situation and fair bidding can take place.
   - Price quoted by the contractor should have some rational basis such as previous work of similar nature whose pricing data is available.
   - Information regarding the cost for completing a similar project helps in the precise evaluation of performance cost.
   - Since a fixed price is quoted, the contractor should carry the uncertainty analysis beforehand to identify some possible risks and add the additional cost incurred to the quoted price to nullify the effect of risk.

2. **Firm-Fixed-Price With Economic Price Adjustments (FPE)**: A fixed-price contract with economic price adjustment is a special type of fixed-price contract. It provides scope for changing the quoted price. This is specifically useful at the beginning of work if some discrepancy or construction cost is higher than the expected cost. Economic price adjustments are of three general types:

   (1) Modifications based on quoted price: If there is a change in construction cost, then alteration from quoted price is possible.
   (2) Modifications based on the cost incurred on labor and material: This price alteration is based on changing labor and material costs that the contractor may face while completing the project.
   (3) Modifications based on labor or material cost indexes: These price adjustments are based on variation in labor costs and material price standards, or indexes explicitly acknowledged in the contract.

3. **Fixed-Price-Incentive-Fee Contract**: Firm-fixed-price contracts refer to contracts where the customer or purchaser pays a fixed price. The amount may vary if the seller meets some predetermined criteria related to their performance.

These type of contracts can benefit both the customer and the seller. It is advantageous to the seller because it typically allows the seller to charge a reasonable base fee and commit to excellent service. It also benefits the buyer as he will be paying a reasonable base fee in the beginning. However, there are chances that prices may increase in the future. When the price increases, the buyer is only liable to reward the excellent performance of the seller. During ordinary circumstances, no additional costs are incurred.

4. **Cost-Plus-Incentive-Fee Contract**: A cost-plus-incentive-fee contract is a distinct type of fixed-price contract that provides contractors and sellers with extra financial incentives to keep the venture's expense as low as possible. An agreement of this nature may likewise offer an incentive when the vendor meets other criteria in the agreement, before work starts. The cost-plus-incentive-fee contract is a compensation mechanism wherein the contractor is given a particular incentive to complete the prescribed budget or an incentive to complete work ahead of the calendar.
5. **Cost-Plus-Award-Fee Contract (CPAF)**: One of the contractual agreements used by most establishments today is the cost-plus contract (cost-reimbursement contract). This type of contract gives compensations to sellers for the actual legal costs sustained for a particular finished work. This type of contract allows financial enticements based on definite objectives. It also provides elasticity to the seller every time there is work that cannot be defined from the beginning. This type of contract works fine for high-risk ventures. One of the forms of cost-reimbursement contracts is the cost-plus-award-fee contracts (CPAF). This type of contract includes compensating the seller for all the legal costs that they have sustained. However, most of the payment earned is based on fulfilling the subjective performance criteria specified in the contract. The fee is evaluated based on the consumer's valuation of the seller's performance. The fee earned in the cost-plus-award-fee contracts (CPAFs) will be proportionate with the consumer's overall schedule, cost, and technical performance and meet the standards of the award-fee plan. This means that the buyer will not be paid if the performance is below a particular threshold. This type of contractual agreement is very elastic, and the contractor usually delivers better products and services. It also promotes an encouraging relationship between the consumer and seller, as good performance is rewarded. On the other hand, this is not a favored contracting method when there is a need for regulating the seller or contractor. It also has a higher cost and increases the obligation when compared to the traditional fixed-price contract.
6. **Cost-Plus-Fixed-Fee Contract (CPFF)**: A cost-plus-fixed-fee contract is a specific type of contract wherein the contractor is paid for the typical costs of the venture, in addition to an extra settled charge for their administrations. These enable the contractor to benefit from the venture, and they allow monetary creation in different businesses. In general, the expenditures in a cost-plus-fixed-fee are evaluated according to market values. However, the "fixed fee" share of the contract is subject to negotiation between the seller and buyer, so it varies based on the project's requirements.

7. **Cost-Sharing Contract**: A cost-sharing contract is a cost compensation contract in which the contractor obtains no fee and is reimbursed only for an agreed-upon share of its allowable costs. A cost-sharing contract may be used when the contractor decides to absorb a share of the costs in the hope of significant compensating benefits.
8. **Cost-Plus-Percentage-Of-Cost Contract (CPPC)**: The cost-plus-percentage of a cost is a type of contract that needs the buyer to repay all legitimate project costs to the seller. Aside from repaying costs, the buyer also needs to pay a percentage of cost as specified in the contract. This type of contract raises the supplementary fee as the cost of the contractor increases. Moreover, many sellers seldom use it because it provides little or no incentive to keep costs low.

This type of contract ensures long-term relationships and good performance of the seller, and the concluding cost may be less than the fixed-price contract agreement. However, it is essential to provide extra supervision to ensure that only the allowable costs are paid. The project manager needs to make sure that the project management activities are exercising satisfactory overall costs.

This type of contractual agreement is suitable when it is necessary to shift some risk of the contract performance from the contractor to the consumer. This type of contract is used when the procured item cannot be defined or when the data to evaluate the cost is not suitable. In most cases, this type of contract is appropriate for research and development (R&D) projects.

12.5 Balancing Contractor's Risk with Different Types of Contracts

Various models have been developed by researchers for helping contractors to calculate their risk in the tendering stage itself. Understanding how contractors arrive at estimated bid prices in real life; would allow us to appreciate how the risk associated with a project influences pricing. It is a general trend in the industry that contractors often do not have a systematic risk identification system, so risk accountability is more than the cost to the contractor. It is often difficult for them to determine how much allowance they should get for risk in a specific construction project. The various types of contracts to balance contractors' risk are explained in Table 12.4.

Table 12.4 Types of contracts to balance contractor's risk

Design and build	• This is the riskiest type of contract for the contractor • The responsibility for designing and constructing the project on-site lies solely in the hands of the contractor • The project is even riskier when the contract is set up on a lump sum basis where the contractor is required to quote a price for completing the project while bidding. Still, the risk is somewhat balanced as most preparatory work is done well before starting the project
Traditional contract	• Traditional lump sum contracts are meant to provide a fair allocation of risk between client and contractor. In this arrangement, the client designs the project, and the contractor is responsible for executing the construction work on site • The risk allocation can be adjusted as per the requirement. Still, too much risk cannot be allocated to the contractor because, with an increase in risk, the risk premium associated with the project also rises, increasing the tender amount quoted by the contractor
Management contracting	• The client takes up the majority of the risks. Management contractors are appointed during the design process to help choose the best possible design using their experience • The various work required for project completion is allocated to different contractors by the management contractor • Different work contractors help in starting the project even before the design is finalized. For example, work on the foundation of a building can begin before completing the detailed plan for the superstructure. This practice saves time, but the cost remains uncertain • The client has to administer just one contractor. The project's risks regarding delay and defect are allocated between client and contractor as per the contract between them

12.6 Popular Tools and Techniques in Project Procurement Analysis

To make the best use of the available monetary resource, the company or client needs to perform a regular procurement analysis. Procurement analysis is the study of what goes on in the company's purchasing department to keep up with the purchasing policy, and the inventories are kept at the optimal levels. Concerning suppliers, it makes sure if the best suppliers providing the best prices are chosen. Companies need to analyze costs and inventories. As organizations grow larger and their purchasing needs become more complex, it is essential to analyze delivery times, maverick spending, and whether the best use of suppliers is made or not. The techniques used in project procurement analysis are explained in Table 12.5.

The key performance indicators (KPIs) for supplier selection are explained in Table 12.6.

Table 12.5 Tools and techniques in project procurement analysis

Cost analysis	• The first thing that we have to see is the actual cost to procure the goods. Delivery, handling, and storage of material also have a cost associated with it • The company may incur higher costs for delivery and installation when required to pay all the people involved • If the company stores goods in its warehouse, then inventory cost is also taken into account. An exhaustive procurement analysis is required to understand the total cost of purchase from every supplier
Inventory analysis	• It is essential to identify the difference between the cost of storing a particular item and the cost of not having the item available for sale or manufacture. The item should be stored only when the cost for the latter case is more • A precise procurement analysis of the company's stocks and production capacity can help in predicting the number of goods to be stored in the inventory to obtain maximum profit • It is a fact that people and machines are a liability for companies, if left idle. Hence, proper utilization of available resources should be made
Best use of supplier analysis	• A company cannot produce everything required to build an item, so it depends on its suppliers for different raw materials and subparts, necessary for production. Identifying which suppliers provide them with what is a must • Suppose it is found that the company relies only on one supplier, and the risk associated with the item in terms of its criticality for production is very high. A monopoly like this would not be good, and the company should find a way to diversify its source of supply • On the other hand, if too many suppliers are involved in the supply chain, the company should try to reduce the number of suppliers so that many items can be purchased from few suppliers in order to get the benefits of quantity discounts
Maverick spending	• In some cases, if the organization utilizes purchase requisition, things may be requested that are not appropriate or of a higher cost or lower quality than they typically buy. Similarly, when one seller offers an item at a lower price, another would have been more suitable • A procurement examination of all your purchasing needs is essential. If procurements are done through maverick purchases, then it may cost the organization significantly higher
Delivery time analysis	• Delivery times can be significant to some big companies, mainly when perishable, consumable, and large goods are concerned • Payments for a large delivery made at the last minute can result in chaos with a tight cash flow. The handling of large goods requires a large storage space as well as personnel • It may be necessary for a company to carefully control its inventory needs since some perishable goods may need special storage facilities, such as freezer areas, which may not be readily available

12.7 Selection of Type of Contract Under Risk and Uncertainty

Procurement is one of the essential steps in the project life cycle. It helps choose the right contractor for completing the work that the client himself cannot do due to lack of expertise, capabilities, time, or because it is cheaper for the client to outsource the work. An infrastructure project, for example, usually has a considerable cost

Table 12.6 KPIs for supplier selection

Satisfaction level of the stakeholders with the quality of the arms supplied	• Let us say the supplier has previously been involved in some military contracts. It can be determined if a product is a quality one by getting feedback from their operators (such as soldiers, pilots, technicians) about how well the product performs and other suggestions (e.g., adding another option to the equipment) • Suppose the government has no experience in dealing with the supplier in the past, then the market reputation of the supplier should be evaluated • Any supplier who is entirely new to the field should be avoided
On-time delivery	• The supplier must make on-time deliveries without delay in timely military exercises and operations • Even early delivery can be detrimental as it increases the inventory and maintenance costs of the equipment when not in use • It can be measured by comparing the frequency of on-time deliveries to early and late deliveries
Purchase order accuracy	If a supplier has a history of not providing all the features promised at the agreed-on price settled at the time of contract, it should be avoided
Cost savings by improving the suppliers' competition	Allowing more competitors to participate in the tender process would help receive the best bid, thereby reducing costs
Willingness to invest and manufacture in India	A manufacturers' willingness to invest and produce the weapons indigenously can significantly reduce the cost
The political influence of the supplier	A supplier's political affiliations and influence on the global power balance should also be evaluated. Firms backed by significant political power often engage in lobbying to get the best deal for themselves and offer equipment at an exorbitant price
Degree of obsolescence of the products offered	It has been often observed that equipment offered at a cheaper price quickly becomes obsolete. This might increase the costs in the long run and risk the project. Suppliers who provide highly obsolescent equipment should be avoided, no matter how lucrative the bids are

associated with it. The project's viability depends upon its ability to create value for the client or customer, and its price should be more than the cost incurred. We cannot always predict and quantify the risks associated with various phases of the project. In this regard, the risk is said to be high if there are uncertainties around the project. Time, quality, scope, and cost decide the project's performance, and unforeseen events or risks negatively impact the performance. According to PMBOK, there are three types of contract which can offer better protection against risks.

1. **Fixed price**: This is also called a lump sum contract. In this contract, the seller quotes a fixed price to the buyer for completing the project. So the buyer knows the amount of payment required to get the work done. Thus, liability for the buyer is fixed. It can be said that the buyer has low-cost risk and uncertainty. On the other hand, the seller does not know precisely how much money will be needed to complete

the project, and the price quoted by him is based on the time required for completion and his experience with similar projects. These things are not absolute but project-specific and may change. In this type of contract, if the seller can complete the project within the allocated budget, he will be in profit. Otherwise, the seller has to sustain losses. Thus, it can be concluded that risk is low for buyers and high for the seller.

In fixed-price contracts, goods and services are provided to the project at an agreed-on price with little to no scope for any deviation. In case of any requirement of change in the contract due to unavoidable circumstances, the associated price is relatively high. The quality, timing, and price of goods and services to be delivered are included in the contract. For the type of work where the scope is very clearly defined and unlikely to change, fixed-type contracts are most desirable. Managing the work to meet the needs of the project lies with the seller/contractor is done. In contrast, the buyer/project team tracks the quality of work and schedule progress to ensure that the contractor is keeping the sanctity of the contract intact and the needs of the project are met. Hence, such contracts place the maximum risk on the contractor or seller and minimum risk on the buyer.

2. **Cost-reimbursable contract**: When a particular project scope is not well defined and the risk associated with the project is high, then a cost-reimbursable contract is used. The buyer pays the seller for all costs incurred during the completion of the project, and the seller gets an additional fee as profit. Here, the buyer bears all the risks associated with the project. The major problem related to the project is that the seller can elevate the construction cost, as the buyer will pay this money. This can be prevented if the total profit earned by the seller should be capped, for example, at X% of the total cost. Thus, it is evident that most of the risk is with the buyer, and the seller has shallow risk.

A cost-reimbursable contract or cost-plus contract is used when the work scope and the work's cost requirements are not well known. The buyer agrees to pay the seller or contractor for the cost of supplying the goods and services as part of such a contract. It is stipulated in the contract to pay the allowable expenses related to the seller or contractor for completing the job or transporting the goods on time.

Thus, cost-reimbursement contract types place minimum risk upon the seller and impose higher risk and administrative burden on the buyer. It is superficially agreed upon by the seller/contractor to use their best efforts to accomplish the undertaken task within the time and budget constraints, as established in the contract. In cases where the work is not entirely done, or the goods are not delivered within the original agreed-on contract cost, the seller/contractor has no obligation for further performance unless the contract is changed to increase costs. In such contracts, the seller/contractor is less motivated to manage its costs to keep its expenditure within the estimated budget specified in the contract unless some incentives are promised in the contract to do the same. Good documentation of the costs is vital in cost-reimbursable contracts to ensure that the seller/contractor is paid for all the work performed. The buyer does not pay anything for unfinished work or unallocated products.

3. **Time and material contract**: In a time and material contract, also called a T&M contract, the risk associated with the project is shared between both the parties involved. Equal risk allocation is done to both seller and buyer. When the project

starts, the buyer does not know how long the project may extend or how much material, machinery, and other resources are required to complete the project. In this case, the buyer has some degree of uncertainty associated with the project. On the other hand, the seller does not know how the cost of labor and construction material will change along the construction span of the project. Labor costs may be increased during construction, and hence the profit earned by the seller would fall. Thus, the seller also has some degree of risk.

12.8 Requirements for the Formal Acceptance and Closure of the Contract

The last phase of a project life cycle is the closure of the contract after project completion. The contractor is responsible for satisfying all the project stakeholders and completing the contract as per norms in code and local federal laws. When that occurs, the project can end. Despite being neglected, the project closure phase is an essential phase of the project. When a project is finished, people tend to hurriedly move on to the next project without executing proper closure procedures.

The key activities in project consummation are gathering venture records; spreading data to formalize acknowledgment of the item, administration, or undertaking; and performing project conclusion. The project manager is expected to audit project records to verify their credibility. For instance, some changes proposed in the last phase of the project might have impacted the quality of the deliverables. The deviations need to be appropriately documented.

It is a common belief among contractors that as the project ends, the contract also gets terminated. In actual terms, the contract ends only after the implementation of contract closure; the closure procedures are run to see if all contract agreements are satisfied by the contractor or not and check the quality of work.

Contract closure supports the project conclusion process because the contract closure procedure determines if the work defined in the contracts was completed precisely and agreeably. Every project is not done under contract, and in such cases, contract closure is not required. These procedures apply only to those stages, deliverables, or portions of the project that were done under contract. The various requirements for the formal acceptance and closure of the contract are summarized in Table 12.7.

12.9 Make or Buy Decision

Make or buy decision involves determining whether a particular product or a component of the same shall be manufactured in-house or purchased from a third party. If

Table 12.7 Requirements for acceptance and closure of the contract

Final payment	• The last installment of money paid to the contractor usually is more than work that remains to be completed. In projects, it is always seen that challenging parts in the execution are left for the last time. Thus, the previous payment to be released is also significantly larger regarding the percentage of work left • This motivates the contractor to finish the work on time • If the contractor has met the stipulated expectations, the client can approve the final payment
Post-project evaluation	• Before dissolving the project team, an audit is done to understand the learning from the project, as every project is unique in itself, and not everything goes according to plan. Hence, the project team can document many essential learnings from the project once the project is completed • It should be examined how learning from one project can be utilized in other projects
Schedule and budget management	• Here, the actual event calendar of the project is compared with the proposed network diagram developed at the beginning of the project. This helps analyze the events that could not be done on time and the events whose delay increased the project completion time • Carrying schedule and budget management are necessary to create skills required to evaluate schedule components in future projects. These things are not done to blame someone but to improve upon scheduling management in future projects
Customer satisfaction	• The reviews of customers are sought for improving the execution of future projects. A report containing an outline of the project, its goals, and objectives is created • The report communicates how the goals were achieved and objectives were met • This report can be shared with the customer for feedback

the capability to produce a product in-house at a cost lower than the market price exists, the choice should be to make the product within the organization's facilities. Otherwise, it should be outsourced to some external agency if they provide reasonable offers. However, the overall cost of procuring the products, including direct and indirect costs, should be considered before deciding. Budget constraints play the most crucial role in any such decision. But there are other factors like in-house production capabilities, availability of funds, and the product's strategic importance that play a crucial role in the decision-making process.

Make or buy analysis can decide whether a critical product should be purchased if it is available at a lower price or made in-house. If we keep the manufacturing in-house as an essential product, we need to make several comparisons, qualitative and quantitative, before making a judgment. Typically, the make or buy analysis for a critical product includes three crucial steps, as summarized in Table 12.8.

Table 12.8 Steps in make or buy analysis

Step 1: Quantitative analysis	• A quantitative analysis can be carried out by comparing the approximate overall expenses incurred in each option. Although the critical product might be available at a lower price in the market, we must compare it with the opportunity cost of time required to manufacture such product in-house • The time and money spent on in-house production can also be diverted to select other project activities. This is typically called an opportunity cost. If the opportunity cost of manufacturing the product in-house comes out to be greater than the cost at which the product can be purchased from the market, one can consider buying it
Step 2: Qualitative analysis	• The qualitative factors must be considered that may influence the decision to manufacture or buy the critical product of the project. All the factors that cannot be expressed in numbers for the sake of comparison are included here. For example, the quality of the product available in the market, the company's previous experience in manufacturing critical products in-house, learning curve effects if manufactured in-house, etc. If the product's quality is low, it is of no value, even if it is available at a lower price • Suppose one has a core competence in knowledge, skilled manpower, and technology in manufacturing such critical products in-house. In that case, it might be better to keep the production in-house and retain the expertise
Step 3: Integrating qualitative factors with quantitative analysis	• The next step involves factoring the qualitative aspects into the quantitative assessment to complete it. Even if it is cheaper to buy the product, there are grounds to believe that it would be more expensive to do so in the long run • With time and practice, a company can acquire the necessary competence in manufacturing the product and avail the advantage of economies of scale by producing it for its consumption and other industries in the market

12.10 Opportunism in Project Procurement

Opportunism is the policy and practice of consciously taking advantage of circumstances for selfish interests, politics, business, and even personal relationships, without any regard for principles and consequences. Opportunistic actions are convenient and practical actions that are governed by self-interested motives. The term can be applied to individuals and groups, behaviors, organizations, styles, and trends. It can capitalize on the opportunities created by the mistakes, errors, weaknesses, or competitors' distractions. It is advisable to be wary of such partners or collaborators, typically in an industry setting. For example, a supplier may act opportunistically and violate the terms and conditions of the contract by disclosing the design of the component to other OEM manufacturers. The critical attributes of opportunism in a project are summarized in Table 12.9.

Table 12.9 Key attributes of opportunism in project

Understanding what is going on	• It is not that rare for people to be misunderstood, and the same might be the case with the supplier or contractors. One should try to look into the background and market reputation of the suppliers and contractors to investigate the cause behind opportunistic behavior • It might be the case that suppliers or contractors have experienced heavy losses in their business and get prompted to opportunistic behavior
Talking in private	• It is sometimes necessary to informally discuss with your suppliers or contractors on select confidentiality issues and sensitize them • The members might be aware of the consequences of the opportunistic behavior, and a timely confidence-building exercise can help them get adopted to the correct behavior
Not saying "yes" all the time	• It is not necessary to be aggressive with business partners such as suppliers and contractors; however, if they tend to transfer inventory, monetary issues, warranties, changes in scope, etc., to you, one must be assertive about the boundary of work • A project manager should also know to stand his ground in the workplace and not budge all the time. Though one must know cooperation, it must not turn into an opportunity of opportunism for others. If your colleague tries to hog the team's or your accomplishments, do not be afraid to be vocal against it
Staying in the loop	• A project manager needs to examine why his team members or suppliers and contractors seem to be getting more information from the top management than himself • If they try to bypass you and do not keep you informed, exercise your authority as a leader and make yourself visible
Escalating the problem	• If you feel that you are being taken advantage of by your peers and are overworked but undervalued, speak with your boss about it • If your boss is your issue, talk with HR • Get to the root of things by passing through the proper channels so that you can effectively plot your next move • If things look entirely out of hand, and you are not getting any support from anywhere and feel completely powerless, then it is better to end it with the organization and look for an exit plan from the project

12.11 Monitoring of Contract Performance Cycle

The contract performance cycle is kind of a continuous procurement process to check if all contractual obligations between the contractor and client are met or not. The key steps involved in the monitoring of the contract performance cycle are summarized in Table 12.10.

Table 12.10 Key steps in the monitoring of contract performance cycle

Plan and scope	• Identifies the objectives to be met and quality standards to be followed. This is done considering the availability of resources and project priorities • For a construction project, this helps define the role of the client as a consultant and the role of the contractor as an executor
Stakeholders	• Different external and internal stakeholders may exist for a project depending upon its size. Collaboration between these stakeholders is essential for the successful completion of the project • Different stakeholders in a typical construction project can be the client's top management, project manager, team members, customers, government, etc.
Relationship management	• It is essential that every person in the team clearly understands their role in the project • Due to unforeseen reasons, the change in construction events should be communicated effectively to the project manager on time to ensure timely action
Performance management	• Before starting the project, a framework should be in place, including service level agreement between stakeholders, key performance indicator, problem resolution mechanism, and mechanism to judge different suppliers and provide quality and efficient delivery incentives • Construction project contracts often have provisions for incentives on the timely completion of the project
Risk and uncertainty	• As the size of the project increases, the uncertainties around it also increase. It becomes challenging to know about every expected problem from the beginning of the project. Thus, proper risk allocation is essential • Risk allocation between different parties varies according to contracts, but in the infrastructure sector, generally, the PPP model is used where both government and contractor share risks
Monitoring	• As the work on a project proceeds, monitoring becomes essential, as it ensures that proper quality work is done and safety measures are implemented. This also helps in checking if goals set in the timeframe are met or not • In a typical construction project, the client agency generally must oversee the contractor's construction work
Supplier relationship management	• The purpose is to develop the right kind of environment between client and contractor to facilitate the work smoothly. The aim behind this is to improve the efficiency of the contractor • In a construction project, improvement activities are carried out to find cheaper and more sustainable materials with similar strength, for example, fly ash bricks in place of standard bricks. This creates a long-term value for both contractor and client
Contract closure and termination	• It is often overlooked as people rush to work on another project. Still, it is an important point where the completed project is handed over to the client, who then checks whether all contractual agreements are met and all objectives have been achieved • After successful handling of the project to the client, the contract can be terminated
Asset management	• It helps to understand the requirements of the business • Lessons learned from previous projects are used in enhancing the efficiency of future projects • The cycle then begins again

Project Management in Practice

Procurement and Contracts Issues in Construction of Multi-specialty Hospital: A Hypothetical Scenario

In August 2017, a tender was invited to construct a multi-specialty hospital on five acres of land in a district in Gujarat, India to cater to the ever-growing healthcare demands of the rural and urban areas of the state. The "request for proposal" involved an invitation on Engineering Procurement Contract mode for design, planning, and construction of a 700-bedded hospital before the occasion of World Health Day on 7th April 2019.

The tender was attended by leading developers. According to the tender requirement, 60% weightage was given to technical specifications of the proposal and 40% on the financial aspects. After analyzing the proposals submitted by different companies in the bidding, ABC Ltd was finally selected. It won the bid under a fixed-cost contract for an amount of 300 crores. The contract defined the roles of the parties involved, the schedule of compensation, clauses of penalty on non-performance by the contractor, and the client's right to terminate the contract in case of delay of more than 50 days on the schedule sketch proposed by the contractor. It also bears the cost overheads on the contractor in case of delays due to dereliction by the same. The hospital had already acquired the land in advance so that delays were avoided. The payment for work was to be done in phases by the hospital quarterly every year from the commencement of the work, subject to completion of at least 90% work scheduled up to that quarter, emphasizing quality construction. As this was an EPC contract, ABC Ltd was responsible for planning, design, and construction.

The contract also made the contractor liable for litigation in non-compliance with the clauses stipulated in the contract. The responsibility of procuring materials and equipment for construction lay with the contractor. The production of materials, plumbing, wiring, and HVAC installation of HVACs in the facility was outsourced by M&T and given to local subcontractors on a pay-on-work basis. Out of the many plans and designs ABC Ltd offered, the hospital selected a four-story building with two acres of green space. The work commenced from September 2017. The schedule proposed by ABC Ltd involved the details given in Exhibit 12.1.

The materials were to be procured weekly, and a cost-reimbursable-type contract was given to a local supplier. The supplier had a storage warehouse close to the site. As the supplier worked with ABC Ltd in the past with satisfactory results, no special precautions were taken before awarding the tender. No bidding was done for the tender since the supplier offered a proposal at a discounted rate. This caused a significant issue in the monsoon season. When the construction work of the third floor's columns was to be started in July 2018, it was delayed due to heavy rain in western India, which affected some parts of Gujarat severely. The supplier's failure in delivering the material on time had created a problem at the construction site. However, ABC Ltd, a giant construction firm, had contacts and long-term relationships with some other suppliers in the state. The material was procured directly from a nearby supplier on an urgent basis. ABC Ltd had to bear an additional overhead cost of about seven lakhs and pay a total amount of 23 lakhs to the supplier immediately

Exhibit 12.1 Project schedule proposed by ABC Ltd.

Construction activity	Duration (days)	Material	Equipment	Skilled worker
Site setting and site clearance	7	Steel Fencing	Excavator, dump truck, dozer, chain saw, survey equipment	10 General workers, 4 truck/dozer operators
Excavation and earth works	20	Natural soil	Excavator, payloader, six wheel truck, bob-cat, shovel, meter tape	10 General workers, truck driver, excavator operator
Pile boring and reinforcement installation	10	Steel bore, reinforcement bar, bar bender	Rebar bending machine, boring machine	12 General workers, boring machine operator, bar bending machine
Pile filling	5	Concrete	Concrete mixer, concrete dispenser, concrete vibrator	5 General workers
Pile cap construction	7	Concrete, reinforcement, timber formwork	Bar bending machine, saw, vibrating machine, concrete mixer	6 General workers, 3 bar benders, 2 carpenters
Ground beam construction	15	Concrete, reinforcements, timber formworks	Concrete vibrator, rebar bending machine, table saw, concrete mixer	12 General workers, 5 bar benders, 4 carpenters
Ground floor slab construction	3	Concrete, reinforcement bars, timber formworks, Damp Proof Membrane (DPM), rigid floor insulation	Concrete vibrator, rebar bending machine, table saw, concrete mixer	10 General workers, 3 bar benders, 2 carpenters
Ground floor columns	12	Concrete, reinforcement, timber formwork	Concrete vibrator, rebar bending machine, table saw, concrete mixer	10 General workers, 5 bar benders, 4 carpenters
First floor beam	15	Concrete, reinforcement, timber formwork	Concrete vibrator, rebar bending machine, table saw, concrete mixer	12 General workers, 5 bar benders, 4 carpenters
First floor slab	3	Concrete, reinforcement, timber formwork, DPM	Concrete vibrator, rebar bending machine, table saw, concrete mixer	10 General workers, 3 bar benders, 2 carpenters
Ground floor wall	15	Concrete, reinforcement, timber formwork	Concrete vibrator, rebar bending machine, table saw, concrete mixer	10 General workers, 4 bar benders, 2 carpenters

(continued)

Exhibit 12.1 (continued)

Construction activity	Duration (days)	Material	Equipment	Skilled worker
Ground floor stairs	3	Concrete, reinforcement, timber formwork	Concrete vibrator, rebar bending machine, table saw, concrete mixer	8 General workers, 3 bar benders, 2 carpenters
First floor columns	12	Concrete, reinforcement, timber formwork	Concrete vibrator, rebar bending machine, table saw, concrete mixer	10 General workers, 5 bar benders, 4 carpenters
Second floor beam	15	Concrete, reinforcement, timber formwork	Concrete vibrator, rebar bending machine, table saw, concrete mixer	12 General workers, 5 bar benders, 4 carpenters
Second floor slab	3	Concrete, reinforcement, timber formwork, DPM	Concrete vibrator, rebar bending machine, table saw, concrete mixer	10 General workers, 3 bar benders, 2 carpenters
First floor wall	15	Concrete, reinforcement, timber formwork	Concrete vibrator, rebar bending machine, table saw, concrete mixer	10 General workers, 4 bar benders, 2 carpenters
First floor stairs	3	Concrete, reinforcement, timber formwork	Concrete vibrator, rebar bending machine, table saw, concrete mixer	8 General workers, 3 bar benders, 2 carpenters
First floor wall	10	Concrete vibrator, rebar bending machine, table saw, concrete mixer	Concrete vibrator, rebar bending machine, table saw, concrete mixer	8 General workers, 3 bar benders, 2 carpenters
Second floor columns	12	Concrete, reinforcement, timber formwork	Concrete vibrator, rebar bending machine, table saw, concrete mixer	10 General workers, 5 bar benders, 4 carpenters
Third floor beam	15	Concrete, reinforcement, timber formwork	Concrete vibrator, rebar bending machine, table saw, concrete mixer	12 General workers, 5 bar benders, 4 carpenters
Third floor slab	3	Concrete, reinforcement, timber formwork, DPM	Concrete vibrator, rebar bending machine, table saw, concrete mixer	10 General workers, 3 bar benders, 2 carpenters
Second floor wall	15	Concrete, reinforcement, timber formwork	Concrete vibrator, rebar bending machine, table saw, concrete mixer	10 General workers, 4 bar benders, 2 carpenters

(continued)

Exhibit 12.1 (continued)

Construction activity	Duration (days)	Material	Equipment	Skilled worker
Second floor stairs	3	Concrete, reinforcement, timber formwork	Concrete vibrator, rebar bending machine, table saw, concrete mixer	8 General workers, 3 bar benders, 2 carpenters
Second floor wall	10	Concrete vibrator, rebar bending machine, table saw, concrete mixer	Concrete vibrator, rebar bending machine, table saw, concrete mixer	8 General workers, 3 bar benders, 2 carpenters
Third floor columns	12	Concrete, reinforcement, timber formwork	Concrete vibrator, rebar bending machine, table saw, concrete mixer	10 General workers, 5 bar benders, 4 carpenters
Fourth floor beam	15	Concrete, reinforcement, timber formwork	Concrete vibrator, rebar bending machine, table saw, concrete mixer	12 General workers, 5 bar benders, 4 carpenters
Fourth floor slab	3	Concrete, reinforcement, timber formwork, DPM	Concrete vibrator, rebar bending machine, table saw, concrete mixer	10 General workers, 3 bar benders, 2 carpenters
Third floor wall	15	Concrete, reinforcement, timber formwork	Concrete vibrator, rebar bending machine, table saw, concrete mixer	10 General workers, 4 bar benders, 2 carpenters
Third floor stairs	3	Concrete, reinforcement, timber formwork	Concrete vibrator, rebar bending machine, table saw, concrete mixer	8 General workers, 3 bar benders, 2 carpenters
Third floor wall	10	Concrete vibrator, rebar bending machine, table saw, concrete mixer	Concrete vibrator, rebar bending machine, table saw, concrete mixer	8 General workers, 3 bar benders, 2 carpenters
Fourth floor columns	12	Concrete, reinforcement, timber formwork	Concrete vibrator, rebar bending machine, table saw, concrete mixer	10 General workers, 5 bar benders, 4 carpenters
Fifth floor beam	15	Concrete, reinforcement, timber formwork	Concrete vibrator, rebar bending machine, table saw, concrete mixer	12 General workers, 5 bar benders, 4 carpenters
Roof construction	3	Concrete, reinforcement, timber formwork	Concrete vibrator, rebar bending machine, table saw, concrete mixer	5 General workers, 3 bar benders, 2 carpenters

(continued)

Exhibit 12.1 (continued)

Construction activity	Duration (days)	Material	Equipment	Skilled worker
Fourth floor wall	15	Concrete, reinforcement, timber formwork	Concrete vibrator, rebar bending machine, table saw, concrete mixer	10 General workers, 4 bar benders, 2 carpenters
Fourth floor stairs	3	Concrete, reinforcement, timber formwork	Concrete vibrator, rebar bending machine, table saw, concrete mixer	8 General workers, 3 bar benders, 2 carpenters
Curtain walls installation	15	Glass, steel, screws	Wrench, level, screw driver	8 General workers, 4 special worker
Railing installation	7	Aluminum handrail (staircase), aluminum guardrail (balcony)	Wrench, level, screw driver	5 General workers, 2 special worker
Door framing and installation	20	Hollow single rabbet metal frame	Wrench, level, screw driver, measuring tape, marker	8 General workers, 4 special worker
Window framing and installation	20	Aluminum casement frame, glass window, screws	Wrench, level, screw driver, measuring tape, marker	8 General workers, 4 special worker
Electrical wiring	7	Insulated copper wires, electrical tape, connector	Test pen, voltage meter	12 Electricians
Plumbing	10	Assorted dimensions of PVC pipe, connectors, PVC glue	Saw, measuring tape	8 Plumbers
Sanitary fittings	12	Ceramics	Saw, measuring tape, drill	8 Special worker, 2 general workers
Tiling	15	Mortar, grout, homogeneous ceramic tiles	Putty knife, spacers, level	10 Tilers
Plastering	20	Plaster board	Ramset concrete nailer, Makita concrete driller, Laser Level machine, Plumbs, Basic hand tools, Makita steel bench cutters, Ceiling 20 gun, Ladder, Steel square ruler, Drilling Machine	50 General workers
Painting	15	Roller brush. Brush extension pole, masking tape	large bucket, wheelbarrow, trowel, garden hose	6 General workers

(continued)

Exhibit 12.1 (continued)

Construction activity	Duration (days)	Material	Equipment	Skilled worker
Hardware installation	5	Stove ventilation hood, cabinets, counter, etc.	Measuring tape, level, screwdriver	5 General workers
Exterior finishes	5	Lime, sand, water	Large bucket, wheelbarrow, trowel, garden hose	3 General workers
Cleaning	2	Water, cleaning liquid	Mop, cloth, vacuum cleaner	5 General workers

for the same materials worth sixteen lakhs. It could not blame the supplier because of the nature of the contract being a cost-reimbursable type. In the late delivery of materials, such losses may have been avoided if the contract terms included penalties for the materials supplier.

There was another incident that ultimately led to the delay of the overall project. The project was scheduled to be completed by the last week of Jan 2019 but was still delayed by fifteen days, and the facility was handed over in the second week of February 2019. Thankfully, it was still before the deadline mentioned in the contract, because of which ABC Ltd did not have to face any problems related to payments by the hospital. In December 2018, when plastering was ongoing, workers went on a strike demanding a raise on their daily wage. They demanded an increment of Rs. 300 per day. Keeping in view the cost quoted by ABC Ltd in the contract, it was difficult to accept this demand immediately. Work was halted on-site for fifteen days. The strike even took a political angle, and the construction firm could not fire the existing workers and hire new skilled ones.

This created a significant difficulty for the project manager on site. ABC Ltd generally does not hire workers on a contract basis with fixed pay. Most workers are hired through brokers and paid daily. The work to be done was crucial, and the company had to settle with the decision to increase the workers' daily wages by Rs. 200 per day. For the remaining period of the contract, it had to bear extra labor costs. Work resumed as usual after the strike was called off, but ABC Ltd could not catch up with the schedule, and ultimately the whole project was delayed by fifteen days. Thankfully, this delay was within the margin of 50 days as allowed by the contract between the two parties involved, and ABC Ltd had to bear no penalties from the side of the hospital. But it did incur significant losses. This tells us the value of establishing contracts with our suppliers/workers and on doing so can lead to some increment in the work cost. If the workers would have been hired initially on contract basis, such a situation would not have occurred. Ultimately, incorporating a delay of fifteen days, the facility's construction was completed on February 13, 2019, and handed over to the hospital for inauguration on April 7, 2019, on the World Health Day.

This narration highlights the importance of procurement and contract management in the construction industry. The readers may identify a set of viable strategies to help project manager.

Summary

During the project life cycle, project procurement management becomes indispensable to establish and maintain relationship with vendors of numerous goods and services.

It ensures that the availability of materials and services needed for the project to run smoothly and achieve its objectives within the budget and time.

Repeated problems with procurement can lead to some long-term issues, including money loss, diminishing profitability, bad reputation, and negative publicity.

Having poor procurement strategies means making redundant purchases, paying too much for transportation, suffering from a lack of quality control or developing problems with contract scope.

Procurement is one of the most important steps in project life cycle. It helps in choosing the right contractor for completing the work which the client himself cannot do due to lack of expertise, capabilities, time or because it is cheaper for the client to outsource the work.

Make or buy decision involves determining whether a particular product or a component of the same shall be manufactured in-house or purchased from a third party. If the capability to produce a product in-house at a cost lower than the market price exists, then the choice should be making the product within the organization facilities. Otherwise, it should be outsourced to some external agency if they provide reasonable offers.

Opportunism is the policy and practice of consciously taking advantage of circumstances for selfish interests, in politics, business, even in personal relationship, without any regard for principles and consequences. The opportunistic actions are convenient and practical actions that are guided by self-interested motives.

Questions for Discussion

1. Why is project procurement a critical issue? What are the problems a project firm may face if the procurement management is not executed correctly? Explain with some real-life case examples.
2. What are the different phases in project procurement as per PMBOK guidelines? Explain each phase in detail. Explain the significance and criticality of each phase for a typical construction and R&D project in a tabular format.
3. What is the importance of contracts in Project Management? What are the contract selection criteria? Explain each one in detail with its significance for a typical construction project.
4. What are the contractor's risks? What are the different types of contracts in project management [For example, firm-fixed price contract (FFP), firm-fixed price with economic price adjustments (FPE), firm-fixed price with economic price adjustments (FPE), incentive contracts—fixed-price-incentive-fee contract (FPIF), cost-plus-incentive-fee contract (CPIF), cost-plus-award-fee contract (CPAF), cost-plus-fixed-fee contract (CPFF), cost-sharing contract (CS), cost contract (C), cost-plus-percentage-of-cost contract (CPPC)]? Explain each one in detail.
5. How do you balance contractor's risk with different types of contracts? Consider a case of a construction project and explain your response in detail.
6. What are the various popular tools and techniques used in project procurement analysis?
7. Consider yourself as a VP (global purchasing) of a Medical Equipment Manufacturing Company. How can you select reliable suppliers to procure the advanced technology and exact components for your medical equipment manufacturing? What are the KPIs you will consider?
8. How would you select a contract type considering the risk and uncertainty involved in the project? Explain from the perspective of both the parties (i.e., buyer and seller).
9. What are the various requirements for the formal acceptance and closure of the contract? Explain each one in brief.
10. What do you determine to make or buy decision? Shall you go for buying (outsourcing) your entire research thesis if you can get it at a lower price? Explain with proper justifications.
11. What is opportunism? How do you deal with the situation if you find your future life partner as an opportunistic person? What are the managerial lessons you will derive from it to implement in your business/ organization?
12. Briefly explain the process of monitoring the entire contract performance cycle. Describe each phase for a typical construction project.

Web-Based Exercise

Visit the Web sites of megaprojects and go through their online tenders for selection of vendors and purchase. Enlist at least twenty essential requirements covering financial and technical aspects of the tender. Identify the critical criteria megaprojects apply for vendor selection and evaluation.

Chapter 13
Project Quality Management

Critical Questions
What is the role of quality in managing projects?
What are the key quality concepts useful for project management?
How do we apply TQM principles for quality management in projects?
What is the relevance of Deming's principles and Juran's Triology in managing projects?
How do we execute Six Sigma—DMAIC cycle for an effective project management?

13.1 Definitions of Quality and Relevance for a Project

> Project quality management encompasses the processes and activities used to figure out and achieve the quality of the deliverables of a project. However, quality can be an elusive word.

> Project Quality Management includes the processes & activities that determine quality policies, objectives & responsibilities to ensure that the project satisfies the needs for which it is undertaken.

The International Organization for Standardization (ISO) defines **quality** as "the degree to which a set of inherent characteristics fulfills requirements" (ISO 9000:2000). The present competition and increasing customer expectations do not permit the businesses to accept a single definition of "quality." The perspectives on quality are summarized below.

1. **Fulfillment of Purpose**
   A quality of a product or project is defined in terms of the "fulfillment of the purpose." This definition applies to various processes, services, or products. However, its measurability is difficult.

J. J. Thakkar, *Project Management*, Management and Industrial Engineering,
https://doi.org/10.1007/978-981-15-3695-3_13

For example, a new racing car has the purpose of being fast, efficient, secure, sound braking and controlling system, comfortable and safe, coupled with flying, which defines it as quality.

2. **Fulfillment of Requirements**
   Quality is multi-dimensional and the degree to which a set of inherent characteristics fulfill requirements.
   For example, an IT sector client X defines its application requirements to be developed by an IT company Y. The quality of the developed application needs to be measured against the client's requirements. This understanding is essential for the quality assurance team to validate processes, systems, services, and product quality. The only problem with this definition is that various requirements may offer a biased and subjective view of quality from customer to customer.
3. **Cost as an Indicator of Quality**
   Traditionally, it was the perception that product quality is directly proportional to the cost of the material.
   For example, a watch made of gold is considered to be of higher quality than one made of plastic. Sheets of higher quality have a thread count of 180 or higher. A quality product costs higher to produce. It is valid for only select products. It is inapplicable to technology, art, and culture.
   For example, a construction project can use low price but better strength material for creating an infrastructure using modern construction techniques rather than using a costly building material with low strength and durability via traditional construction methods.
4. **Quality is Price**
   Quality is a vital part of economic models; economists have given various definitions of quality. Willingness to pay for a product or service is a way that economists tend to judge its quality. In other words, "Quality is the price consumers are willing to pay for a product or service."
   For example, in infrastructure projects like introducing a new expressway, people are generally willing to pay high Toll-Tax. It provides the users with a higher level of service, i.e., high-speed mobility with comfortable, safe, and fast road journeys.
5. **Quality in Terms of Standard**
   In various projects, the products and services must adhere to standards and specifications, which ultimately validate their quality. Quality must be maintained for both the final product and processes involved during its entire production.
   For example, construction projects strictly adhere to IS codes that help and guide the various activities to maintain quality as prescribed by the standards, thus ensuring its quality in terms of safety, durability, and level of service to the customer during its proposed design life period.
6. **Quality in Terms of Value for Performance**
   The cost-benefit value of a product defines its quality for a consumer. The product or service with the best performance for the least cost receives the highest value.

For example, in an R&D project, the self-cleaning toothbrush is to be developed, which will cost Rs. 500 to the customers. The regular toothbrush costs Rs. 30 to the customer. Suppose the performance gap between a self-cleaning toothbrush and a regular toothbrush is not significant. The value associated with the self-cleaning toothbrush is marginal and may not motivate the customer to purchase this product.

7. **Quality in Terms of Experience**
   The competition has forced businesses to focus more on better services. Quality offers a satisfying experience to the targeted customers.
   For example, a customer never purchases a flat to satisfy his living requirements. He would like to improve the quality of his life by selecting a reputed builder, good quality construction, safe locality, nearness to the market, airport and other facilities, etc. A project manager needs to understand the explicit and latent expectations of the customers and develop a product that can improve their overall experience.

13.2 Importance of Quality for the Project

Quality is the driving seat of a successful project. Quality must be planned at the design stage of the project. A misconception exists about the deployment of quality. It is perceived that the project managers should only be concerned about the quality during manufacturing, construction, and installation. It is inappropriate. A systematic deployment of quality at each stage of the project is vital for resolving the conflicts arising out of time, cost, and performance trade-offs. It is "quality" that sets the foundation of a healthy and long-term relationship between a customer/client and the company.

The concept of project quality management consists of a plan, quality control, and quality assurance, as shown in Fig. 13.1. The planning phase includes the development of quality policies, standards, metrics, checklists, etc. Quality control ensures

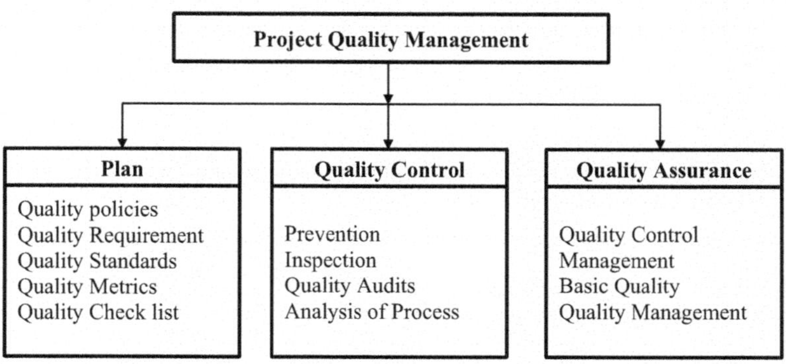

Fig. 13.1 Project quality management

that the project's quality requirements are adequately met during the execution stage of the project. Country-specific standards regulate the assurance of quality in all engineering projects. In India, it is held by the Bureau of Indian Standards (BIS). BIS is a National Standard Body of India established under the BIS Act 2016 for the harmonious development of standardization activities. International standards like ISO, QS, etc., are also well-practiced. Quality assurance ensures that the quality of the product or service fulfills the client's requirements. It is essential to execute the project with an appropriate quality management charter. This helps the project manager and team to minimize quality lapses and ensure client satisfaction at various stages in the project life cycle.

Implications of a Poor Quality on Overall Execution and Planning of the Project

Successful execution and planning can never be achieved when there is poor "quality" management. The major contributor to poor quality is management's inability to ensure and implement quality assurance, control systems, and procedures. This is most prevalent in scenarios where both the senior management and project team fail to understand the value of quality for the various phases of the project life cycle. Problems can be effectively resolved in any project, but it depends on teams of experts who can quickly adapt, organize, and troubleshoot. Most of the time, "quality" is not adequately funded or implemented. The solution is to educate consumers and clients about the "cost of quality."

Example

A typical construction project can be timely completed, and the client requirements can be met if the quality deployment at various phases of construction is adequately planned. While the contractors or consulting engineers accept the ultimate responsibility for quality and meet the design requirements, the whole purpose of quality assurance plans and related construction monitoring is to move the project toward fulfilling the client's needs smoothly.

It is estimated that the global contribution of infrastructure projects will reach US$ 1 trillion by the year 2022, and organizations with an effective quality system will demonstrate success in projects by reducing cost and risk. A compromise on quality adversely affects the success of a project by affecting its safety, schedule, and cost. It is evident from the industry experience that a sound quality management system can reduce the total construction costs by 25%.

Project Management in Practice

Quality Cautions for Project Managers

- Meeting the customer requirement by overworking the project team may lead to adverse consequences in employee turnover
- Meeting project schedules by rushing planned quality inspections may produce negative consequences when errors go undetected

13.3 Deming's Principles with Its Relevance for a Project

Quality is often mistakenly related to the final product. In his book, Out of the Crisis,[1] Edward Deming introduced 14 critical principles for management for creating the constancy of purpose toward improvement of product and service. He advocated that organizations increase quality and simultaneously reduce costs (reducing waste, rework, staff attrition, and litigation while increasing customer loyalty). The principles are universal and can be successfully applied to various projects for better quality management. The 14 fundamental principles of Deming are explained below.

1. **Create constancy of purpose**
   This reflects the need for continuous improvements. Most projects are run with a reactive mind-set. Managers react to the current problems than avoiding them proactively. Very stiff long-term plans are difficult to implement. It is necessary to develop both short-term and long-term goals with quality management practices adopted at various phases. Previous pitfalls in the same or similar projects can be used for the improvisation of plans.
2. **Adopt the quality philosophy**
   Management should try to discard the existing philosophy of being on par with the competitors. Recognize the changes in the competitive environment and accept your new responsibilities. With the existing policies of the organization, this cannot be accomplished. Organizational change requires leadership acceptance.
3. **Stop dependence on inspection for quality**
   Quality cannot be ensured by the surveillance of the workforce alone. Inspection can be used as a tool to improve the strategies of workmanship planning. This mass inspection is costly, ineffective, and unreliable. Preventive measures have to be adopted to minimize the wastage of resources. It is crucial to create a project culture in which the manager acts as a leader, and the workforce is equipped with the necessary skills, knowledge, and experience. The quality must be built into the product right from the design stage. Inspection of an employee using cameras or surveillance of their Web history will not do well in any industry. It creates uncertainty and limits freedom.
4. **Move toward single supplier**
   Suppliers should not be selected just based on the price. Choosing a single supplier for specific items can improve the long-term relationship by building loyalty and trust. This can also facilitate the changes in procurement due to the constant improvements in the work plan.
5. **Improve constantly and forever**
   Problems with systems can continue to cause problems in the project until they are resolved. Continuous improvements for constantly decreasing the cost must be adopted as the core strategy of the organization.
6. **Institute training on the job**

[1] Deming, W. E. (1982). Out of the Crisis. Massachusetts Institute of Technology.

Workers should be trained as per the necessities of the organization. They must be made aware of the objectives and quality criteria. Training can help to orient the workmanship to the common goal of quality.

7. **Institute leadership**
   A manager should not supervise but lead the activities. The manager or the leader must be well aware of the tasks carried out under him. He must be able to recognize between the common variation and unique variation of results. A leader translates the vision of a project to some activities that can excite and motivate workers.
8. **Drive out fear**
   Deming believes that fear of punishment or penalty cannot create an environment to effectively yield a worker's potential. It is counterproductive as well. A friendly policy of open-door approach, attentiveness to their problems, faster troubleshoots, etc., can encourage the workers to admit their mistakes or report their problems more freely, improving quality.
9. **Remove the barriers between departments**
   Competitions and barriers due to the organizational structure within any project can affect the quality of the project. This hampers the smooth working of the organization as a whole to execute a common goal. Each department sets its goals and works to achieve that without paying much attention to the common objective. An attempt to remove such barriers can cause short-term productivity loss, which can be beneficial in the long run.
10. **Eliminate slogans**
    Eliminate slogans and numerical targets to increase productivity. This can affect the quality of the workforce. In a free and friendly environment, quality can be improved. Deming thinks that the process is responsible for more of the mistakes than the people.
11. **Eliminate management by objectives**
    Setting production targets can only cause people to achieve it by compromising the quality. Reactively setting targets based on immediate problems can cause quality problems. Milestones have to be placed with proper planning so that they are achievable and flexible to unforeseen circumstances.
12. **Remove barriers to pride of workmanship**
    People are interested in doing a good job. Management should be responsible for providing them with the equipment and good working conditions. Recognizing the team or individuals for their achievements can also enhance their involvement. It is possible to unlock the value even from failures. It is possible to motivate workers by making them feel important.
13. **Institute education and self-improvement**
    To improve quality, it is essential to educate the workforce and identify the project's goals. Additionally, this can improve the living conditions of the workforce and thus the quality of the work.
14. **Transformation is everyone's job**
    The workers cannot do the transformation alone, nor can the managers. Every part of the organization must undergo a complete transformation.

For example, a hypothetical IT company, XYZ Ltd. is developing a new version of anti-virus software and a publicly funded defense research organization, NDRO is handling a R&D project of an intercontinental ballistic missile. The application of the 14 principles of Deming is demonstrated for these two different types of project organizations in Table 13.1.

13.4 Juran's Trilogy with Its Relevance for a Project

> Quality" means freedom from deficiencies – freedom from errors that require doing work over again (rework) or that result in field failures, customer dissatisfaction, customer claims, and so on.
>
> *Joseph Juran*
>
> *(Juran, J. M., & Godfrey, A. B. (1998). Juran's quality handbook. New York: McGraw Hill.)*

Joseph Juran has advocated his quality improvement model based on three universal processes: quality planning, quality control, and quality improvement, popularly named Juran's trilogy[2]. The lessons learned during the quality improvement phase are primarily utilized for the quality planning phase. The planning addresses the issues that affect the product and service design, such as the definition of a project, identification of customer needs, development of the product and process according to customer's need, and establishing the quality objectives. Furthermore, quality improvement is related to improving the level of performance of the process to meet the project's stated objectives. The quality control phase is established to assure and sustain the realized gains. It measures the actual operating performance and compares this with the defined goals. Figure 13.2 indicates a sudden increase in the cost (maybe more than 40–50%) named "sporadic spike," which happens due to unplanned events such as power failure, critical equipment or machinery breakdown, or human error. This can be brought under control with timely measures and policy decisions. An application of Juran's principles for R&D and IT project organizations is presented in Table 13.2.

[2] *The Juran Trilogy*. Juran. (2018, March 1). Retrieved from https://www.juran.com/blog/the-juran-trilogy-/

Table 13.1 Application of Deming's principles for IT and R&D project organizations

Principle	XYZ Ltd.	NDRO missile project
Create constancy of purpose toward improvement	Undertake a detailed study on customer requirements, security threats, and possible future developments	While planning the research of a new missile, they need to think about developing missile tracking and dismantling systems in various parts of the world
Adopt the quality philosophy	An Antivirus needs to be superior to its fellow software in terms of safety	Since this deals with the nation's security, quality and safety at any stage cannot be compromised
Stop dependence on inspection for quality	Robust monitoring systems of a camera can be a financial burden. Checking the web histories and restricting some websites can affect the morale of workers. Instead, they need to be made aware of the importance of the project	Instead of supervising or monitoring, employees can achieve quality by being motivated and understanding the national importance of the project
Move toward a single supplier for any one item	While outsourcing some parts of the project to freelancers or other companies, reliability is ensured by a warm relationship	Some parts of the missile may need to be manufactured in other companies or labs. Also, raw materials are to be brought from a single supplier to ensure the timely supply and quality of raw materials
Improve constantly and forever	Constant improvements are required to satisfy customer needs. New security risks arise constantly. A regular upgradation of features is essential	Constant improvement has to be initiated while choosing raw materials and technologies since the costs of these can vary over time. Also, advancement in the electronics field may help to improve the scope and features of the project
Institute training	Software developers should be updated with the competitors' products and offerings	Like manufacturing, designing, etc., relevant training must be imparted for necessary skill building in every department

(continued)

Table 13.1 (continued)

Principle	XYZ Ltd.	NDRO missile project
Institute leadership	The manager of each team that develops various parts of the software, like user interface, data clouds, development, etc., should act as a leader and implement the company's quality philosophy within his team	The functional managers should lead and motivate his team to timely achieve expected research and development targets. He must not merely act as a supervisor
Drive out fear	Employees should experiment and feel free to admit their mistakes. This enhances group learning	A small error in the design process can negate a considerable quantity of work done for some weeks. But, un-identifying it can cause a substantial loss in money and even the nation's security
Break down barriers between departments	Experts in other departments may be helpful for the development of specific features	Functions should work in a symphony, keeping a healthy competition and collaboration among the employees
Eliminate slogans	Projects must be scheduled realistically, and managers need to encourage and motivate workers to accomplish the stipulated objectives	Unachievable targets in high tech R&D projects reduce employee morale and quality of work
Eliminate management by objectives	Management should motivate the team and push them toward the successful accomplishment of the project. Too much emphasis on meeting deadlines may increase the errors and compromise the quality of the software	The development of a new missile has to override the short-term or myopic functional objectives, which can contradictory at times
Remove barriers to pride of workmanship	Good workplace, computers installed with latest versions of relevant software, amenities etc. helps	Good working environment, good quality equipment, employee-friendly work culture
Transformation is everyone's job	Quality depends on the works of both the beginner level developer and top-level management	Involvement at all levels (top management to workers) is essential

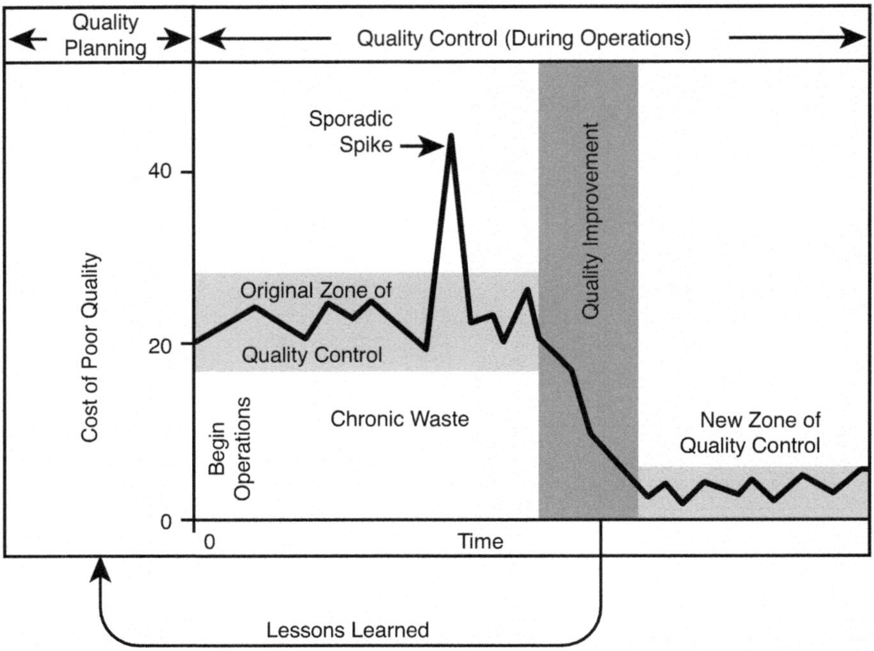

Fig. 13.2 Juran's trilogy (adapted from: *The Juran Trilogy*. Juran. (2018, March 1). Retrieved from https://www.juran.com/blog/the-juran-trilogy-/)

13.5 Total Quality Management (TQM) and Its Relevance in Project Management

Total Quality Management is a customer-centric approach, which aims for continuous improvement of business processes using fact-based decision making. It ensures that employees work toward the common goals of improving product quality or service quality.

Total quality management is a strategy to improve the processes that continue to meet the ever-changing quality criterion. It ensures the involvement of employees and management for the continued improvement of products and services aiming at long-term benefits. The five pillars of TQM are product, process, system, people, and leadership which helps the organization to realize continuous improvement in the various processes and functions.

Principles of TQM:

1. **Focus on Customers**
   Quality is defined as meeting or exceeding customer expectations. The level of quality depends on the customer's aspirations. Therefore, this TQM process is customer-driven. The "focus on customers" is significant in avoiding scope creep and keeping the project manager and team members committed to the client's requirements.

Table 13.2 Application of Juran's trilogy for R&D and IT project organizations

S. No	Juran's trilogy	Relevance to R&D project	Relevance to IT project
1	**Quality Planning**: According to Juran's trilogy, quality planning is a simultaneous exercise that involves all the dramatic events related to the product and services to provide ideas and give early cautions during the planning processes **Steps involved**: (1) Defining the project (2) Identification of the key customers (3) Detection of customer needs (4) Drafting and proposing the product and processes to meet the customer's requirements (5) Instituting the quality objectives (6) Preparing the plans to achieve these objectives	XYZ Limited is a leading international mobile company. Its R&D team is developing an enhanced version of a mobile camera that can record 3D videos. This idea is finalized based on studying the market trend and customer requirements, and behavior survey. The product is targeting a premium class customer	ABC Limited is a leading IT Company in India. The team is developing software to analyze criminal activities and locate probable crime spots. This software will benefit police departments in checking and reducing criminal activities, thus ensuring the safety of the citizens
2	**Quality Control:** As per Juran's trilogy, quality control involves formulating and maintaining operational methods to assure performance of work as they are designed and that the target levels of performance are being achieved. Quality control solely concentrates on the execution of plans. It is primarily to control that occasional spike in error in the process. Quality control is a short-term process to check that spike **Steps involved:** (1) Evaluate actual performance (2) Compare actual performance with quality goals (3) Act on the difference	The company will evaluate the quality of the videos and their 3D effects. The team will focus on optimizing the camera's performance for varying weather conditions	The company will evaluate the quality of the software performance, i.e., the accuracy of the result. The team runs many trails to validate the software performance. The various coding platforms can be tested to improve the versatility and compatibility of the software

(continued)

Table 13.2 (continued)

S. No	Juran's trilogy	Relevance to R&D project	Relevance to IT project
3	**Quality Improvement:** This aims to enhance the performance of the process through breakthrough improvements. An innovation or an innovative solution is implemented to improve the current performance levels. This ensures that the new performance levels are achieved, and then quality control mechanisms are in place to sustain **Steps involved:** (1) Identify the improvement projects (2) Establish project teams (3) Diagnose the causes (4) Establish controls to hold the gains	When the team kept working on improving the quality of the camera and its 3D effect, they revealed that the camera was effective in capturing 3D photos. This has a significant impact on sales and enhancing customer value. The result is a new feat in the mobile industry, with delighted customers adding a new value to the company's brand	With a continuous effort to improve the accuracy of the results, it was revealed that the software was also helpful in catching criminals around the predicted criminal spots. This additional capability significantly increases the demand

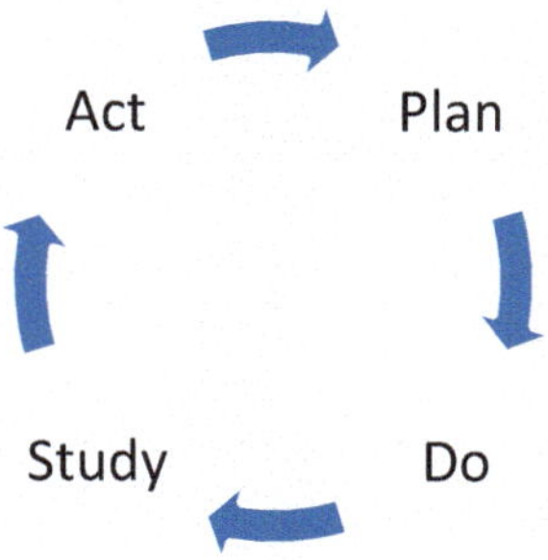

Fig. 13.3 Plan-Do-Study-Act cycle

2. **Continuous Improvement**
   TQM focuses on the continuous improvement of systems. This reimagines the achievement of quality as a continuous process. Plan-Do-Study-Act cycle depicts how the continuous improvement principle is followed (Fig. 13.3). Managers need to study the current status of the processes and the problems prevailing. This data can be used to evaluate and propose new plans to improve the quality. This plan is implemented and studied whether this change could improve the process. The P-D-S-A cycle keeps repeating.
3. **Leadership**
   Strong and visionary leadership is essential to align the organization's purpose, focus, and internal environment. Leadership must set measurable and achievable targets and involve the employee to realize the organization's overall goal. Motivating the workforce is an essential function of leadership.
4. **Involvement of People**
   The success of a project significantly depends upon the involvement of its employees and their effective skill utilization. The involvement of employees is vital to value their critical suggestions and create a culture of ownership. This cultivates a culture of openness and improves coordination among various functions participating in the project.
5. **Process Approach**
   Processes are to be defined carefully by determining objectives, function, inputs, and outputs. The control at the process stage helps achieve the project's desired outcomes without compromising the project's time, cost, scope, and performance.
6. **System Approach to Management**
   Interrelated processes can be converted into systems to achieve the objectives of the project effectively. Multiple processes managed together as a system can improve the effectiveness. Processes and functions are combined to form business plans. Sub-goals of different processes are aligned with the larger organizational objective. Overlapping competencies of different processes are avoided by proper management. This can be well addressed while developing the work breakdown structure (WBS) of the project.

7. **Factual Approach to Decision-Making**
   Intelligent decisions are based on a scientific analysis of the facts. Decision-making must involve gathering inputs from various sources, identifying facts, objectively organized data. This is essential for complex projects which involve multiple stakeholders and their conflicting objectives.
8. **Mutually Beneficial Supplier Relationships**
   Good relationships with suppliers largely contribute to the success of a project. A well-designed relationship between supplier or contractor and client helps improve the project's overall efficiency and effectiveness. It is mutually beneficial.

Example: Application of TQM principles for IT and R&D projects (Table 13.3).

- IT Project: IT company XYZ Ltd. executes a project to build a website for food delivery
- R&D Project: Cement company ABC Ltd. researches on developing a low-cost construction material to replace cement

Table 13.3 Importance of supplier relationship for IT and R&D projects

Principle	IT project (Food delivery website)	R&D Project (Low-cost construction material)
Focus on customers	Exhaustive research about the customer needs is necessary. Already there are several food delivery websites. So, the project should set its objective keeping the requirements of the customers. From the survey, it has been found that long delivery time is one of the problems. This project aims at a faster delivery	The cost of construction is skyrocketing due to the increase in the prices of cement. A low-cost binding material to reduce construction costs is the objective of this project. Also, cement production is causing high carbon emissions, which also needs to be addressed
Continuous improvement	While deciding the features to be included in this new website, continuous improvement is essential. There will be a new feature to order food daily. PDSA cycle is carried out, and modifications are made	Research should accommodate technological advancements and market expectations
Leadership	To achieve the project's goals, the project manager must motivate and lead the other functional managers. This leadership must extend up until the stage of the delivery boy	Continuous motivation from the leadership is required as failures in experiments can reduce the employee's morale

(continued)

Table 13.3 (continued)

Principle	IT project (Food delivery website)	R&D Project (Low-cost construction material)
Involvement of people	Innovation and new ideas can germinate at various levels. A delivery boy can report the delivery problems, which provides the basis for improvement	Research cannot succeed by simply following the unanimous management's decisions. It must include suggestions and contributions from employees at various levels
Process approach	The project is divided into different processes: planning of features, coding, graphics, testing, etc.	This R&D project can be classified into different processes, such as chemical composition, manufacturing, testing, procurement…etc. Each process needs to be organized and aligned with the project's objective
System approach to management	Graphics and coding can be combined as a user interface system. This can improve the customer experience	It is necessary to link research and development with existing processes and capabilities
Factual approach to decision-making	Information about the restaurants, frequent buyers, potential zones or markets, etc., need to be evaluated	Information regarding sources and cost of procurement should be examined for achieving the least-cost mix
Mutually beneficial supplier relationships	Restaurants are a sort of supplier in this project. A good relationship with them is required to ensure the quality of food for delivery The company might be outsourcing some parts of graphics design. A good relationship with the vendor can expedite the development and improve the quality of the website	Best quality raw materials are essential for the research. Quality checks may not be possible every time. The reliability of the supplier is necessary. A good relationship with the supplier can help to expedite the project

13.6 Six Sigma-DMAIC for Project Management

Six Sigma is a quality improvement methodology introduced by Motorola. Its focus is on improving customer satisfaction, cycle time reduction, and reducing defects to 3.4 per million. It has evolved from a quality improvement policy to a company-wide philosophy. It uses a considerable amount of data and statistical tools to predict the sources of defects, methods to eliminate them, and choose the best alternative.

Six Sigma has three different elements:

- *Measure:* to quantify the deviation Target: 3.4 defects per million
- *Philosophy:* reduction of cost through the reduction of variability

DMAIC, which stands for "Define, Measure, Analyze, Improve and Control" is one of the essential methodologies for adopting Six Sigma (Fig. 13.4).

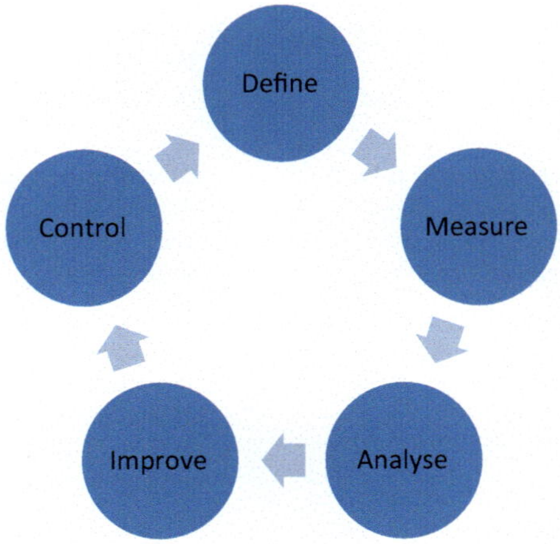

Fig. 13.4 DMAIC

Define	• It is required to define the customer requirements, objective, and scope of the project precisely • Supplier, Input, Process, Output, and Customer (SIPOC) tool is commonly used in this step to define the problem and scope • Do not initiate the process in this phase. The initiation can be done after informing leaders and other employees about the objective and the scope of work
Measure	• Identify and collect the data related to the expected problems and the processes which need improvement and measure the present situation for the data obtained • The variations are observed over some time
Analyze	• The data obtained is compared with the current processes, and the deviation is measured. The root cause for this deviation is identified by analysis • Assumptions that may create bias in the analysis must not be taken into account • Data must be verified carefully before decision-making. Critical reasoning should happen at this stage
Improve	• Improve the process by eliminating the problems. Processes that cause issues are identified and rectified. A potential solution is developed. Involve operators and fix the practical issues, if any
Control	• This step ensures that the processes causing undesired variations were being rectified. A new process is to be implemented. It checks whether the new approach could eliminate the variations and enhance the quality of the output

Project Management in Practice

Challenges in Managing Six Sigma Projects

by

Mr. Ciby C. James, Director, American Society for Quality (ASQ) India

The power of the project-based approach to improvement associated with Six Sigma is that it unleashes motivated individuals and teams within the organization and facilitates them in executing critical projects. If this process is managed well, many large and small improvement projects are being run across the company, each bringing some vital change. This is a cultural facet that an organization's management would do well to establish, ensuring intrapreneurship and leadership development from the grassroots.

The greatest challenge to project success in such a scenario lies in the management's understanding and commitment to this process. Middle management members who do not appreciate the power of this process or do not want to cede power or control to their subordinates may put roadblocks in the way of such an initiative. Therefore, this calls for top management to put their full weight behind this movement and back it up by ensuring a sustained review process and recognition process. Projects by young professionals tend to be successful only when there are clear expectations out of them. Review and periodic monitoring by managers are essential requirements to ensure the success of projects.

The results can often be surprising! Some projects can bring value over 10 times the annual salary drawn by all the team members, and collectively, the projects can bring in savings or revenue opportunities that rivals the profits earned through a normal business. It is a good practice to have the Finance Department formally vet the financial benefits achieved by the projects so that all claims are fully credible.

The other challenge is the capability of the individuals leading the project teams. If they are working on a project for the first time, they may face issues in problem-solving and people management that can seem insurmountable. This is where formal training and project support through a coach become effective. Once a project leader completes one or two projects, confidence in one's own ability follows. Such individuals then become resources for the organization that further help to sustain and grow the project practice by helping to initiate and coach others in doing new projects.

Project Management in Practice

Why do we need quality management for construction projects?

Project managers in the construction industry encounter a great deal of challenge for smooth conduct of site operations, maintaining safety and meeting the stipulated schedule and budget requirements. The key challenges for project managers are:

- Managing the conflicting objectives of stakeholders
- Educating managerial staff and site workers for quality control procedures
- Ensuring satisfaction of client
- Dealing with government regulatory norms and political disturbances
- Identifying and developing an appropriate quality control method for the project

Project managers in the construction industry must appreciate the role of quality management and develop proactive strategies to deal with following key challenges.

Risk Management

- Do not ignore the short-term issues and put the entire focus on long-term risk management.
- For example, unreliable subcontractors, scheduling conflicts, changing expectations of stakeholders can hamper the performance of the project.
- Contingency plan is a necessity!

Structure

- It is essential to have target setting for the functional managers and their team.
- A clear goal setting improves accountability of the people.

Forecasting

- It is difficult for the project manager handling a typical construction project to deal with escalating and fluctuating stakeholders' and market expectations. This severely hampers the productivity and performance of the project.
- A forecast should not only focus on long-term issues but adequate accommodate short-term requirements. This should be timely communicated to the stakeholders.

Cash Flow

- The construction business relies on invoicing, which can sometimes be an outdated system. And if payments fall behind, it could negatively impact a company's cash flow. This can in turn dry up a well of funds for other projects and cause delays.

- A scientific IT-based invoicing system is a necessity !

Communication

- Construction project involves many work packages and demands a strong coordination among functional managers.
- A clear and effective communication improves the synchronization across work packages.

Skills

- Construction industry is governed by market reputation and trust. It is necessary to examine a skill gap at managerial and worker levels and devise a suitable mechanism for skill upgradation.
- A skill gap and lack of knowledge on advanced practices can cause significant delay in construction projects.

Sources:

(a) Chaturvedi, S., Thakkar, J.J. and Shankar, R. (2018) "Labor Productivity in Construction Industry: An Evaluation Framework for Causal Relationships", Benchmarking: An international journal, Vol. 25 No. 1, pp. 334–356.
(b) Tanmay, M., Thakkar, J.J., and Maiti, J. (2017) "Modelling critical risk factors for Indian construction project using Interpretive Ranking Process (IRP) and System Dynamics (SD)", International Journal of Quality and Reliability Management, Vol. 34 No. 9, pp. 1451–1473.

Summary
"Quality" in today's world cannot be defined in a single sentence as it all depends upon the frame of reference with respect to which we want to define quality.

The final products and services in various projects from different field must adhere to certain specified standards and specifications which ultimately validates it quality.

Quality must be maintained for both the final product and processes involved during its entire production.

Deming has advocated fourteen principles for quality management.

Joseph Juran has explained his model of quality improvement on the basis of three universal processes: quality planning, quality control, and quality improvement which have been popularly named a Juran's trilogy.

Six Sigma is a quality improvement methodology introduced by Motorola. It focuses on improving customer satisfaction, cycle time reduction, and reduction of defects to 3.4 per million.

Questions for Discussion

1. What are the various definitions of quality? Explain each one with a suitable example specifically to construction or R&D or IT projects.
2. What is the importance of quality for executing a project? What are the implications a poor quality may have on the overall execution and planning of the project? Explain this for a typical project executed by an infrastructure company.
3. Enlist and explain the principles of Deming's with its relevance for a typical R&D and IT project (in a tabular format).
4. Enlist and explain Juran's trilogy and principles on quality with relevance for a typical R&D and IT project (in a tabular format).
5. What is TQM (Total Quality Management)? What are the pillars of TQM?. Explain the relevance of TQM principles for a typical R&D and IT project (in a tabular format). How can it help R&D and IT projects to improve their quality? – Explain in detail.
6. What is Six Sigma? What is the DMAIC cycle, and what are the various issues considered in each of the phases of DMAIC? What is the utility of the DMAIC cycle for controlling/ improving the quality of a project? Illustrate an application of the DMAIC cycle for a typical construction project. Explain each phase in detail.

Group Project

Based on the reading of this chapter, develop a detailed list of quality indicators for measuring the performance of an infrastructure project. Undertake a visit to ongoing government infrastructure projects such as road construction in your vicinity and appreciate the various challenges specific to the project life cycle. Evaluate the performance against a set of quality indicators on a scale of 0–10. Draw the detailed recommendations for the government agencies to improve upon the cost, quality, and schedule of the ongoing road construction project.

Chapter 14
Human Issues in Project Management

Critical Questions

Why considerations to human issues are important in project management?
What are the consequences of ineffective human resource utilization in project?
What are the key strategies for an effective in human resource utilization?
What is HDBI profile and its importance for project management?
What are group maturation phases?
How do we keep people motivated in the project environment?
What are the leadership models?
What are the key time management principles?
What are the key principles of personal mastery? Why is it important for project leaders?

14.1 Importance of Human Resource in Project

Project Human Resource Management includes the processes that organize, manage, and lead the project team. The project team comprises people with assigned roles and responsibilities for completing the project. Project team members may have varied skill sets, may be assigned full or part-time, and may be added or removed from the team as the project progresses.

Human Issues Are Critical in Managing Project Because

- Project management is teamwork, and the presence of many people tends to give rise to issues that disturb the project's environment.
- Managing people's issues well can help things going smoothly despite abrupt changes.
- More often, the project fails because the organization did not sufficiently focus on the project's effect on people.

J. J. Thakkar, *Project Management*, Management and Industrial Engineering,
https://doi.org/10.1007/978-981-15-3695-3_14

Consequences of Neglecting Human Issues in Project Management

- This crucial neglect tends to keep people under stress, making people more likely to make mistakes, and performance ceases to be at an optimal level.
- When people's issues are given less importance, it brings resistance, apathy, lack of commitment, and lack of enthusiasm.
- Issues and differences, if not taken care of, may not help in keeping people together.

Human Issues in Typical Construction Projects

1. Recruiting qualified workers.
2. Ensuring safety and controlling construction worker's costs.
3. Coping with changing labor laws.
4. Temporary workers and high employee turnover.
5. Construction companies often employ workers from other companies.
6. There is a risk that performance concerns subjugate the needs of employees.
7. Construction projects are given short notice because of tender bidding, so this
8. Leads to insufficient time for strategic planning.
9. Multiple organizations are working together on a project.
10. Internal temporary appointment of employees because of the increasing tendency of construction companies to appoint sub-consultants and subcontractors.
11. The workforce may work long hours.
12. Staff turnover tends to be high enough.
13. Different skills and experiences of employees are required.
14. There are many health and safety risks in the construction organization.
15. The great number of legal requirements to be satisfied.
16. Managing employees' payroll, benefits, and compensation.

Human Issues in Typical R&D Projects

1. R&D projects involve geographically dispersed participants and diverse teams.
2. R&D projects involve people of different cultural and professional backgrounds, which may lead to tensions.
3. Recruiting experts as R&D projects demand it.
4. Employees need to be given proper training; hence, it requires enough training period.
5. Abrupt changes due to frequent technical changes may lead to enthusiasm reduction.
6. Changes in contracting authority break the rhythm of the project.
7. Changes in in-laws affect the expenses of the project, which indirectly affects employees.
8. The R&D leader has to deal with technical professionals who may have various distinctive characteristics regarding their goal orientation, need structure, and behavioral patterns.
9. Technical experts tend not to accept hierarchical control.

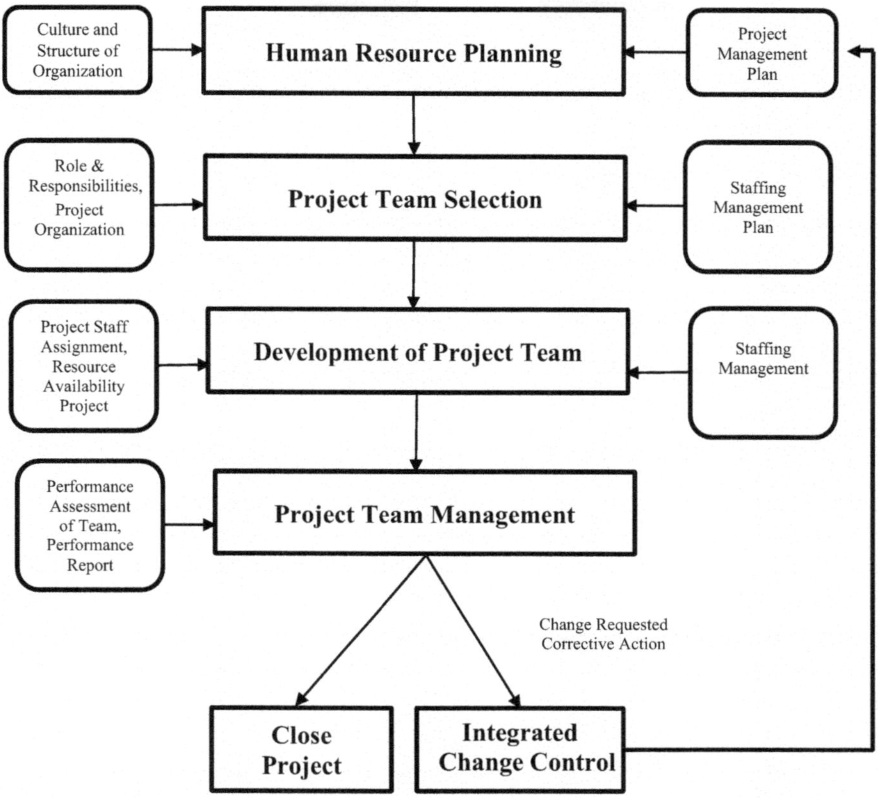

Fig. 14.1 Project human resource management

10. Since R&D projects attempt to conquer the unknown, it creates scheduling issues for employees.

Project human resource management is a highly complex affair as the project demands different skills at different phases in executing the project. Any mismatch in the human resource requirement planning cycle causes a server delay and cost overrun of the project. Figure 14.1 illustrates the critical elements of project human resource management: project team selection, development of the project team, and management of the team.

14.2 Whole-Brain (HDBI Profile) Model

Importance of Recognizing "Human Potential" in Executing Project

- Requirement of core competencies for the specificity of the projects.

- It ensures the effective utilization of human resources to enhance the project's overall productivity.
- It significantly impacts human motivation if the task accommodates their skills, interest, and potential.

Whole-Brain Exercise (HDBI) Profile

- Herrmann Brain Dominance Instrument (HDBI) is a system to measure and describe thinking preferences in people often compared to psychological assessments.
- It is a 120-question highly validated diagnostic survey where a person's answers indicate his thinking style preferences.
- HDBI model shows how different people approach problems and challenges differently.
- The way humans communicate, work together, deal with conflicts, or are creative is primarily determined by the HDBI index.
- The four-quadrant model of Ned Herrmann, responsible for our behavior, is presented in Fig. 14.2. This is as per *Project Planning, Scheduling and Control: A Hands-On Guide to Bringing Projects in On Time and On Budget by James P. Lewis, Third Edition, McGraw-Hill, 2001.*
- HBDI is very valuable in understanding the complex nature of the project.
- HBDI helps individuals understand preferred thinking styles.

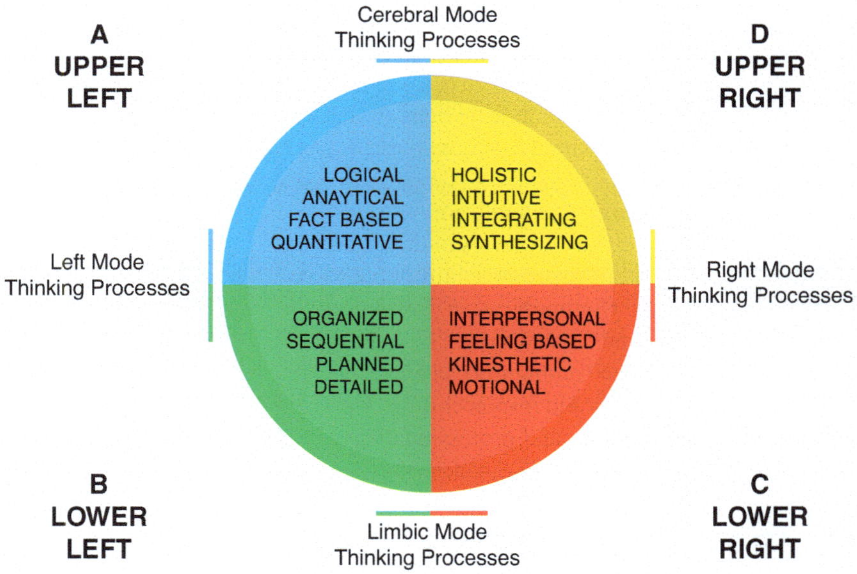

Fig. 14.2 Whole brain model of Ned Herrmann. *Source* Project Planning, Scheduling and Control: A Hands-On Guide to Bringing Projects in On Time and On Budget by James P. Lewis, Third Edition, McGraw-Hill, 2001

- HBDI helps develop the skills to broaden their thinking beyond their preferences and communicate across thinking styles for better results.

  Organizations introduce HBDI to spark motivation and interest among employees.

- It helps a manager explain the importance of team building to the employees and allows individuals to understand their thinking styles.
- It helps employees working in different departments enabling their team fit.
- HBDI help team members to appreciate diverse thinking styles, thought, and logic process as summarized in Table 14.1. It is also essential to understand the differences between team and group as listed in Table 14.2.

Table 14.1 Relevance of HBDI elements

Element	Qualities	Relevance
A	• Logical • Analytical • Fact Based • Quantitative	• The person with this trait is highly logical, interested in technical issues of the project, analyze the situation critically and undertake problem solving • If the person with predominant "A" profile (i.e., he or she has very little preference for thinking in the other quadrants, particularly the C quadrant), this person may be seen as cold, uncaring and interested only in the problems presented by the project
B	• Organized • Sequential • Planned • Detailed	• This is the preferred thinking of many managers, administrators and planners, bookkeepers, supervisors and manufacturers • Individual who have single-dominant profiles in the B quadrant could be concerned with the detailed plans of a project, and with keeping everything organized and controlled • Individuals with financial interests who are dominant in quadrant A will probably be financial managers, whereas those with dominant B quadrant profiles may be drawn to cost accounting
C	• Interpersonal • Feeling based • Kinesthetic • Emotional	• The people with "C" profile are "touchy-feely" and "people oriented" • The people with this trait are interpersonal, emotional, musical, spiritual, and talkative • Such individuals are often teachers, social worker, counsellors, and ministers
D	• Holistic • Intuitive • Integrative • Synthesizing	• The person with this profile is highly useful for coordination of project activities and building relationships with various stakeholders. He can coordinate well with people both inside and outside the team • For highly political projects, this trait is extremely useful

Left Mode Thinking Processes: Language, Mathematics, General analytic, logical type of thinking used in constructing network diagrams
Right Mode Thinking Processes: Synthetically perceptual, realizing the whole, simultaneous relations and dependencies

Table 14.2 Difference between team and group

Team	Group
The team refers to the number of people working together in an activity	Group refers to the number of people connected by some shared interest
Work for a common goal	Not necessarily work for a common goal
Specific tasks are assigned to each individual	Specific roles and duties are not assigned to individuals
Members of the team are interdependent	Members of the group are independent
Members know each other and are aware of each other's strengths and weaknesses	Members may not know each other

Importance of team formation:

Employees are the critical asset for any organization, and team building boosts employees' morale and increases the project's overall success. The reasons for team formation include:

1. Team formation enables open communication among employees.
2. It motivates employees to take on new challenges.
3. Working together as a team promotes creativity and ignites new ideas.
4. It also increases the trust factor among employees.
5. In a team, lots of suggestions and discussions raise the performance.
6. It helps in developing mutual trust among employees.

Various Roles People Can Play in a Team

- Shaper role: Performed by people who are dynamic and fight challenges.
- Implementer role: Performed by those who are practical and efficient and convert ideas into an actual plan.
- Finisher role: Those who have a complete eye on a project.
- Coordinator role: They guide the activities of the team.
- Team worker role: Works toward resolving issues and keep the team united.
- Monitor evaluator role: Cautious and does not hurry in decision-making.
- Specialist role: A person with expert knowledge in a specific area.

The importance of team roles in construction and R&D projects is presented in Tables 14.3 and 14.4, respectively.

Table 14.3 Importance of team role in construction projects

Type of role	Its importance in project
Construction manager	• Oversees entirety of the project from beginning to end • Coordinate among different construction managers • Responsible for planning, budgeting & scheduling
Estimator	• Responsible for estimating costs, labor needed & material required • Has to be well versed with project
Architect	• Responsible for overseeing client's need • Develop creative plans
Supervisor	• Acts as middlemen between field workers & office • Takes care of any conflict between field workers & office
Construction worker	• Follow the instructions of managers • Use heavy machinery • Give realistic completion to the project

Table 14.4 Importance of team role in R&D projects

Type of role	Role importance in project
Technical expert	• Suggest problem solution • Generate new ideas • Work as a technical consultant • Involved in basic research, new product development
Team builder	• Conflict solver • Cohesiveness builder • Considerate and empowers the team
Champion	• Risk-taking • Leads into coalition building for support
Gatekeeper	• Contacts with information sources • Circulate technical and market information
Strategic planner	• Deals with planning, • Oversees scheduling, task coordination, and goal setting

14.3 Group Maturation Process

Before we get into the group maturation, we must get to know why the group is formed. In simple terms, it is done to achieve the common goals set up by the leader. This can be accomplished through a group maturation process. Group maturation can be defined as changes that occur in the group over time. The model developed by Wheelan (1990) has described that the group maturation evolves through four stages which include:

Stage 1: dependency and inclusion
Stage 2: counter-dependency and fight
Stage 3: trust and structure
Stage 4: work and termination.

Although many such models have been defined that elucidate group maturation, it is no secret that very often, a group maturation model developed by Bruce Tuckman (a renowned psychologist) in 1965 is widely used. This model comprises the five phases of group maturation—forming, storming, norming, performing, and adjourning. The key characteristics of the phases are summarized in Table 14.5.

Table 14.5 Key characteristics of group maturation phases

Stages	Relevance
Forming	• The group is uncertain about their purpose, structure, and leadership, and herein they decide all those factors which can influence the project. Individuals comprising groups make an effort to know each other, and their motivation comes from within • This is the stage where creating a team charter is critically essential. In this stage, members "test the water" to see what type of behavior is acceptable • This stage primarily concerns the group individual with accepting their fellow members rather than achieving the team goal. Of course, they avoid conflict and controversy at this stage • The team leader has a vital role in guiding the group for team goals and providing clear direction
Storming	• This stage is characterized by tension, completion, and conflict among group members • Very often, questions arise about who is responsible for what and what the rules are. This stage is where many teams fail and sail on, and it needs a prudent leader who uses his leadership style to resolve the issues • If the leader cannot find a way to lead together, resolve differences and problems productively, individuals in the group won't be influential and lead to the group's collapse
Norming	• Storming fructifies to this stage only when trust grows among members comprising the group • Individuals start appreciating each other skills and knowledge • Group shows cohesiveness, members of the group start to understand that there should be a common goal and leader who should lead us • Individual lacking in specific domain asks for help and guidance from other members
Performing	• This is the ripe fruit of the stages of group development, where the structure is now fully functional and acceptable. In this stage, the group operates with little friction due to developing a high level of trust, respect, and motivation • Team energy moves for performing the intended task • The role of the leader is to communicate effectively and assist the team if it attempts to revert to the prior stage
Adjourning	• The last stage of group development is preparation for disbanding as the intended task is done or at last leg is left behind. Sometimes, even the permanent team is dissolved if there is significant change through an organizational structure • The leader has a vital role in play here as some members may find it difficult to form strong bonds or if their future looks uncertain • It is the role of the leader and manager to identify this stage and use their leadership style along with the specific tools and techniques to maximize the team's effectiveness

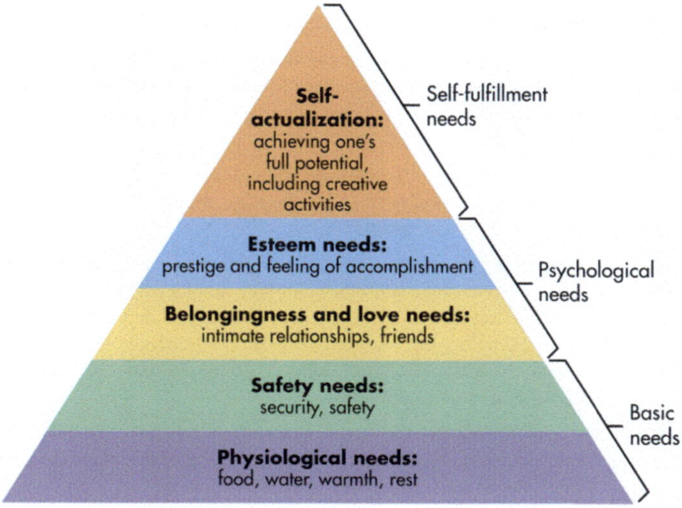

Fig. 14.3 Pyramid classification of Maslow needs

14.4 Role of Motivation in Project Management

At the starting of the project, people are most motivated to work toward goals and achieve good results. As time passes and people get exposed to various challenges, members tend to fade the motivation and forged relations they shared. At this stage, one needs to enthuse motivation among the members to achieve their common goals. This helps the team to boost its productivity and efficiency. It also has a positive impact on the morale and quality of work of the people.

Although there are many theories of motivation to increase productivity, such as Herzberg's motivation theory, Maslow's motivation theory, Hawthorne effect, to name a few, the first two have received the maximum attention. Maslow's motivation theory (Fig. 14.3), also known as "Hierarchy of Needs," was coined by psychologist Abraham Maslow in his paper "A theory of Human Motivation".[1] It comprises the five-tier model of human needs, often denoted with the pyramid. It implies that lower-level needs in the hierarchy must be satisfied before individuals can attend to higher needs.

It comprises physiological, safety, love and belonging, esteem, and self-actualization (see Fig. 14.3). The five-stage model is divided into deficiency and growth needs; the first four levels are often referred to as deficiency needs (**D-needs**), and the top-level is termed as growth or being needs (**B-needs**). Maslow stated that D-needs must be met before the individual desires for the secondary or B-needs. A detailed narration of all the five needs is summarized in Table 14.6.

[1] Maslow, A. H. (1943). "A Theory of Human Motivation". In Psychological Review, 50 (4), 430–437.

Table 14.6 Maslow's hierarchy of needs

Physiological conditions	It is the physical requirement for human survival. If this is not met, the next four needs of the pyramid are hard to realize. This means all other needs become secondary unless the physiological needs are met. It includes food, water, warmth, sleep, shelter
Safety needs	It is logical that once the survival is secured, the person strives for security. It includes protection from law, policy and should provide stability and fearless working conditions
Love and belongingness needs	It includes social and feelings of belongingness. According to Maslow, humans need to feel a sense of belonging among social groups. Many people experience loneliness, social anxiety, and clinical depression in the absence of love and belonging. As seen in children who cling to abusive parents, this need can override safety in childhood. These particular needs include friendship, intimacy, and family
Esteem needs	The need for respect and status, recognition, fame, and attention is the "lower" version of esteem. The "higher" sense of esteem is the satisfaction of one's needs for self-respect and can also include self-confidence, competence, mastery, and independence. The "higher" version takes guidelines into account, so the hierarchies are interconnected rather than sharply divided. Therefore, esteem and the next levels are not strictly distinct; instead, they are closely related. Maslow classified it into two categories: (1) esteem for oneself and (2) the desire for reputation and respect from others. Maslow indicated that the need for respect or reputation is most important for children and adolescents
Self-actualization needs	One of the B-needs comes at last when all the D-needs are fulfilled. It refers to the need for personal growth and discovery that is present throughout a person's life. In self-actualization, a person comes to find a meaning to life that is important to them. This includes needs such as partner acquisition, parenting, utilizing and developing talents and abilities, and pursuing goals

Herzberg's Motivation Theory: Also known as the two-factor theory and Herzberg's motivation-hygiene theory, it was proposed by Frederick Herzberg in 1959 (Herzberg, F., 1959). Mausner. *SNYDERMAN., The motivation to work, New York.*). According to theory, some job factors result in satisfaction while other job factors cause dissatisfaction and act independently. Herzberg developed this theory based on the research which he conducted by asking a group of people about their good and bad experiences at work. He advocated that people's job satisfaction depends on factors for satisfaction (motivators/satisfiers) and factors for dissatisfaction (hygiene factors/dissatisfaction).

Motivators/satisfiers include performance, recognition, job status, responsibility, opportunities for growth, etc., while all about salary, secondary working conditions, and relationships with colleagues, physical workplace, etc., are included in hygiene factors/dissatisfaction.

As per the theory, he claims that these factors function on the same plane, which means that satisfaction and dissatisfaction aren't opposites. Taking away an employee's dissatisfaction with some kinds of perks does not mean the employee will be satisfied. The employee is just no longer dissatisfied. According to this theory, there are four possible combinations:

High hygiene + High motivation	This is an ideal situation that employees are highly motivated and have few complaints
High hygiene + Low motivation	They have few complaints but are not highly motivated. They see their job as payback
Low hygiene + High motivation	Employees are motivated but have a lot of complaints. The job is exciting and challenging, but salaries and work conditions are not up to par
Low hygiene + Low motivation	This is the worst situation where employees are not motivated and have many complaints

Unlike Maslow, who had little data to support his ideas, Herzberg presented considerable empirical evidence to confirm the same but has been criticized many times. Behavioral scientists have surfaced several inadequacies. The most basic of such criticism is that both of these theories contain the relatively explicit assumption that happy and satisfied workers produce more ideally might be far from the truth.

For example, the relevance of the theory can be explained for a typical construction project. As per Maslow's motivational theory, the basic structure of pyramid includes physiological need, which is abused in most specific construction projects. Regardless of their physiological needs such as food, water, sleep, or shelter, the workers all live under the same construction roof or build tin shades to keep them warm. Coming to the safety needs, they are provided with almost nothing or bare minimum on these safety needs. Dingy work conditions enthuse no motivation, but the run for money makes them go along their needs. Similarly comes the abusing or denying of social belongings and esteem needs which they need to forgo as construction sites are far away from their home.

14.5 Leadership and Importance for Project Management

Leadership is an essential ingredient of project management. It is the fuel that propels the projects forward toward delivery. Leadership involves motivating a group of people to take action toward a common goal. In a business setting, this can mean providing workers and colleagues with a strategy to meet the company's needs. Leadership includes creating a vision, strategizing to achieve the vision, making decisions and solving problems, building and empowering teams, and influencing and communicating with stakeholders or clients. Without an influential leadership, a project manager may underperform, miss strategic opportunities, stifle innovation, underutilize team members, and ultimately may lead to a shortfall of project goals, quality, performance, and productivity. The leadership traits a project manager must possess primarily includes motivating and inspiring others. Other leadership skills

Table 14.7 Relevance of leadership traits for R&D projects

Trait	Importance for R&D projects
Ability to organize and delegate	R&D projects require an extensive study of cluttered information, samples, and specified objectives. The given traits help in avoidance of macro-management and embracing delegation
Flexibility and quick understanding of the situation	Due to the high uncertainty associated with the R&D project, it may collapse at any stage. Flexibility and quick understanding of the situation provides patience, stamina, and a strong vision for handling difficult situations
Capability to balance logic with creativity	This helps the leaders to create a bigger picture, comprehend the individual elements, and discern the subtlest issues of the project
Communication mastery	R&D involves many stakeholders with conflicting requirements. This trait helps a leader to better negotiate and pursue the matter to the satisfaction of key stakeholders
Leadership, competence, and experience	R&D projects are excruciating, challenging, and seemingly difficult to obtain the desired result very often. This trait enthuses leaders to go on with what they had started, face the difficulties with a bold move, and experience nudges to elude the failures
Inspiring leadership	R&D project demands confidence building among various agencies and seeking their consensus on conflicting matters. This trait helps the leader to augment and consolidate for the common goals to be achieved

include negotiations, communications, influencing skills, and team building. Overall, the leadership traits a project manager must possess are as follows:

1. Ability to organize and delegate
2. Flexibility and quick understanding of the situation
3. Capability to balance logic with creativity
4. Communication mastery
5. Leadership, competence, and experience
6. Leadership that inspires

Complex projects such as R&D are tough to execute and complete within the stipulated deadlines. This demands leaders to perform with unique traits. Table 14.7 summarizes the relevance of various leadership traits for R&D projects.

14.6 Leadership Theories/Models for Project Management

There are three types of valuable leadership for project management, as explained in Table 14.8.

Table 14.8 Types of leaderships

Situational leadership	• Adaptable leadership as per the situation and maturity of the project team to perform the task • The project manager directs the tasks and monitors the project progress or sometimes even sharing the decision-making aspect, whichever suits the situation • A manager finds and provides what the project team needs to perform the task • The manager expects to have a short relationship with the project team till the completion of the project
Contingency leadership	• It claims that there is no best way to get a project done • A manager may get stressed out because he has to assign jobs to all the employees and meet the project's deadline, which in other cases brings all blame to him • The success of the project demands control of the behavior of the employees through strict job descriptions • Easy to meet tight project deadlines but raises unique issues in the future as the primary focus of team members is on current problems
Servant leadership	• It starts with a natural feeling to serve by sharing power, putting the needs of the employees first and helping people develop and perform • Customer service associates as the main priority, i.e., project team comes before short-term goals related to project completion • The employee identifies himself with the project leading to high loyalty and corporate culture development • Ideal for highly uncertain projects that require an adaptive approach

14.7 Personal Mastery: Must for Project Leaders

Importance of Personal Mastery in Project Management

- Guided by passion, an individual can work with an underlying sense of purpose.
- Sharp focus on goal achievements and objectives.
- It helps in becoming initiators acting despite having to be fearful.
- Volunteering for owning responsibility is a prominent characteristic.

Key Dimensions in Personal Mastery

Personal Vision

- Ability to picture the best leader one can be and work toward that with focus.
- Expresses one's underlying purpose into the domain of acts and commitments.

Organizational Culture

- Describes the fundamental assumptions of an organization's values, beliefs, norms, symbols, language, rituals, and myths that guide people's behavior.
- Personal mastery characterizes open, trustworthy, collective, and empowering organizational culture.

Competence

- Competence is the capability to perform well in a project.
- Includes emotional intelligence, interpersonal skills, and systems thinking.

Personal Values

- Shapes and influences the general nature of an individual's behavior.
- Self-respect, self-fulfillment, security, sense of belonging, warm relationships with others, fun, and enjoyment of life are common positive personal values.
- It can also be defined as the degree to which individuals express pride, loyalty, and cohesiveness.

Motivation

- Explain why humans are inspired to do certain things.
- Willingness to commit themselves to be personal and professional development.
- Leads to better performance and individual happiness.

Training and Development (T&D)

- T&D enables organizational transformation.
- Effective T&D will lead to professional development, which will benefit the project.

14.8 Time Management in Managing Projects

Time management is one of the critical components of driving a project completion effectively. Good time management leads to a project completed at the desired time efficiently and productively. The primary goal of a project manager is to ensure that the resources are used most efficiently to complete the project on time with desired quality. Efficiencies could be hampered, and, in turn, project success might be hindered in the absence of time management. The pickle jar theory for time management is illustrated in Fig. 14.4.

The pickle jar theory for time management has three layers:

(1) The rocks fill up most of the jar. These are the essential tasks in the project, and they should be completed with utmost urgency and priority.
(2) The small pebbles that fill up the voids between the rocks represent the essential tasks but not as urgent as the rocks that comprise a majority of our project. However, once the majority of the project is done, the pebbles should be on priority.
(3) The water that fills up the rest of the space comprises the last 5–10% of the project and should be done with the least priority and urgency. Without time management, the project might not finish successfully if these tasks are done before the previously mentioned tasks.

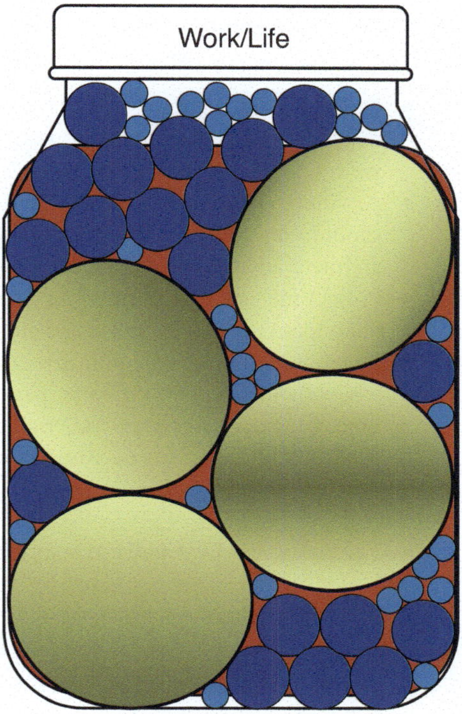

Fig. 14.4 Pickle jar theory for time management

For example, a construction project can be well managed using the pickle jar theory:

- The most important and urgent tasks should be prioritized: Planning the project, site clearing; raw materials import; task delegation.
- The second layer comes on top of the most urgent tasks, which are execution: General excavation, placing formwork for concrete, installing sewer lines.
- The third and final layer should come last in this priority: Pouring concrete; support for the project; post-project cleaning (if any).

Project Management in Practice

Human Issues in "Construction Project"

A typical construction project has three phases: Feasibility or planning, preconstruction, construction, after which the handover to the client occurs. For each of these phases, various stakeholders and human factors are involved, which could cause intricacies in the project's development. One of the project manager's roles is to manage human resources efficiently to get them completed in due time with the best quality. The processes and activities involved in a typical construction project can also be laid down into these three categories. As a result, we will break our case study into these sub-cases and analyze them individually:

1. **Planning/feasibility phase**:
   The various stakeholders or people involved in this phase are:
   - Owner/Client: Negotiations with them regarding finances, time schedules, or resources required can be tedious and time-consuming and should be planned well.
   - Architect: He designs the construction plan. Again, the project manager should convey to him his requirements in the most lucid manner, and the architect and the project manager should be on the same page.
   - Project management team: There should be many members in the management team to delegate tasks and sub-tasks that could be done individually.
   - Financial advisor: A significant entity, the project manager, and financial advisor should be on the same page for the project to be completed in due time. Cost is one of the most crucial attributes in a construction project, so dealing with finance is needful.
2. **Preconstruction**:
   - The client, architect, and PM team all have the same responsibilities as before. Major human issues could happen if the project manager is not comfortable with any of them
   - Local authorities: The people in the local area are vital factors that might drive/hinder the project, and hence, the project manager should manage the regulations (economic, cultural, social, and ecological) and should be on the same page with the locals.
3. **Construction**:
   - Suppliers: Many suppliers come into the picture to supply the raw materials necessary for the project. Cost, quantity, quality, reliability, sustainability, delivery time, etc., are factors that the project manager needs to discuss with the suppliers to succeed successfully.
   - Contractors: One of the essential entities in the project, and the most problematic situations are usually experienced between the project manager and the contractors. Most of the time, they are outsourced by an external agency, so the project manager needs to have good ties with such agencies.
     But he should also manage the contractors and see to it that the work done meets the prescribed quality standards and holds up to the schedule.

Question for Class Discussion

1. What are the critical human management issues in a typical construction project?
2. What is the role project manager in dealing with conflicting human relationships in such a project?
3. What is the conflict resolution mechanism a project manager should adopt?

Summary

Project management is a teamwork and the presence of large number of people tends to give rise to issues which disturbs the environment of the project.

Managing people issues well can help things going smoothly in spite of abrupt changes.

More often project fails because organization did not sufficiently focus on the effect that project brings on people.

Herrmann Brain Dominance Instrument (HDBI) is a system to measure and describe thinking preferences in people often compared to psychological assessments. It is 120-question highly validated diagnostic survey where answers of a person indicate his thinking style preferences. HDBI model shows how different people approach to problems and challenges differently.

Group maturation can be defined as changes that occur in the group over time.

Maslow's motivation theory comprises of five-tier model of human needs and often denoted with pyramid. It implies that needs lower down in hierarchy must be satisfied before individuals can attend to needs higher up. It comprises psychological, safety, love and belonging, esteem and self-actualization.

Leadership is an essential ingredient of project management. It is the fuel that propels the projects forward toward delivery. It is actually the ability to get things done through others. It consists of focusing the efforts of a group of people toward a common goal and enables them to work as a team.

Time management is one of the key components of driving a project completion effectively.

Good time management leads to a project completed at the desired time, efficiently and productively, without any delays.

Questions for Discussion

1. Why are Human issues critical in managing projects? What are its consequences on project execution and planning? Enlist at least 15 human issues for typical construction and R&D projects.
2. What is the importance of recognizing "Human Potential" in executing a project? What is Whole-Brain Exercise (HDBI) profile? Explain each element of the HDBI profile in detail with suitable examples of various projects like R&D, IT, Construction, etc.
3. What is the difference between team and group? What are the various roles in a team? Check the relevance of such roles for R&D and construction projects in a Tabular format.
4. What is the process of group maturation? Why is this important? Explain the Group Maturation Process (like Forming, Storming) in detail with suitable examples.

5. What is the need for motivation in managing projects? Explain the widely used motivation theories such as Maslow's Motivation theory and Herzberg's Motivation theory in detail. Discuss the importance of both approaches for a typical construction project environment.
6. What is leadership? What are the leadership traits a project manager must possess? What are the positive effects of such features in executing a complex project such as an R&D project (explain in a tabular format—Column 1: Trait, Column 2: Importance for R&D project)?
7. Explain the three most important and widely used leadership theories/models—(i) Situational Theory; (ii) Contingency Theory; and (iii) Servant Leadership Theory with a specific focus on its relevance for the project environment. Compare the critical feature of all the three theories on the set of criteria for three different project environments such as (i) Construction Project; (ii) IT Project; (iii) R&D Project in TABULAR FORMAT (Be creative in constructing a Table which can present the comparison in a concise format).
8. What is the importance of "PERSONAL (INDIVIDUAL) MASTERY" for a project manager to effectively execute the project? What are the critical dimensions of "PERSONAL MASTERY"? Explain each one in detail and how it will positively or negatively affect the management capability of a project manager.
9. What is the importance of "Time Management" for a project manager? Explain the scientific model/theory for Time Management in detail with suitable examples of managing a construction project.

Chapter 15
Project Communication Management

Critical Questions
Why communication management is most critical in managing projects?
What are the consequences of an ineffective communication?
What are the communication models and what is its relevance for project management?
How opportunism and effective communication are linked?

15.1 Importance of Project Communication

Project communication is an essential pillar for the success of a project.

> Project Communications Management includes the processes that are required to ensure timely and appropriate planning, collection, creation, distribution, storage, retrieval, management, control, monitoring, and the ultimate disposition of project information.

There are three prime reasons for the project team to manage project communication.

- Meet the information needs of your project stakeholders (communication planning and information distribution)
- Track and report on project performance (performance reporting)
- Formally document project results (administrative closure).

Project communication helps in establishing commonness among the people working on the project. We can refer to project communication as "Project-Life Blood," as the performance of a project heavily depends on it. It has been frequently reported by the project managers delivering exceptionally high standards of technical work that communication lapses have led to the stakeholders' dissatisfaction.

J. J. Thakkar, *Project Management*, Management and Industrial Engineering,
https://doi.org/10.1007/978-981-15-3695-3_15

Equal importance to both technical features of the project and communication is vital. There are two issues associated with the need for communication for project managers. In general, they communicate with the stakeholders when any unforeseen event occurs. This is very much essential for risk mitigation. However, steady communication with the stakeholders helps the project managers to evolve a holistic understanding of various issues and minimize the possibilities of critical issues being escalated. This steady and regular communication with stakeholders falls under the project communications management knowledge area of PMBOK. The project communication involves three processes:

1. Plan communications management
2. Manage communications
3. Monitor communications

The details of these three processes are summarized in Table 15.1.

A project manager spends significant time in project communication.

- Project communication helps answer 5Ws (Why, What, When, Where, Who) and 1H (How) related to the project.
- Project communication helps notify the project status and discussions and makes sure new ideas will come from the members. Good project communication helps in the establishment of purpose. This answers the question "Why" behind everybody is doing the project.
- Project communication helps to set the goals/objectives that are to be achieved through the project. Objectives will be the primary indicators of success.
- Project communication helps in determining the key players of the project. Which task should be assigned to which team or person; who can perform it efficiently and effectively; how much time would the team need; what will be the level of involvement of the team members in meetings and standpoint?; is everybody clear about their roles?, etc.
- Task dependencies and how will they meet objectives are monitored through project communication. It keeps everybody updated about the needs of the project.
- It helps team members to be realistic on time and scale.
- It cultivates a culture of honesty, flexibility, and adaptability among team members.

The critical aspects of project communications include:

- **Mapping Potential Conflicts with Stakeholder Preferences**: When stakeholders come from different fields, backgrounds, organizations, operations, cultures, and states, it is more likely to result in conflicts. Stakeholders might misinterpret the directives. Therefore, it is most important to adopt a professional and scientific way of communicating for highly complex projects. This helps avoid the differences in opinion and conflicts in the various phases of the project life cycle.
- **Clear Default Protocols**: Communication needs are different for different projects, so we need to establish a set of default protocols that are understandable

Table 15.1 Project communication management processes

Process	Relevance	Inputs	Tools and techniques	Outputs
Plan communications management	• A controversial project can be managed well by communicating with stakeholders, such as open houses, town hall meetings, and meetings • The project might not even change significantly from its original plan, but the mere communication with stakeholders can often be the deal breaker that allows the project to proceed • Even if the project is not controversial, the PMBOK dictates that everyday communication needs such as progress updates, investor circulars, and the like are identified and their content planned out in advance in the communications management plan	1. Project charter 2. Project management plan • Resource management plan • Stakeholder engagement plan 2. Project documents • Requirements documentation • Stakeholder register 3. Enterprise environmental factors 4. Organizational process assets	1. Expert judgment 2. Communication requirements analysis 3. Communication technology 4. Communication models 5. Communication methods 6. Interpersonal and team skills • Communication styles assessment • Political awareness • Cultural awareness 2. Data representation • Stakeholder engagement assessment matrix 3. Meetings	1. Communications management plan 2. Project management plan updates • Stakeholder engagement plan 3. Project documents updates • Project schedule • Stakeholder register

(continued)

Table 15.1 (continued)

Process	Relevance	Inputs	Tools and techniques	Outputs
Manage communications	• Managing project communications can be the difference between a successful project and an unsuccessful one during the project execution phase • The communications with stakeholders identified within the communications management plan are put into practice, and any spontaneous communications are executed as necessary • Creating, distributing, and storing communications are an integral part of project management and require the constant attention of the project manager	1. Project management plan • Resource management plan • Communications management plan • Stakeholder engagement plan 2. Project documents • Change log • Issue log • Lessons learned register • Quality report • Risk report • Stakeholder register 3. Work performance reports 4. Enterprise environmental factors 5. Organizational process assets	1. Communication technology 2. Communication methods 3. Communication skills • Communication competence • Feedback • Nonverbal • Presentations 2. Project management information systems 3. Project reporting 4. Interpersonal and team skills • Active listening • Conflict management • Cultural awareness • Meeting management • Networking • Political awareness 5. Meetings	1. Project communications 2. Project management plan updates • Communications management plan • Stakeholder engagement plan 3. Project documents updates • Issue log • Lessons learned register • Project schedule • Risk register • Stakeholder register 4. Organizational process assets updates

(continued)

Table 15.1 (continued)

Process	Relevance	Inputs	Tools and techniques	Outputs
Monitor communications	• Because project communication is so important to overall success, strong project control should be present • This process involves asking whether the project communications at the current point in the project have been adequate • As part of the monitoring and controlling process group, the project manager must include project communication control at regular project status update points • At the same time, as Earned Value Analysis, quality control, scope validation, and the other project control items take place, the project manager must ensure that each stakeholder has received the appropriate communication and whether anything should change	1. Project management plan • Resource management plan • Communications management plan • Stakeholder engagement plan 2. Project documents • Issue log • Lessons learned register • Project communications 3. Work performance data 4. Enterprise environmental factors 5. Organizational process assets	1. Expert judgment 2. Project management information system 3. Data analysis • Stakeholder engagement assessment matrix 2. Interpersonal and team skills • Observation/conversation 3. Meetings	1. Work performance information 2. Change requests 3. Project management plan updates • Communications management plan • Stakeholder engagement plan 4. Project documents updates • Issue log • Lessons learned register • Stakeholder register

Source A guide to the project management body of knowledge (PMBOK Guide), Sixth Edition (https://www.projectengineer.net/project-communications-management-according-to-the-pmbok/) accessed on 1st July 2021) and clear to everyone. Use simple default protocols that are used internally and modify them according to the different project requirements.

- **Match Preferences to Work Styles**: Ensure that no team uses just one method to communicate effectively, especially in today's IT-intensive era when there are many digital resources available. It depends on the characteristics of the project and people's convenience to see which communication method will enhance the effectiveness of coordination. Minimize conflicts with stakeholders using two communication channels, one for the internal purpose and the other for the stakeholders.
- **Transfer Ideas into Information**: Daily data should be converted into pieces of information that are easy to understand and analyze. The key issues and decisions should lead to executive summaries that are easy for stakeholders and team members to understand. The key components necessary for ensuring effective project communication at various phases in the project life cycle are presented in Fig. 15.1.

A project can be affected in many ways in the absence of an appropriate communication channel. Implementation of the key attributes of project communication depends upon the organizational structure, nature of the project, level of complexity, etc. The consequences of inadequate communication can be summarized as below.

1. **Socioeconomic Consequences**:

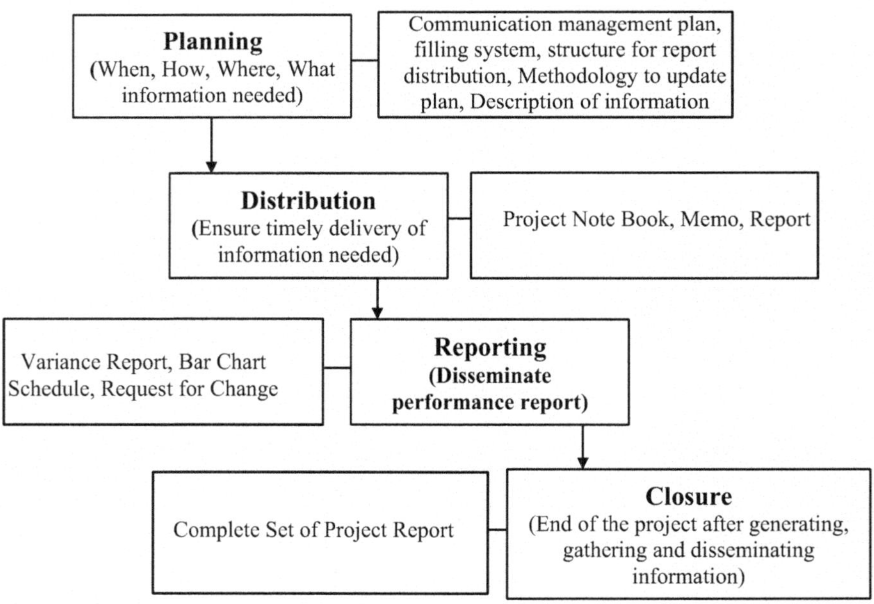

Fig. 15.1 Project life cycle and communication requirements

- **Environmental Protection**: This will inflate the uncertainty because there may be a lack of clarity on how long it will take to get approvals from authorities or regulatory agencies. Lack of definitive criteria and continued revaluation of project requirements also result in additional costs.
- **Public Safety Regulation**: Safety for people's life and health evolves with better project communication channels.
- **Economic Instability**: Budget and economic viability should be discussed during project meetings to improve transparency and trust among team members.
- **Exchange Rate Fluctuations**: This may lead to a financial crisis. New ideas and suggestions from team members help to hedge against such fluctuations.

2. Organizational Relationships Consequences

- **Contractual Relations**: Risks related to internal and external organizational relations are also significant and quite tangible. Sometimes, strained relationships may develop between various organizations involved in the design and construction/production process.
- **Attitudes of Participants**: Bad communication among teams affects the attitude of groups and team members. Teams lose their confidence and interest in a project due to a lack of understanding during meetings and discussions.
- **Communications**: A communication barrier arises from ill-conceived relationships. This needs to be addressed through appropriate contract terms.

3. Technical Problems

- **Design Assumptions**: Assumptions discussed for the designed phase are vital. The design phase should be accurate. A cross-functional communication can significantly enhance the quality at the design stage.
- **Site Conditions**: These are sub-surface conditions that always represent some degree of uncertainties during the construction or production phase.
- **Production/Construction Procedures**: If the procedures are not deliberated in detail during the design phase, the design may have to be modified after the commencement of construction. Here, communication plays a vital role in accommodating later stage modifications.
- **Occupational Safety**: The design phase should give the utmost importance to the safety of employees and workers. A little carelessness can result in severe accidents.

Project Management in Practice

How can project managers communicate effectively?

Communication Methods

- Interactive: meetings, phone calls, video conferencing
- Push: letters, memos, reports, e-mails, faxes, voice mails, press releases
- Pull: internet sites, e-learning, acknowledge repositories

Communication Types

- Formal written: complex problems, project management plans, project charter, communication over long distances
- Formal verbal: presentations, speeches
- Informal written: memos, e-mail, notes
- Informal verbal: meetings, conversations

Key Characteristics of Good Communication

- Understand and follow the communication model (sender, receiver, message, noise)
- Take an informal approach when communicating with stakeholders outside status meetings
- Develop effective listening skills
- Appear interested
- Make eye contacts
- Put your speaker at ease
- Recap what the speaker said in your words
- Do not interrupt

Communication Management Plan Main Contents

- Stakeholder communication requirements
- Information to be communicated
- Reason for distributing the information
- Time frame and frequency
- Responsible person/party to prepare and/or communicate
- Responsible person/party for authorizing the release of confidential information
- Persons who will receive the information (distribute to)
- Communication method, type, and technology
- Allocated resources to perform communication, time, and budget
- Escalation process
- Updating and refining the communications management plan
- Glossary of common terminology
- Project information flowcharts

- Communication constraints.

15.2 Opportunism of the Players and Communication in Project Management

Opportunism is a self-interest-conscious policy of taking advantage of present circumstances during the project. The opportunism of the players is a crucial managerial issue in project management. A lack of professionalism of the project managers leads to poor execution of the project. For example, select clients demand unreasonable acceleration in the project activities. Schedule delays may arise in the projects due to settlement price differences, materials or unqualified engineering acceptance, the low quality of workers, and many other reasons. The various entities in a project try to maximize their gains using opportunism. This damages the confidence fostered in cooperation and weakened the internal and external collaborations. A selfish motive to reap profits at the expense of others evokes opportunism in the project. Opportunism can induce disputes among colleagues, teams, and participants, leading to poor quality and job delays. It is essential to restrain opportunism and develop trust among the stakeholders for achieving the intended project performance. The components of communication structure in project management are illustrated in Fig. 15.2.

An effective communication structure can help to build transparency and coordination in the project, as explained below.

- Employee trust from effective communication structure: Effective communication structure creates transparency, which helps build trust across organizational levels. Open communication can reduce feelings of uncertainty and cluelessness about the company's state, making for a more positive work environment and staff who feel secure and safe.
- Relationships: Effective communication structure helps build professional and social relationships at various levels in the organization. An atmosphere of open communication makes the people comfortable to express their ideas and explore innovative solutions. Communication prevents employees from feeling isolated; it builds teamwork and creates a more collegial atmosphere.
- Clarity: A confusion and ambiguity can create negative feelings and a tense atmosphere. By making roles and responsibilities clear to the team members, a culture of matured understanding and professionalism can be evolved. Communication reduces misunderstandings and, hence, the cost associated with duplication of efforts.
- Collaboration: Effective communication improves cooperation among employees and streamlines the organizational processes. This enhances the overall performance of the team.

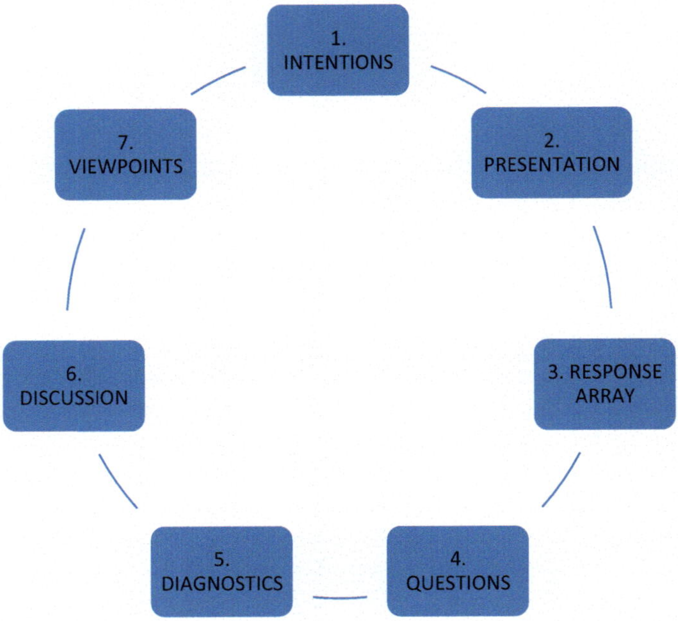

Fig. 15.2 Components of communication structure

15.3 Models of Communication and Relevance for Project Management

The communication models with their relevance to "Construction" and "R&D" projects are summarized in Table 15.2.

15.4 Strategies to Improve Communication in Project

The strategies to improve communication are summarized in Table 15.3.

Project Management in Practice

Role of Communication in Project Management

We consider a project of constructing a new student hostel for an educational institute. The institute's planning department identified the hostel requirements by collecting inputs from the director, dean, and other stakeholders. The institute opens a tender and awards the contract for constructing the hostel to a private contractor. The contractor hires several sub-contractors to carry out specific tasks as part of constructing the hostel. Due to so many stakeholders involved, confusion and misinformation may

Table 15.2 Communication models

Communication model	Relevance/importance for construction project	Relevance/importance for R&D project
Shannon and weaver model **Principle** The information source is where the information is stored. To send the information, the message is encoded into signals to travel to its destination **Key elements** Information source Transmitter Channel Receiver Destination	For a construction project **Sender**: Project manager **Encoder**: Engineers **Channel**: Notice board/letters **Noise**: Missing information **Decoder**: Instructor **Receiver**: Worker Feedback from workers lets the manager know what changes to be made to make the construction process more effective and productive. It is a two-way process	For an R&D project **Sender**: Project manager **Encoder**: Telephone network company **Channel**: Mobile network **Noise**: Missing information **Decoder**: Mobile phone **Receiver**: Testers/customers Feedback from testers and customers lets the project manager know bugs in new technology and helps in improving it
Laswell's model **Principle** The message flow in a multicultural society with multiple audiences **Key elements** 1. Surveillance of the environment 2. Correlation of components of society 3. Cultural transmission between generations	In a construction project **Sender**: Project manager **Message**: Tasks to be done or schedules or notices **Medium**: Using mail or group notice **Receiver**: Workers **Feedback**: Reports to a manager This model gives power to managers in construction projects and determines the most effective medium for communication with workers. It also helps in deciding audiences for a particular message	In an R&D project **Sender**: Project manager/mentor **Message**: Tasks to be done or schedules or notices **Medium**: Personal or group message **Receiver**: Employees **Feedback**: Work completed and issues or bugs This model helps control analysis, content analysis, media analysis, audience analysis, and effect analysis for an effective communication process

(continued)

Table 15.2 (continued)

Communication model	Relevance/importance for construction project	Relevance/importance for R&D project
Linear model **Principle** The process of one-way communication, whereby a sender transmits a message and a receiver absorbs it **Key elements** 1. Information source 2. Transmitter 3. Channel of transmission 4. Receiver 5. Destination	A linear communication model envisages a one-way process in which one party is the sender, encoding and transmitting the message. Another party is the recipient, receiving and decoding the information A construction project helps to focus on how a design/architecture may be changed or modified for a better design	In an R&D project, it helps to focus on how an idea/design may be altered and influenced by the encoding process of the technology, the effects of the communication channel or medium
Interactive model **Principle** It describes communication as a process in which participants alternate positions as sender and receiver and generate meaning by sending messages and receiving feedback within physical and psychological contexts **Key elements** 1. Two sources: The originator of the message and the recipient of the message are both sources 2. The message: the information being exchanged 3. Feedback: takes place after the first message has been received and is returned to the originating source	In a construction project, an interactive model helps in designing/discussing a concept where more than one team is required. They all can interact at the same time. Doing work together on software using cloud computing is the best example of an interactive model of communication	In an R&D project, feedback is essential for new technology. The internet can be used as the best interactive communication method as the receiver can give feedback even in newspapers and books. Human–computer interaction is also now considered interactive communication as the model is circular where the sender's interchange every time

(continued)

Table 15.2 (continued)

Communication model	Relevance/importance for construction project	Relevance/importance for R&D project
Transactional model **Principle** The model is mainly used for interpersonal communication and is also called circular model of communication **Key elements** 1. Communication evolves from the very first (origin) until the current moment 2. Communication is mainly dependent on its past 3. Concept of time	In a construction project, transactional model helps build positive bonds and ensures the flow of positive ethical energy. This model emphasizes norms, values, and laws related to society during construction. It helps communicate with stakeholders or other teams; when two people from the same caste or culture are there, let them discuss	In an R&D project, the efficiency and reliability of communicated messages also depend on the medium used. The transactional model plays a vital role while wishing and greetings colleagues and stakeholders

Table 15.3 Communication strategies

Communication strategy	Relevance/importance for construction project	Relevance/importance for R&D project
Communication flow creation	• Develops clarity on deliverables among stakeholders • Improves transparency and makes the process smoother • Enhances site productivity	• In a typical R&D project, coordination between teams and understanding each other's work is essential • Communication flow helps in creating transparency and understanding among different groups • It reduces troublesomeness so increases productivity
Meeting regularly	• Regular meetings help in keeping track of workflow • It helps in estimating construction project completion time • It helps in accommodating modifications/revisions through improved coordination among stakeholders	• Regular, frequent, and short meetings are necessary for R&D projects to improve coordination and resolve various emerging technical and non-technical problems • Facilitates brainstorming during the design phase of the project
Being inclusive	• In a construction project, everybody plays a vital role from their end. So, managers must make sure not to leave anyone out when inviting people related to the project • Ask for communication preferences of members of the team. And gain more input from people than limited input from limited team members	• Typical R&D project requires new ideas to work on • New ideas come from unique thoughts and fresh minds, so everybody must participate in discussions related to the project • Each member's contribution is vital for the design phase and later
Be transparent, clear, and concise	• Be clear, transparent, and concise regarding the meeting as it saves precious time for everyone and does not make anyone feel bored • Bring clarity to everyone about the purpose of the meeting and what the desired outcome will be	When discussing the technicalities of an R&D project, a manager should be clear and transparent about the goals and methodology used in the project

(continued)

Table 15.3 (continued)

Communication strategy	Relevance/importance for construction project	Relevance/importance for R&D project
Show some respect	• A typical construction project requires many teams to work together and coordinate with each other • Everybody is put on project teams for a reason. Irrespective of their roles, everyone serves an intended purpose and brings an intrinsic value to the project • Everyone should respect each other during communication and listen to others	• In an R&D project, we need to take care of many aspects of the project like socioeconomic, technical, and economy • Each team plays a vital role in each aspect • Everybody should respect each other's ideas and suggestions. This motivates the
Ambiguity reduction	• Discussions among different teams related to a construction project should be more transparent • There should not be any misunderstanding and ambiguity • The ambiguity can be reduced by documenting the minutes of meetings	• Ambiguity on technical aspects can be hazardous in a typical R&D project • The ambiguity comes from uncleared concepts/talks • The manager should try to reduce ambiguity, and notes of meetings play an essential role in reducing ambiguity • Playback and reformulation of meetings are also important
Assertiveness	• In a typical construction project, a good manager should declare things clearly and state them • If somebody loses temper, the manager should remain quietly assertive, rethink the statement, and understand the position	• R&D projects require detailed discussions and critical thinking. This may involve conflicting ideas • A good manager should remain quietly assertive and should take understanding of the position
Confrontations	• During a construction project, criticism is common from workers and others • A good manager should confront them without losing self-control and temper	• Criticism of new ideas and technology at their initial phase is common during discussions • A good manager should confront

(continued)

Table 15.3 (continued)

Communication strategy	Relevance/importance for construction project	Relevance/importance for R&D project
Let others speak	• The flow of information is significant during the conversation • Construction projects require information on a large scale, so it is essential to listen to others, show interest, and give attention	• If all members of the team participate, then it helps the R&D project to be more productive • Communication must motivate the team members

spread, which cause delays and cost overruns. In such a scenario, communication plays a vital role. We will look at the communication problem, leading to the safety concern in construction and its cost implication. We will also look at project delays due to improper communication and the increased labor cost resulting from that. The project details are given in Exhibit 15.1. The contractor also hires 70 laborers/workers.

Organization Breakdown Structure

The organizational structure of the company is illustrated in Exhibit 15.2.

Exhibit 15.1 Project roles and responsibilities

Management role	Responsibility	Number of engineers	Assistant/staff
Project manager	Complete project	4	10
Senior construction manager	Block A, B, C, D	2	8
Assistant construction manager	Cycle parking and external work	1	2
Safety manager	Safety on-site	–	3
Technical manager	Design and QAQC	2	5
MEP manager	MEP coordination	1	3

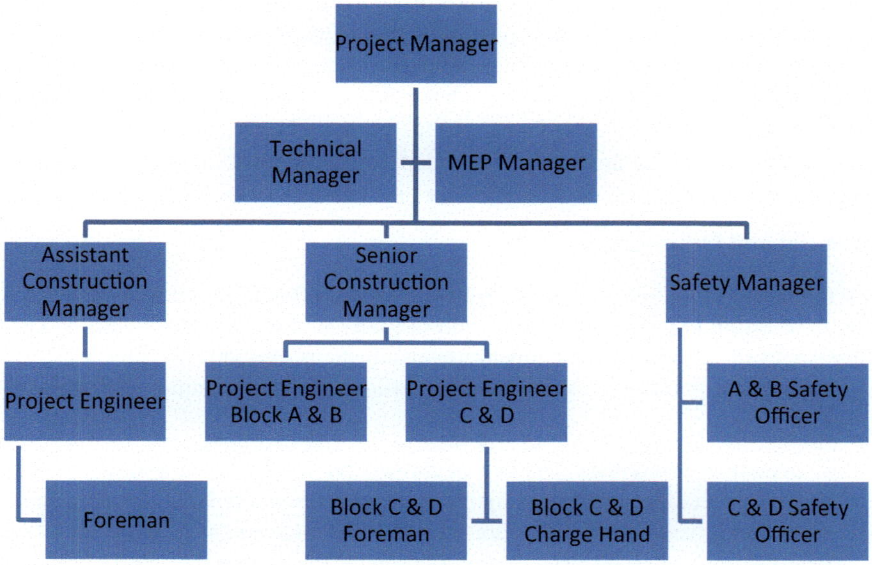

Exhibit 15.2 Organizational structure

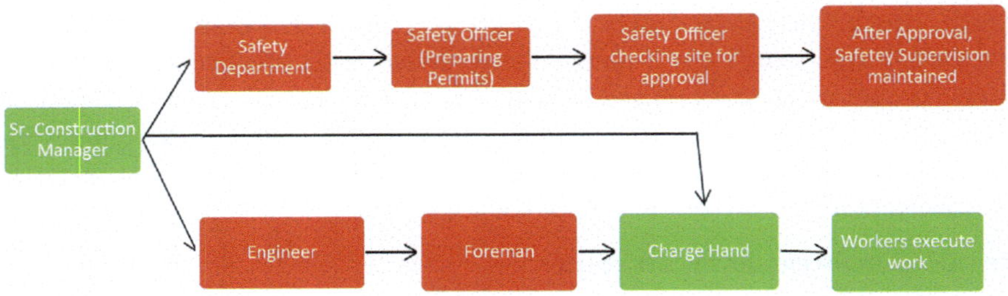

Exhibit 15.3 Order processing sequence

The Accident

The senior (Sr.) construction manager is responsible for a daily work allocation. One of the tasks was to lift a heavy platform and move it to Block D. Project engineer of Block D was on leave that day. This has prompted Sr. construction manager to inform the charge-hand directly to do this task. Six workers started the work the next day without a foreman supervising them. They did not tie the platform correctly, and while lifting, it went out of control. Three workers got crushed under the platform and died, while two were injured.

What is the Cause? Communication Gap

The accident resulted from inappropriate communication by Sr. construction manager when he directly informed the charge-hand to do a dangerous task without notifying the safety officer. Also, he did not consult the engineers for this risky task. In Exhibit 15.3, the red box highlights the people who did not receive the order. Green highlights the receivers of the order. It can be seen that the Sr. construction manager informally communicated with the charge-hand, and the other staff were not updated. As the order was not communicated to the safety department, the work permit reports were never made, and there was no safety officer on-site when the accident happened.

Financial and Other Impact

Three workers died, and two were injured. The project site had to be closed for a month, resulting in paying an extra month of salary to the contractor's staff and the daily wages of 70 laborers. The government authorities also imposed penalties.

The different points which stress the importance of good communication are as follows:

1. Expectation differences: The absence of proper communication may result in varied expectations when moving in the stakeholder hierarchy. The top management sets a deadline of 1.5 years from the commencement of work such that the inauguration takes place on the subsequent foundation day. The communication gap between the different levels may cause laziness in working, which may cause the project to finish behind schedule.
2. Improper communication problems: Improper communication may confuse the system, causing a loss in time and money. The inappropriate information about

the availability of materials used in construction causes loss of time and money that might be saved. The materials are kept in the open, which are unknown to the project managers causing unnecessary procurements and wastage of funds.
3. Confusion: The correct information needs to reach the right people at the right time. The information about the cancelation of previously allotted holidays has caused problems for the laborers or staff.

The communication problems can be minimized using the following recommendations.

1. Multiple communication methods are unnecessary since they cause undue stress to the lower levels of the hierarchy.
2. The system should send reminders at predetermined intervals. This helps people realize that the work is critically important.
3. A database comprising the details of various stakeholders can help to retrieve the necessary information timely.

Questions for Class Discussion

1. What is the role of project communication in efficient execution of the project?
2. What are the key challenges a project manager faces while committing the people for effective communication?

Summary

Communication is an important part of daily life, and the whole world needs it. Project communication is an important pillar for success of a project.

Project communication helps in notifying the project status and discussions and makes sure new ideas will come from other members. Good project communication helps in the establishment of purpose. This answers the question "Why" behind everybody is doing the project.

Opportunism is a self-interest conscious policy of taking advantage of present circumstances during the project. Opportunism of the players is a basic managerial issue in project management.

Questions for Discussion

1. How can communication be utilized as an effective tool to streamline the workflow in complex projects?
2. Why "project communication" is the most critical aspect? What are the consequences of not having an appropriate communication channel in typical projects like construction and R&D? Explain in detail.

3. What is the “opportunism of the players” in project management? How is it linked with communication structure? How can a practical communication structure help to have transparency and coordination in the project?
4. What are the different models of communication most suitable for the project environment? Explain at least three models. Explain the relevance of each of the five models for a typical IT and R&D project in a tabular format.
5. What are the key strategies to improve communication and coordination in the project? Explain the effectiveness of these strategies that are for a set of projects such as road construction, refinery construction, and new product development in a tabular format.

Chapter 16
Project Closure and Termination

Critical Questions
What are the key issues in project closure and termination?
Why it is important to be careful at the project closure and termination stage?
What are the key steps in project closure and termination?

16.1 Importance of Project Closure and Termination

Project closure and termination is an important issue because of the following reasons:

1. **Assurance of Completion of Project Work**:

It is vital to ensure that all the project objectives set during the project planning phase are achieved by the end of the project. A few goals may not be met because of the delayed work or partially completed, perhaps due to resource constraints or low priority work. Therefore, it is necessary to confirm that all project objectives get accomplished, and the project team delivers the intended outcomes.

2. **Communication of Project Conclusion or Termination to Stakeholders**:

Project closure or termination allows the stakeholders to realize the completion of the project. Therefore, the stakeholders must be well aware of the project's completion and not undertake any modifications.

3. **Obtaining Feedback from Project**:

During project closure, meetings are held with clients and the internal team to identify and discuss problems faced during the project and lessons learned to avoid such issues in future projects. It is necessary to have such discussions to reflect upon past mistakes

J. J. Thakkar, *Project Management*, Management and Industrial Engineering,
https://doi.org/10.1007/978-981-15-3695-3_16

and prevent them in the future. In addition, if some strategies or methods worked in the project's favor, they should be discussed and documented to help future projects.

4. **Assessment of Project Management**:

It is essential to ensure the completion of all project management processes and improve upon a set of practices as desired.

5. **Completion of Administrative Tasks**:

Many administrative tasks need to be performed at the end of the project, including re-assigning resources to other projects, completing all contractual obligations, including final payments for the project, and documenting the final project status report. Project closure ensures an appropriate execution of these tasks to avoid any confusion or causing unnecessary wastage of time and resources.

Considering the case of a flyover construction project, the stakeholders are the users (people who will use the flyover), project managers, contractors, suppliers, and social groups, and environmentalists who may oppose the construction. A government organization usually manages a public construction project. For example, in India, the Central Public Works Department (CPWD) executes flyover construction projects. The contractors undertake the construction after a successful award of contract. Table 16.1 summarizes the issues specific to project closure.

At the project closure stage, conflicts may arise among different stakeholders or internally in the management team. For example, there can be conflicts between the finance team and other project teams within the management due to cost overruns. Conflicts may also arise between the administration and contractors concerning financial matters like pending payments or demand for additional money. The strategies which the project managers can follow to minimize conflicts are as follows:

Table 16.1 Issues in project closure

S. No.	Stakeholder	Issue	Solution
1	Users	Due to a lack of knowledge about the constructed flyover, people tend to follow other routes defeating the purpose of the flyover to reduce traffic	The news about the constructed flyover should be disseminated through various forms of media so that more people travel by the flyover, thus reducing the problem of traffic
2	Contractors and suppliers	Contractors and suppliers have not received their final payments	The contractors' and suppliers' payments should be streamlined at the end of the project
3	Social groups	Use of low-quality construction material identified by social groups	Social groups verify the standards followed in the construction. The necessary justification needs to be prepared

1. The documentation of the decisions over the project duration is necessary to avoid any confusion later at project closure or termination.
2. An agreement among the stakeholders is essential for the decisions taken during the project. The dialogue with the concerned stakeholders is necessary to resolve conflicts and avoid any problems further in the project.
3. The information regarding the allocation of resources, finances, and personnel to various parts of a project should be decided and documented simultaneously to avoid ambiguity in the future. The project manager should trace the allocation and utilization of resources closely.
4. The stakeholders should be aware of decisions taken at the various stages of the projects to avoid conflicts toward completing the project. For example, avoiding a dispute between the finance team and project manager is possible if the project managers carefully decide the allocation of funds during planning and document the information. He should frequently check how the funds have been allocated and utilized for different packages of the project. A reallocation of funds is necessary for an underestimated cost.

16.2 Steps Involved in Project Termination Process

1. **Project Closeout Criteria Check**: The project manager confirms the meeting of all the project requirements and completion of work.
2. **Client Closeout/Deliverables**: It helps ensure that the stakeholders are satisfied with deliverables and validate their acceptance.
3. **Organizational Closeout**: This helps to remove the workforce/equipment dedicated to the project. It should release borrowed equipment or employed workforce toward the completion of the project. The given step helps in reducing operational costs.
4. **Subcontract Closeout**: All the subcontracts on the project need to be closed properly. The project manager needs to confirm the adequateness of changes in orders before closing subcontracts.
5. **Risk Assessment**: It helps in uncovering the risk which can derail the project. It is essential to review the project and make a strategy to mitigate risk or liabilities like legal issues, political issues, negative cash flow, and transferring of deliverables, to ensure the final handover of the project easy and efficient.
6. **Write a Final Report**: The lesson learned in the project is documented and utilized to update the organizational processes. This provides necessary guidelines to the project managers working on various projects.
7. **Team Closeout**: Final tour or demonstration of an area or task with the project team helps reveal and document the lessons learned. It minimizes the communication gap among the design team, clients, and associated third parties.

Table 16.2 explains the relevance of the above project closure steps for "construction" and "R&D" projects in Table 16.2.

Table 16.2 Relevance project closure steps for "construction" and "R&D" projects

S. No.	Project closures steps	Construction project	R&D project
1	Project closeout criteria check	• Review legal requirement • Review client requirement • Review contract closure requirement	• Review legal requirement • Review the requirement against various regulations and policies for implementing R&D project • Review technical faultiness • Review technical and market-related objectives
2	Client closeout /deliverables	• Ensure the completeness of work as per client requirements • Conduct customer satisfaction survey to yield feedback from stakeholders	• Verify the effectiveness of deliverables • Determine the probability of success • Verify technical faultiness
3	Organizational closeout	• Compare the money spent on the project with cash budgeted • Pay outstanding invoice • Finance sign off from the project • Removing equipment and employee from the construction site	• Compare the money spent on the project with cash budgeted • Finance sign off from the project after considering all the deliverables • Pay outstanding invoices
4	Subcontract closeout	• Verify the completed work for intended quality standards • Match payment amounts and invoices and submitted to the finance department as needed	• Verify the work for intended quality and regulatory standards • Match payment amounts and invoices and submitted to the finance department as needed • Identify the scope for future R&D requirements
5	Risk assessment	• Review the project and make a strategy to mitigate risk or liabilities like legal issues, political issues, negative cash flow, and transferring of deliverables, to ensure the final handover of the project easy and efficient	• Calculate the probability of market and technical success • Understand market conditions • Review the project and make a strategy to mitigate risk or liabilities
6	Write a final report	• Lesson learned in the project to be documented • Draw insights for the future improvements	• Lesson learned in the project to be documented • Record success and failures encountered during the project • Identify the gaps in R&D and create a plan for successive research in this domain
7	Team closeout	• Make site visits to appreciate the key learnings for future projects	• Communicate project decisions • Conduct meetings with the R&D team for a final walk-through • Discuss success and failures encountered during the project with the team members and receive their feedback

Project Management in Practice

Project Implementation challenges in Central Public Works Department (CPWD).

(Source: Bharti, B. K. and Thakkar, J. J. (2013) "SWOT of Central Public Works Department India: a case study," Journal of Advances in Management Research, Vol. 10 No. 1, pp. 100–121).

Key Challenges:

- "Compliance to envisaged timelines" is one of the essential elements of service delivery, and the same holds good for CPWD in the context of project management.
- The preparation of preliminary design and preliminary estimates is 72 weeks (about 17 months) and 19 weeks (about five months).

Delays in Project Planning and Execution

- A shortage of adequate workforce and lack of coordination creates delays.
- Prime reasons for delay.
  - Inappropriate allocation and organizational resources across departments.
  - Emphasis on "doing the work in-house" with indigenous resources.
- CPWD opts for outsourcing only when technical expertise is not available within the organization and rarely (or never) to ease workload pressure on departmental engineers/architects.

Delays in Project Construction Stage

- Change in scope of work and any specification during the construction.
- A typical problem is in the fixation of rates for the new items with the existing contractor.
- It leads to coordination problems and delays the project.
- The client directly appoints architects/consultants for the preparation of drawings. However, quite often, these drawings are not available in time and are sometimes incomplete. Therefore, it demands CPWD to put in extra time and effort to push the construction activity.
- The contractors do not invest in modern tools and machinery that becomes a bottleneck in timely delivery or reducing the lead time for construction. A related issue is regarding the contractor's unwillingness to recruit full-time skilled labor/technicians. Deployment of semiskilled labor also results in inefficiency and delay in projects.
- Funds from the client department are not made available to CPWD on time resulting in project delay and sometimes stoppage of work.
- There are also procedural delays in funds reaching from the client department to the pertinent contract engineer.
- There is a non-availability of critical raw materials such as steel and cement due to the sudden demand–supply gap.
- Raw material price shoots up, and the contractor tends to delay the construction hoping for a fall in prices.

- Non-availability of labor due to seasonal variations (especially during agriculture sowing season, festivals, etc.)
- Delay caused by external agencies (such as urban local body or the utility concerned) in providing the requisite approval and even the completion /occupation certificate.

Compartmentalization on Cadre/Functional Line

- The department has increasingly compartmentalized on functional lines (cadres), leading to coordination problems and duplicity of command chains, making it difficult to fix accountability.

Career Stagnation

- Several legal suits against the department on promotion issues have led to many vacancies at the Executive Engineer's (EE) level.
- Hampered the department's performance as an EE heads the field unit and carries out the project execution work.
- It reduced the morale and motivation of engineers within the organization.

Flexibility in Operation

- The vigilance cell of CPWD controls the deviation from the laid down procedures. It has made the engineers and engineers cautious about the deliverables of the project. However, an overcautious approach causes delays in the project.

Absence of Knowledge Management System

- It is necessary to create an electronic database of various projects executed by CPWD. This repository can benefit future projects.
- The absence of a computerized knowledge management system available to all the units of CPWD, from central headquarters to field stations, hinders the proper utilization of collective organizational knowledge for executing projects efficiently.

Inadequate Project Monitoring System

- A directorate is responsible for preparing project monitoring reports that comply with a monthly progress report of Central Police Organization (CPO) works costing more than ten lakhs and non-CPO works costing more than one crore.
- The directorate compiles four annual reports, 19 quarterly reports, and eight monthly reports and submits them to concerned officers and clients.
- There is no integrated computerized project monitoring system linked to milestones or stages of project execution.
- Information regarding the status of various projects percolates up the organization through different engineering levels.
- It is critical to have a dynamic IT system that accepts data and provides access to information at different levels (with varying access rights). This brings transparency across the organization and facilitates the development of an effective project execution strategy Exhibit 16.1.

Exhibit 16.1 SWOT of CPWD

Strength • Government involvement and assistance in land acquisition and environmental clearances • India's rising economic status • A strong pool of project management professionals • Employment opportunities	Weakness • Poor use of project management skills in budget estimation and planning • Improper crew formation and resource leveling • Heavy dependence on government funds. Poor financing • Poor career prospects for CPWD employees • Lack of functional autonomy • Inadequate involvement of end-user/client
Opportunity • A chance to leapfrog over the old system and come up with an indigenous and technologically advanced solution • A chance to build Greenfield cities suited and customized to the need of the Indian urban population • An excellent interest by external players to participate in infrastructure projects	Threat • Rising urban consumerism may threaten to swamp transportation and other available resources • The alignment of urban governance as a state subject makes it more challenging to craft a uniform policy • Huge requirements of funds • Massive backlog in the provision of urban services

Critical Risks in Indian Infrastructure Project

- Delays in initial stages
- Communication breakdowns
- Unrealistic deadlines
- Scope changes
- Resource competition
- Absence or improper use of project management techniques
- Failure to manage risk
- Insufficient team skills
- Inadequate engagement of customers and end-users during the project
- Availability of skilled and semiskilled workforce is insufficient

Question for Discussion:

(1) What are the key implementation challenges in the CPWD infrastructure project? Enlist at least five recommendations to overcome such challenges?

(2) How do you compare the execution of infrastructure projects in Government, Private, and Public–Private Partnership (PPP) modes? What are the key advantages and limitations in each case?

(3) What are the issues in the closure and termination of the infrastructure projects executed by the CPWD? How do you relate the implementation challenges with closure issues?

Project Management in Practice

Project Management Complexities at ABC Ltd.

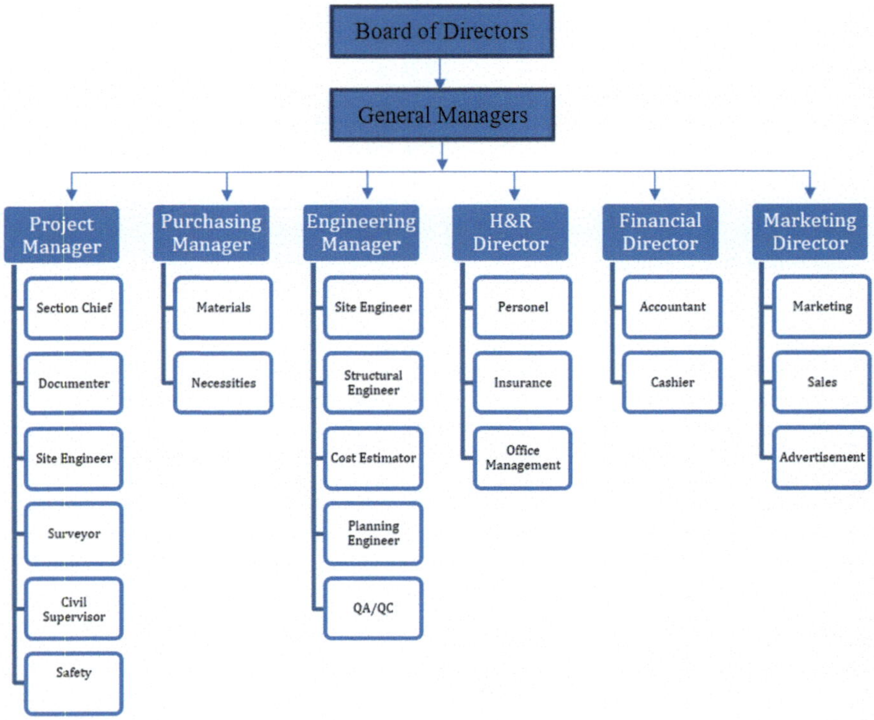

Exhibit 16.2 Organizational structure of ABC Ltd

Exhibit 16.2 illustrates the organizational structure of ABC Ltd., a hypothetical infrastructure company.

Process:

The infrastructure process consists of six distinct stages, as explained below.

1. **Concept:** All infrastructure projects begin with design and planning, also referred to as architectural programming. Several interconnected steps occur during this conceptual or design phase.
2. **Contracts and bid documents:** The builder should supply potential bidders with complete drawings and plans for the projected structure and project specifications for approving construction bids.
3. **Bidding:** After examining the project's feasibility, the owner will solicit bids or proposals from general contractors.
4. **Construction:** It is the actual construction of the project. Fieldwork consists of building permits, subcontractors, scheduling subcontractors, shop drawings, project submissions, and change orders.
5. **Construction payments:** Make the payments for the completed work.
6. **Completion:** If all the requirements are satisfied, the project will be declared completed.

Role:

1. **Business case development:** The scope includes (i) defining project objectives, risk, and budget and (ii) indicating the impact of the project, etc.
2. **Planning:** Successful planning details all the specifications of the project.
3. **Reporting (communication):** It ensures effective communication among all the stakeholders.
4. **Chairing project board:** It requires recording and coordinating of all the actions discussed and agreed upon.
5. **Work package assignment:** Technical team ensures all the technical requirements are adequately satisfied.
6. **Managing and reporting risk:** This involves managing and reporting hazards throughout the project.

Complexities Faced:

Complexities Causing Delays:

1. Lack of coordination and inadequate workforce.
2. The contractors do not readily invest in the modern equipment necessary to expedite the project.
3. The prices of raw materials shoot up, and the contractor tends to delay the construction hoping for a fall in prices.
4. The funding is not received on time, leading to delays in procurement and other activities.
5. Local approval agencies cause delays in providing the requisite approvals.

Operational Complexities:

1. A typical problem is in the fixation of rates for the new items with the existing contractor.
2. The contractor's production schedule does not always match with the production schedules of the project company.
3. The contractor's use of a poorly skilled workforce leads to deviation from stipulated quality standards.
4. Transportation of delicate raw materials like glass and wiring need adequate safety and precautions. Therefore, it is preferred to source such material from nearby locations.
5. The conceived design should be legal as per government regulations.
6. There is an uncertain supply of raw materials such as steel bars, concrete due to fluctuations in market demand.

Human Resource Complexities:

1. Maintaining the motivation and morale of the site engineers
2. Irregularities in releasing funds for salaries
3. Deployment of safety measures.

Questions for Class Discussion

(1) What are the key issues and conflicts ABC should proactively address in the smooth running of the project?
(2) How would ABC take care of T-C-P trade-offs in executing the project within the real-life constraints?
(3) What are the issues and concerns in defining the roles in an infrastructure project?

Summary

It is important to ensure that all the project objectives set during project planning phase are achieved by the end of the project.

A few objectives may not be met because work was delayed or partially completed, perhaps due to a lack of resources or because the work was deemed low priority. It is necessary to confirm that all project objectives are accomplished and all deliverables are completed.

Project manager should try to settle all the conflict at the end of the project and ensure a flow less transfer of ownership of the project.

Question for Discussion

1. Why project closure and termination are essential issues? What are the various critical issues specific to each of the stakeholders in a typical construction project (flyover construction) at the time of project closure or termination? How would you resolve these? What are the strategies to handle or minimize conflicts at the stage of project closure? Explain in detail with a suitable example.
2. Explain each of the steps involved in a scientific project termination process?
3. What is the relevance and importance of each of the steps for a typical R&D and construction project?

Web-Based Exercise

Collect the details on at least three mega-projects executed under the Public–Private Partnerships (PPP) model from the Internet. Compare the execution difficulties in each of the projects. Summarize your insights on problems in the closure and termination of such tasks. What do you recommend as critical measures to minimize such difficulties and conflicts at the stage of project termination or closure?

Group Project

Interact with at least three project managers with more than ten years of experience in different sectors. Based on your interviews with them, identify and enlist the key project termination strategies and challenges. Then, undertake a brainstorming session to relate these challenges with the various phases of the project life cycle. What are measures do you recommend to resolve such issues right at the planning and design stage of the project?

Index

J. J. Thakkar, *Project Management*, Management and Industrial Engineering,
https://doi.org/10.1007/978-981-15-3695-3

The manufacturer's authorised representative in the EU is Springer Nature Customer Service Centre GmbH, Europaplatz 3, 69115 Heidelberg, Germany. If you have any concerns regarding our products, please contact ProductSafety@springernature.com

Printed and bound by CPI Group (UK) Ltd, Croydon, CR0 4YY
30/09/2026
02225315-0013